# The Moths of North America

FASCICLE 9.6

## TORTRICOIDEA

TORTRICIDAE (Part)
OLETHREUTINAE (Part)
EUCOSMINI (Part)

*Epiblema*
*Sonia*
*Suleima*
*Notocelia*

Donald J. Wright
Todd M. Gilligan

2023

THE WEDGE ENTOMOLOGICAL RESEARCH FOUNDATION

**THE MOTHS OF NORTH AMERICA**

**Fascicle 9.6**

*Epiblema* Hübner, *Sonia* Heinrich, *Suleima* Heinrich, and *Notocelia* Hübner of the Contiguous United States and Canada (Lepidoptera: Tortricidae: Eucosmini)



The Wedge Entomological Research Foundation
Bakersfield, California

First printing, 2023
Printed in the U.S.A. by Sheridan, Minneapolis, Minnesota

ISBN 978-0-933003-22-4

Library of Congress Control Number: 2023943939

Layout and design: Todd M. Gilligan

Please cite this work as:

Wright, D. J. and T. M. Gilligan. 2023. *Epiblema* Hübner, *Sonia* Heinrich, *Suleima* Heinrich, and *Notocelia* Hübner of the Contiguous United States and Canada (Lepidoptera: Tortricidae: Eucosmini). Wedge Entomological Research Foundation, Bakersfield, California. 280 pp.

---

Distributors:

*The Wedge Entomological Research Foundation*
9417 Carvalho Court
Bakersfield, CA 93311-1846
USA
www.wedgefoundation.org

*Entomological Reprint Specialists*
2985 E Manzanita Ridge Place
Tucson, AZ 85718-7342
USA
www.ersbooks.com

*Antiquariat Goecke & Evers*
Inh. Erich Bauer
Sportplatzweg 5
75210 Keltern
Germany
www.insecta.de

*Pemberley Natural History Books*
18 Bathurst Walk
Iver, Buckinghamshire
SL0 9AZ
United Kingdom
www.pemberleybooks.com

*Roppon-Ashi Entomological Books*
Sanbancho MY building, Sanbancho 24-3
Chiyoda-ku, Tokyo, 102-0075
Japan
www.kawamo.co.jp/roppon-ashi

## PATRONS



The Board of Directors of the Wedge Entomological Research Foundation is sincerely grateful to the late

BRYANT MATHER

of Mt. Salus, Mississippi who very generously bequeathed money to enable the Foundation to continue its research and publication programs.

## BOARD OF EDITORS

The dates of publication of previous parts of this work:

Fascicle 21 **Sphingoidea** • 30 January 1971
Fascicle 20.2A **Bombycoidea**, Saturniidae (part) • 31 December 1971
Fascicle 20.2B **Bombycoidea**, Saturniidae (conclusion) • 28 April 1972
Fascicle 13.1A **Pyraloidea**, Pyralidae (part) • 31 October 1972
Fascicle 13.1B **Pyraloidea**, Pyralidae (part) • 11 December 1972
Fascicle 20.1 **Mimallonoidea**, Mimallonidae and **Bombycoidea**, Apatelodidae, Bombycidae, Lasiocampidae • 31 July 1973
Fascicle 13.1C **Pyraloidea**, Pyralidae (conclusion of part 1) • 31 January 1974
Fascicle 6.2 **Gelechioidea**, Oecophoridae • 1 July 1974
Fascicle13.2A **Pyraloidea**, Pyralidae (part) • 16 September 1976
Fascicle 13.2B **Pyraloidea**, Pyralidae (part) • 22 December 1976
Fascicle 22.2 **Noctuoidea**, Lymantriidae • 25 March 1978
Fascicle 6.1 **Gelechioidea**, Cosmopterigidae • 29 December 1978
Fascicle 18.1 **Geometroidea**, Geometridae (part) • 25 May 1985
Fascicle 15.2 **Pyraloidea**, Pyralidae (part) • 20 February 1986
Fascicle 7.1 **Gelechioidea**, Gelechiidae (part) • 26 November 1986
Fascicle 27.2 **Noctuoidea**, Noctuidae (part) • 22 October 1987
Fascicle 5.1 **Sesioidea**, Sesiidae • 18 January 1989
Fascicle 15.3 **Pyraloidea**, Pyralidae (part) • 27 June 1990
Fascicle 25.1 **Noctuoidea**, Noctuidae (part) • 23 August 1991
Fascicle 26.1 **Noctuoidea**, Noctuidae (part) • 30 March 1995
Fascicle 15.4 **Pyraloidea**, Pyralidae (part) • 27 March 1997
Fascicle 27.3 **Noctuoidea**, Noctuidae (part) • 24 September 1998
Fascicle 7.6 **Gelechioidea**, Gelechiidae (part) • 29 December 1999
Fascicle 15.5 **Pyraloidea**, Pyralidae (part) • 16 December 2003
Fascicle 27.1 **Noctuoidea**, Noctuidae (part) • 22 December 2004
Fascicle 17.2 **Geometroidea**, Geometridae (part) • 30 December 2008
Fascicle 26.9 **Noctuoidea**, Noctuidae (part) • 9 October 2009
Fascicle 8.1 **Tortricoidea**, Tortricidae (part) • 10 June 2012
Fascicle 9.4 **Tortricoidea**, Tortricidae (part) • 30 April 2015
Fascicle 9.5 **Tortricoidea**, Olethretinae (part) • 1 June 2017
Fascicle 22.1A **Drepanoidea**, Doidae, **Noctuoidea**, Notodontidae (part) • 21 December 2018
Fascicle 25.4 **Noctuoidea**, Noctuidae (part) • 18 February 2020
Fascicle 22.1B **Noctuoidea**, Notodontidae (conclusion) • 30 December 2021

Check List of the Lepidoptera of America North of Mexico • 30 May 1983
The Hawk Moths of North America • 26 December 2007

# Abstract

We review a total of 83 species representing four genera in the tribe Eucosmini that occur in the contiguous United States and Canada. Diagnostic morphological features useful in species identification are emphasized. Included are 51 species of *Epiblema* Hübner, 15 species of *Sonia* Heinrich, 11 species of *Suleima* Heinrich, and 6 species of *Notocelia* Hübner. Twenty new species are described: *Epiblema adusta* Wright and Gilligan, sp.n.; *E. cimarronensis* Wright and Gilligan, sp.n.; *E. wielgusi* Wright and Gilligan, sp.n.; *E. coracinana* Wright and Gilligan, sp.n.; *E. browni* Wright and Gilligan, sp.n.; *E. glaseri* Wright and Gilligan, sp.n.; *E. colfaxana* Wright and Gilligan, sp.n.; *Sonia roseomaculana* Wright and Gilligan, sp.n.; *S. ferruginana* Wright and Gilligan, sp.n.; *S. griseolana* Wright and Gilligan, sp.n.; *S. nigrana* Wright and Gilligan, sp.n.; *S. albimacula* Wright and Gilligan, sp.n.; *S. parva* Wright and Gilligan, sp.n.; *S. hodgesi* Wright and Gilligan, sp.n.; *S. gibsoni* Wright and Gilligan, sp.n.; *Suleima powelli* Wright and Gilligan, sp.n.; *S. leeana* Wright and Gilligan, sp.n.; *S. blanchardi* Wright and Gilligan, sp.n.; *S. flavidana* Wright and Gilligan, sp.n.; *Notocelia lobana* Wright and Gilligan, sp.n. *Eucosma mandana* Kearfott is transferred to *Epiblema*, comb.n. Ten lectotypes are designated. Included are 540 adult images, 612 genitalia drawings, and discussions of several unresolved species complexes.

This book is dedicated to the memory of

Wilhelm Carl Paul Gottlieb Heinrich (1880–1955)

on the 100th anniversary of the
publication of his celebrated monograph

"*Revision of the North American moths of the subfamily Eucosminae of the family Olethreutidae*"

# Contents

# Acknowledgments

We appreciate the cooperation of the following individuals and their institutions in providing access to and/or loans of material in the collections under their care: Robert Androw (CMNH); John W. Brown (USNM); Richard L. Brown (MEM); Charles V. Covell Jr. (MGCL); Jason Dombroskie (CUIC); David Grimaldi, Suzanne Rab Green, and Agnieszke A. Pierwola (AMNH); Charles E. Harp (CSUC); James Hayden (FSCA); Jean-François Landry and Christi Jaeger (CNC); Sangmi Lee (ASUHIC); David Lees and Geoff Martin (BMNH); Chris Looney (WSDAC); Crystal Maier and Rachel Hawkins (MCZ); Tommy McElrath (INHS); Luciana Musetti (OSUC); and Jerry A. Powell and Peter T. Oboyski (EME).

We extend our thanks to many friends and collectors who provided information, advice, fellowship, and specimens for examination, especially: Clifford Ferris, Loran Gibson, Steven Johnson, Eric Metzler, Tomas Mustelin, John Nordin, Steven Passoa, Greg Raterman, Kelley Richers, Michael Sabourin, John Schneider, Brian Scholtens, J. B. Sullivan, James Vargo, David Wagner, and Jerry Wiedmann.

We are especially grateful to John W. Brown and Richard L. Brown for reviewing the manuscript and to the Wedge Entomological Research Foundation and its Board of Directors for supporting the publication of this volume.

# Introduction

*Epiblema* Hübner is a Holarctic genus composed of approximately 90 species (Gilligan et al. 2018). It shares taxonomically important characters with the other three genera reviewed here: *Notocelia* Hübner, *Sonia* Heinrich, and *Suleima* Heinrich. *Notocelia*, with 24 species, is widely represented in the Holarctic and includes a few species from the Afrotropical and Indomalayan Regions. *Sonia* and *Suleima* are strictly Nearctic, both with seven species. These four genera, along with *Eucosma* Hübner and *Pelochrista* Lederer, account for much of the diversity in the North American branch of the tribe Eucosmini, and their taxonomic histories are intertwined. Lacking well defined generic concepts, early tortricid systematists tended to arrange most eucosmine species under *Eucosma* or *Epiblema*, depending on geographic region (North America or Europe, respectively), and despite groundbreaking work by Henrich (1923) and Obraztsov (1946, 1965a, 1967, 1968), uncertainty persists regarding some generic boundaries. Gilligan et al. (2014) conducted the first modern phylogenetic analysis using both morphological and molecular data to refine the definitions of *Eucosma*, *Pelochrista*, *Epiblema*, and other closely related genera. Wright and Gilligan (2015, 2017) followed this study with two volumes describing the species diversity of *Eucosma* and *Pelochrista* in North America. This, the third volume in that series, provides species-level treatments of the North American members of *Epiblema*, *Sonia*, *Suleima*, and *Notocelia*.

Life history information is available for about 40% of the North American species of *Epiblema*, likely because in many of those species the larva develops in a conspicuous gall on the upper part of the host plant. All the recorded hosts are in the family Asteraceae, with most belonging to the genera *Ambrosia* L. (ragweed), *Rudbeckia* L. (coneflower), or *Solidago* L. (goldenrod). Our primary sources for this information are: Heinrich (1923), who reported hosts for reared specimens deposited in the USNM; Miller (1976), who provided detailed life histories of three gall-inducing species; and various species accounts in Bottimer (1926), Decker (1932), and Powell and Opler (2006, 2009). *Epiblema strenuana* has been studied extensively due to its use as a biological control agent for invasive weeds in Australia and China (summarized in Gilligan et al. 2020b). Some species of *Epiblema* bore in stems and terminals without inducing galls, others feed at the base of the plant or in the roots. Adult capture records suggest that many species are bivoltine, especially in southern United States. Some are undoubtedly univoltine, such as *E. macneilli*, which is known only from the alpine zone in the high Sierra Nevada.

Life histories are known for only a few species of *Sonia*, and in each case the larva is a root-borer in Asteraceae, utilizing genera such as *Isocoma* Nutt. (jimmy weed or goldenweed), *Gutierrezia* Lag. (snakeweed), and *Solidago* L. (goldenrod). Powell and Opler (2006, 2009) provide accounts of feeding habits for several western species.

Most species of *Suleima* are associated with the genus *Helianthus* (sunflower), where the larvae bore in stalks, terminals, and seeds. Our accounts rely on label data obtained from museum specimens, published reports by Riley (1881), Putnam (1942), Satterthwait (1948), and Powell and Opler (2006, 2009), and on unpublished field notes taken by J. A. Powell and deposited in the EME. *Suleima helianthana*, widely known as the sunflower bud moth, is a minor pest of commercially grown sunflowers (Knodel et al. 2010).

All the *Notocelia* species treated here are known to utilize *Rosa* L. (cultivated or wild rose) as a larval host. Two of the North American species were introduced from the Palearctic, and their life histories are well documented in the European literature (e.g., Razowski 2003; Hancock et al. 2014), with larvae feeding in buds, twigs, and webbed leaves. One such species, *N. cynosbatella*, also utilizes other plants in the Rosaceae as well as a few tree species (Razowski 2003).

Using forewing length (FWL) as a proxy for specimen size (Miller 1977), we consider adults of the four genera treated here to be small to medium-sized tortricids (FWL: 3.7–13.2 mm, average = 7.4). As in many other Eucosmini, forewing color and maculation can vary within local populations and from one population to another in geographically widespread species. Genitalia character states are substantially stable within species but can be unreliable in distinguishing between closely related species, especially in *Sonia* and *Suleima*. Under these circumstances, species circumscription by traditional morphological techniques can be problematic. Our treatment includes several instances in which a particular taxon may be an unresolved complex of two or more similar species that we are unable to diagnose effectively due to intergradation of character states. In these cases, we discuss the taxonomic issues involved and illustrate the various phenotypes.

**Taxonomic History**

*Epiblema* was described by Hübner ([1825]). A decade later, it was interpreted by Stephens (1834) as a "division" of *Spilonota* with *Phalaena foenella* Linnaeus (1758) as its only species. In his monograph on tortricid genera and their types, Fernald (1908) concluded that the type species for *Epiblema* should be *foenella*, based on Stephens' (1834) treatment, and that position has been accepted by subsequent authors (e.g., Heinrich 1923; Razowski 1989; Brown 2005).

The taxonomic status of *Epiblema* has fluctuated considerably over the years. Early European authors such as Treitschke (1829), Stephens (1834, 1852), and Duponchel (1834) assigned species that would later be recognized as *Epiblema* to various genera, including *Aspidia*, *Carpocapsa*, *Halonota*, and *Spilonota*. A summary of such early arrangements is provided by Obraztsov (1965a). Some consistency emerged in the mid-1800s when authors generally referred these species to *Paedisca*, a taxon that has been treated both as a genus (Walker 1863) and as a subgenus of *Grapholitha* (Heinemann 1863; Wocke 1871). Meyrick (1895) included *Epiblema* amongst 24 genera comprising his family Epiblemidae, a name that is no longer in use but might be considered a first approximation to what would eventually become known as the tribe Eucosmini. In the second edition of the Staudinger catalogue, Rebel (1901) assembled under *Epiblema* a variety of species that today would be placed in other eucosmine and non-eucosmine genera. His organizing principle seems to have been the presence of a costal fold at the base of the male forewing.

The first two species of *Epiblema* discovered in North America were described by Clemens (1860) as *Hedya scudderiana* and *Monosphragis otiosana*. Fernald (1882) transferred these taxa to *Paedisca* and later (1908) treated *Paedisca*, *Epiblema*, and 23 other generic names as synonyms of *Eucosma*. Thus, in the early part of the 20th century, most eucosmines in North America were referred to *Eucosma*, while their counterparts in Europe were placed in *Epiblema*. Walsingham (1914) followed Fernald's (1908) consolidation with a listing of 38 generic names, including *Epiblema,* in the synonymy under *Eucosma*. A decade later, Heinrich (1923) reinstated *Epiblema* as a distinct genus with 34 North American species.

The description of *Notocelia* Hübner ([1825]) appeared in the same publication as that of *Epiblema*. Westwood (1840) has been credited with designating *Phalaena uddmanniana* Linnaeus (1758) as the type species (e.g., Razowski 1989), but Stephens (1834) included *uddmanniana* as the only species in his genus level treatment of *Notocelia* and was thus considered by Fernald (1908) to be the original designator. As with *Epiblema*, *Notocelia* was treated by early European authors as synonymous with a variety of other genera, including *Eupoecilia* Stephens (1828) and *Spilonota* Stephens (1829). Walker (1863) transferred *uddmanniana* to *Paedisca*; Meyrick (1895), Rebel (1901), and Fernald (1908) treated *Notocelia* as a distinct genus; and Walsingham (1914) listed *Notocelia* as a synonym of *Eucosma*.

The first two species of *Notocelia* discovered in North America were described by Walsingham (1879) as *Paedisca culminana* and *Paedisca illotana*. Fernald (1882, 1891) concurred with Walsingham's generic placement of these two taxa, but later authors transferred both species to *Eucosma* (Fernald 1903; Barnes and McDunnough 1917) or *Epiblema* (Heinrich 1923). Brown (1979) was the first to recognize that *Notocelia* has representatives in North America, and he listed four names under that genus that previously had been placed in *Epiblema* by Heinrich (1923).

In the 1920s, the use of genitalic characters produced fundamental changes in tortricid taxonomy. Pierce and Metcalfe (1922) published the first comprehensive study to include genitalia descriptions and illustrations for the Tortricidae of the British Isles. They divided the family into groups (based on the structure of the female signum) and treated *Epiblema* and *Notocelia* as distinct genera in a group they called "Epiblemidii." Those authors noted that *Notocelia* is distinguished by "the presence of two spines at the orifice of the aedeagus," a feature that is still considered the primary diagnostic character for that genus. Heinrich (1923) followed Pierce and Metcalfe with a revision based on male genitalia of the North American eucosmines. He interpreted *Epiblema* as a distinct genus that is separated from *Eucosma* by the presence of a "small rudimentary clasper" on the valva, but he treated *Notocelia* as a junior synonym of *Epiblema*. He also described two new genera: *Sonia*, with type species *Paedisca constrictana* Zeller (1875); and *Suleima*, with type species *Semasia helianthana* Riley (1881). He considered *Sonia* to be derived from *Epiblema* (agreeing with the latter in male genitalia but differing from it in forewing venation) and *Suleima* to be derived from *Sonia* (agreeing with it in forewing venation but lacking a clasper on the male valva).

Approximately 20 years after Heinrich's (1923) revision, Nicholas Obraztsov began a comprehensive review of the Palearctic Tortricidae based on wing venation and genitalia. He tried to reconcile prevailing differences between the North American and European classification systems. In the Eucosmini, he separated out numerous non-eucosmine taxa from previously proposed groups such as Epiblemidii (of Pierce and Metcalfe 1922) and effectively formalized many of the generic concepts in use today. Obraztsov's (1946, 1965a) definition of *Epiblema* was based on wing venation, the presence of a forewing costal fold in the male, and structures on the male valva. He also divided it into three subgenera (*Epiblema*, *Cacochroea*, and *Notocelia*). A decade later, Kuznetzov (1978) elevated *Notocelia* to genus status and synonymized *Cacochroea* with *Epiblema*, producing the arrangement since adopted by authors such as Powell (1983) and Razowski (1989, 2003).

Obraztsov's revisions (1946, 1965a, 1967, 1968) were, at the time, the most thorough attempts to distinguish between the hard-to-define eucosmine genera *Epiblema, Pelochrista,* and *Eucosma*. His diagnoses of these taxa relied heavily on the presence or absence of a lobe (that he called the pulvinus) at the distal margin of the basal excavation of the male valva, but he did not distinguish between several possible interpretations of that structure. The term "basal excavation" refers to an emargination of the medial surface of the valva. In Eucosmini, it extends from the base nearly to the neck. Its distal margin is often straight or weakly curved (e.g., Figs B12, B13) but can have one or more of the following modifications: 1) a basally directed tablike projection (e.g., Figs. B1, B11); 2) a basally directed bulge, producing a somewhat sinuous distal margin (e.g., Figs. B14, B15); or 3) a strongly developed basally directed lobe (e.g., Obraztsov 1967, figs. 16–17 and 21–24). Obraztsov seems to have included under the purview of "pulvinus" an outwardly projecting structure on the medial surface of the valva located at or near the distal margin of the basal excavation (e.g., Figs. B1, B11, B15). Heinrich (1923) referred to this last feature as a "rudimentary clasper."

Obraztsov (1967) used the presence of a pulvinus (sensu modification 3) to characterize *Pelochrista*, which he divided into two subgenera (*Pelochrista* and *Pseudeucosma*). He characterized *Epiblema* by the presence of a clasper and proposed three subgenera (*Epiblema*, *Cacochroea*, and *Notocelia*) based on the presence or absence of a second pulvinus (sensu modification 1) and properties of the cornuti (deciduous vs. non-deciduous). Horak (1984) defined the term "pulvinus" in her discussion of taxonomically significant characters in Tortricinae and pointed out that the structure she referred to by that name does not occur in the Olethreutinae. Ambiguity in the literature regarding the meaning of "pulvinus" prompted Gilligan et al. (2014) and Wright and Gilligan (2015) to use the term "basal process" for the feature described above as modification 1. Moreover, Gilligan et al. (2008) and Wright and Gilligan (2015) followed Heinrich (1923) in referring to the outward projection on the medial surface of the valva as the clasper. We continue these practices in this volume and recognize five primary clasper types in *Epiblema*, *Sonia*, and *Notocelia*. In a few *Epiblema*, the clasper is reduced to a weakly developed bulge on the medial surface of the valva, a feature that is found in many species of *Eucosma*. Razowski (1989) noted that this reduction might be the result of the clasper being lost in *Eucosma*, raising the prospect that *Epiblema* and *Eucosma* might not be distinct genera. Gilligan et al. (2014) tested this hypothesis with a robust molecular phylogenetic study and determined that *Epiblema* is a well-supported monophyletic group. *Epiblema* species in which the male clasper is reduced can be separated from *Eucosma* by features of the female sterigma and sternum 7 (Type II vs. Type I, as defined by Gilligan et al. 2014).

The four genera treated in this volume can be separated from one another by combinations of wing venation and male genitalia characters, as presented in the following key:

**Key to separating *Epiblema*, *Sonia*, *Suleima*, and *Notocelia***

1a. Forewing with 4 radial veins ($R_4$ and $R_5$ united) .......... 2
1b. Forewing with 5 radial veins ($R_4$ and $R_5$ separate) .......... 3

2a. Male valva with clasper absent .......... *Suleima*
2b. Male valva with clasper present .......... *Sonia*

3a. Male vesica with two non-deciduous cornuti .......... *Notocelia*
3b. Male vesica with all cornuti deciduous .......... *Epiblema*

We treat 51 species of *Epiblema*, 15 species of *Sonia*, 11 species of *Suleima*, and 6 species of *Notocelia* that occur in the contiguous United States and Canada. Included are descriptions of 20 new species, a new combination (*E. mandana*), discussions of several unresolved species complexes, and designations of 10 lectotypes. The species accounts are accompanied by 540 adult images and 612 genitalia drawings. We do not treat *E. sticticana* (Fabricius), recently reported from Gros Morne National Park, Newfoundland, Canada (Gilligan et al. 2020a); an account of this species can be found in Razowski (2003: species 321).

## Materials and Methods

We examined more than 4,100 adult specimens and 1,330 associated genitalia preparations from the following institutional and private collections:

AMNH = American Museum of Natural History, New York, New York, USA
ANSP = The Academy of Natural Sciences of Drexel University, Philadelphia, Pennsylvania, USA
ASH = Aaron S. Hunt Collection, Block Island, Rhode Island, USA
ASUHIC = Hasbrouck Insect Collection, Arizona State University, Tempe, Arizona, USA
BMNH = The Natural History Museum, London, United Kingdom
CAS = California Academy of Sciences, San Francisco, California, USA
CMNH = Carnegie Museum of Natural History, Pittsburg, Pennsylvania, USA
CNC = Canadian National Collection of Insects, Arachnids and Nematodes, Ottawa, Ontario, Canada

CSUC = C. P. Gillette Museum of Arthropod Diversity, Colorado State University, Fort Collins, Colorado, USA
CUIC = Cornell University Insect Collection, Ithaca, New York, USA
EME = Essig Museum of Entomology, University of California, Berkeley, USA
FSCA = Florida State Collection of Arthropods, Gainesville, Florida, USA
INHS = Illinois Natural History Survey, Champaign, Illinois, USA
JBS = J. Bolling Sullivan Collection, Beaufort, North Carolina, USA
JJD = Jason J. Dombroskie Collection, Ithaca, New York, USA
JLW = Jerome L. Wiedmann Collection, Painesville, Ohio, USA
JV = James Vargo Collection, Mishawaka, Indiana, USA
LDG = Loran D. Gibson Collection, Florence, Kentucky, USA
MCZ = Museum of Comparative Zoology, Harvard University, Cambridge, Massachusetts, USA
MEM = Mississippi Entomological Museum, Mississippi State University, Starkville, Mississippi, USA
MGCL = McGuire Center for Lepidoptera, Florida Museum of Natural History, Gainesville, Florida, USA
MS = Michael Sabourin Collection, Marshfield, Vermont, USA
OSUC = C. A. Triplehorn Insect Collection, The Ohio State University, Columbus, Ohio, USA
SJ = Steven Johnson Collection, Sunbury, Pennsylvania, USA
TMG = Todd M. Gilligan Collection, Loveland, Colorado, USA
TMM = Tomas M. Mustelin Collection, Seattle, Washington, USA
UCMS = University of Connecticut, Storrs, Connecticut, USA
USNM = National Museum of Natural History, Smithsonian Institution, Washington, D. C., USA
WSDAC = Washington State Department of Agriculture Collection, Olympia, Washington, USA

Morphological terminology follows Gilligan et al. (2008), except "phallus" is used in place of "aedeagus." Descriptive comments regarding forewing appearance utilize as much as possible the nomenclature for the reputed ancestral fasciate tortricid forewing pattern proposed by Brown and Powell (1991) and modified by Baixeras (2002).

## Wing Characteristics and Terminology (Figure A)

Two wing venation characters play a role in defining the four genera treated here. One involves the radial vein of the forewing, which can have either four or five branches. In the case of five branches (Fig. A1a), $R_1$–$R_4$ terminate on the costa and $R_5$ extends to the termen just below the apex; in the reduced case (Figs. A2a & A3a), $R_5$ is absent (a condition usually described as $R_5$ being "united" with $R_4$). We found that the easiest way to distinguish between the two conditions is to count the veins that terminate on the distal margin between the apex and $CuA_2$, which often can be accomplished by brushing a few scales from the ventral surface of the wing. In the reduced case there are four: $M_1$–$M_3$ & $CuA_1$; otherwise, there are five: $R_5$, $M_1$–$M_3$, & $CuA_1$. The second venational character pertains to the hindwing, where $M_3$ is either stalked with or united with $CuA_1$ (Figs. A1b & A2b or A3b). All species in a particular genus have the same combination of these states.

Most species recognized here have a semifasciate forewing pattern (Figs. A4–A7) which includes at least a portion of a subbasal fascia (the **subbasal mark**) and one to three fragments of a median fascia. The largest of the latter fragments (the **pretornal mark**) is located on the inner margin (also referred to in the literature as the dorsum) adjacent to the proximal margin of the ocellus; the smallest (the **mid-**

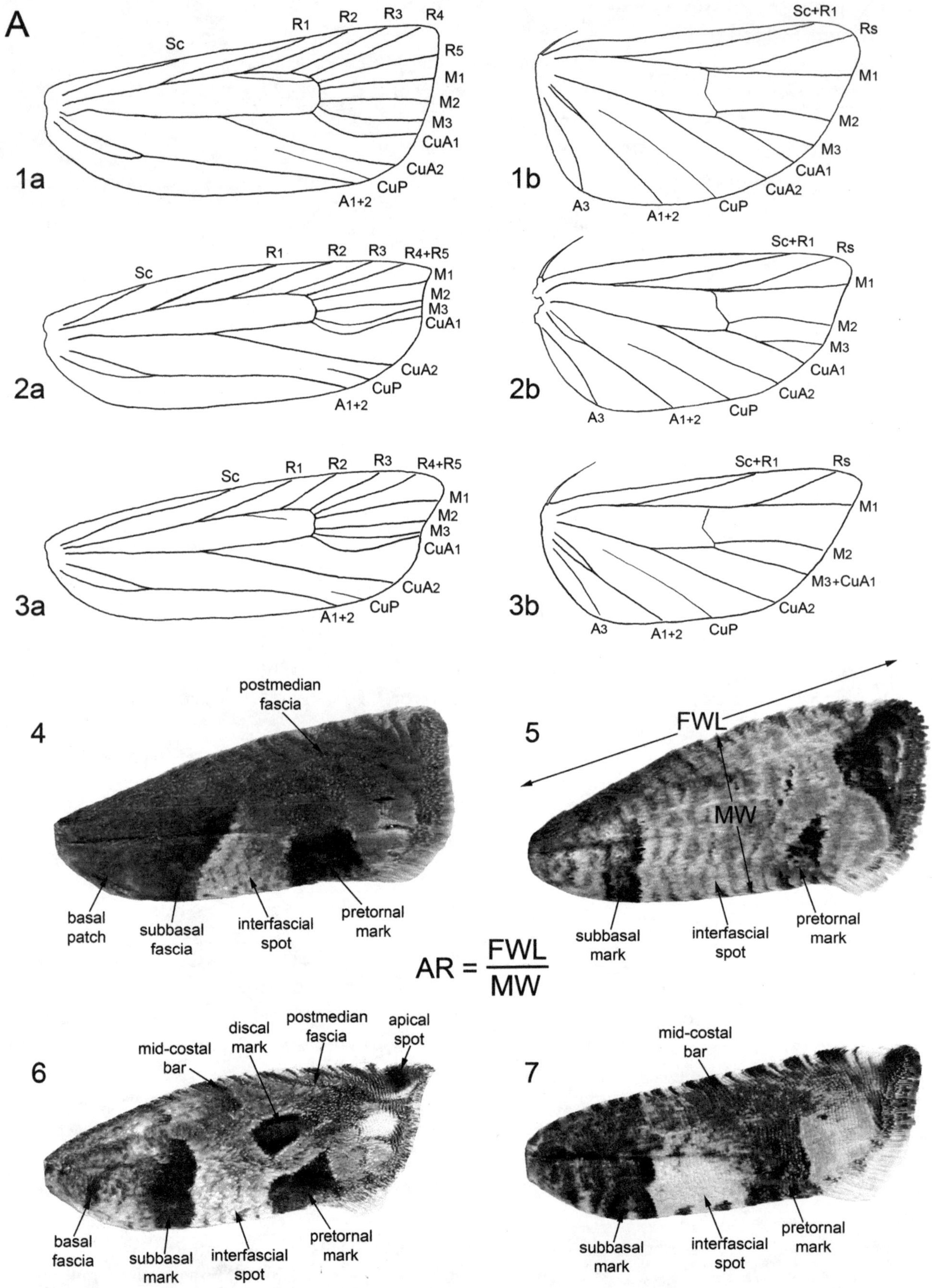


Figure A. Morphological terms used to describe forewing venation, geometry, and markings; abbreviations and ratios are explained in the text.

**costal bar**) is a short, usually narrow, segment emanating from the costa between strigulae 4 and 5; and the third fragment (the **discal mark**) is an irregularly shaped spot at the distal end of the discal cell. In some instances, the basal portion of the wing is concolorous with the subbasal fascia, forming a prominent **basal patch**. Many species have a thin **postmedian fascia** that arises on the costa between strigulae 6 and 7, arches over the apical margin of the ocellus, and extends to about mid-termen. The area between the subbasal and pretornal marks is referred to as the **interfascial spot**. It is almost always paler than the adjacent fasciate markings, often is conspicuous, and usually extends from the inner margin to the discal cell. Occasionally, the apex has a dark circular spot (the **apical spot**).

Forewing length serves as a proxy for specimen size (Miller 1977). It is measured to the nearest 0.1 millimeter and is reported for each species as a range of values and as an average value, once in Appendix A and again in the captions for the adult images. Numerical characteristics pertaining to the forewing are defined and abbreviated as follows (Fig. A5):

FWL = forewing length = distance from base to apex including fringe
MW = medial forewing width, measured perpendicular to the inner margin
AR = forewing aspect ratio = forewing length divided by medial forewing width
CFR = costal fold ratio = length of costal fold divided by forewing length

**Genitalia Characteristics and Terminology (Figures B & C)**

In Nearctic *Epiblema*, the shape of the male valva is variable (e.g., Figs. B1, B11, B12); in *Sonia*, *Suleima*, and *Notocelia*, it is substantially uniform (e.g., Figs. B13, B14, and B15, respectively). We utilize two morphometric characters, **neck ratio** and **saccular angle**, to describe the geometry of the valva. They and their constituent measurements (Fig. B1) are defined as follows:

BVW = basal valva width = perpendicular distance from the costa to the ventral margin of the sacculus, taken near the distal margin of the basal excavation
NW = neck width = width of the valva at the narrowest point of the neck
NR = neck ratio = neck width divided by basal valva width
SA = saccular angle = angle formed at the distal end of the sacculus by linear approximations to the ventral margins of the sacculus and the base of the neck

The measurements were taken on projected images of the genitalia produced by a Ken-A-Vision X1000-1 microprojector and were recorded to the nearest degree (SA) or nearest 0.5 millimeter (NW, BVW). The values of SA and NR for a particular specimen were obtained by averaging the values associated with the left and right valvae of that specimen. A species value for SA and NR was then calculated as the average of the specimen values for several representative specimens. The species values are recorded in Appendix A, with SA reported to the nearest degree and NR rounded to two decimal places.

Males of *Epiblema*, *Sonia*, *Suleima*, and *Notocelia* have a cluster of deciduous cornuti in the vesica. Within a given species, the number of deciduous cornuti can vary considerably from one specimen to another. The range of those numbers for each species is reported in Appendix A. Frequently, the counts were based on an examination of the sockets. In some individuals, a precise count was virtually impossible due to the large number of cornuti (as many as 88) and/or complications in interpreting the slide preparations. Males of *Notocelia* also have two non-deciduous cornuti at the distal extremity of the vesica (Fig. B10). Among 43 male *Notocelia* genitalia examined, we found a few exceptional cases: two specimens had three non-deciduous cornuti; four specimens had one. A detailed discussion of cornuti in Eucosmini is provided by Zlatkov (2016).

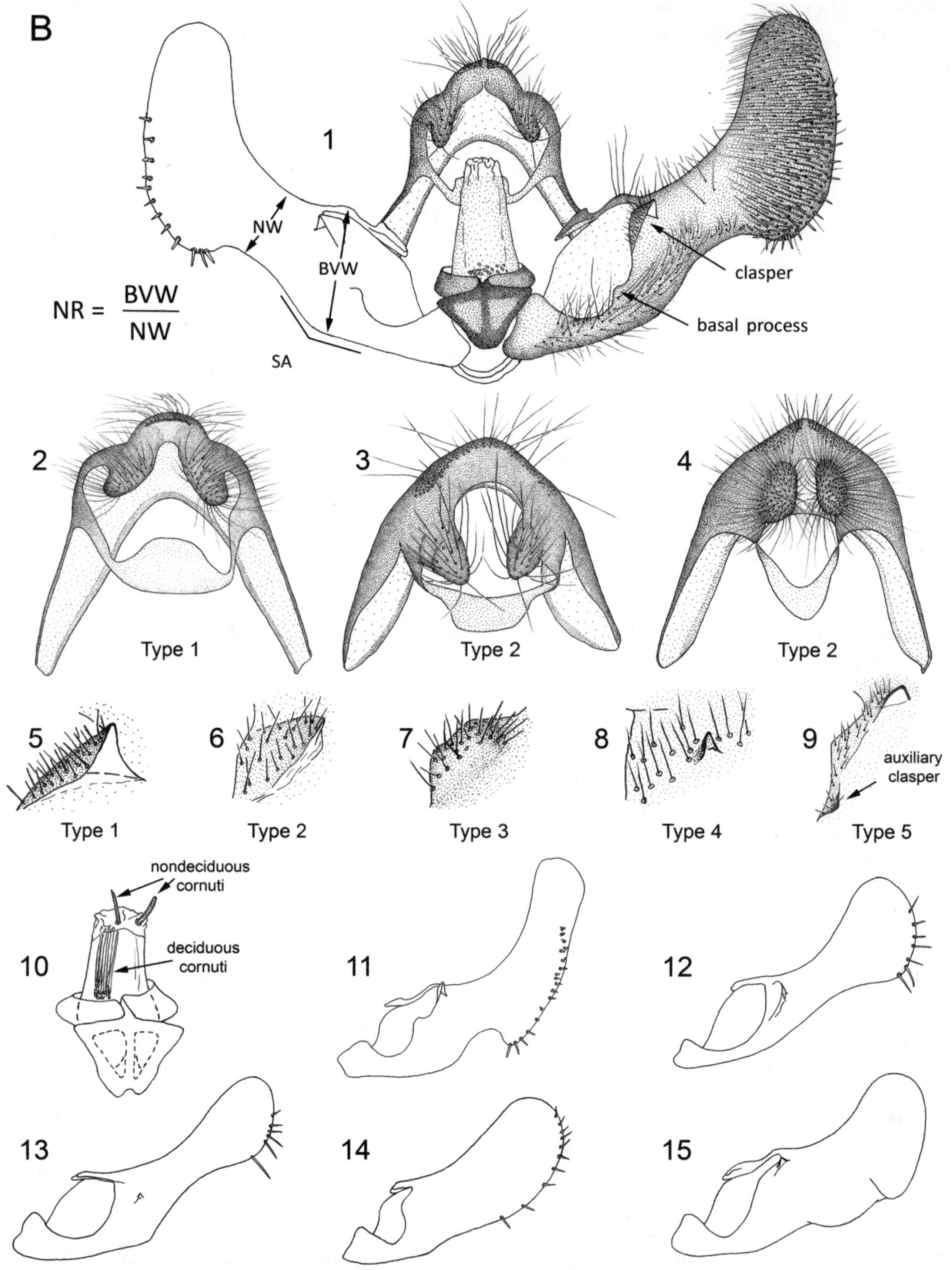


Figure B. Morphological terms used to describe the different types of male genitalia (explained in the text).

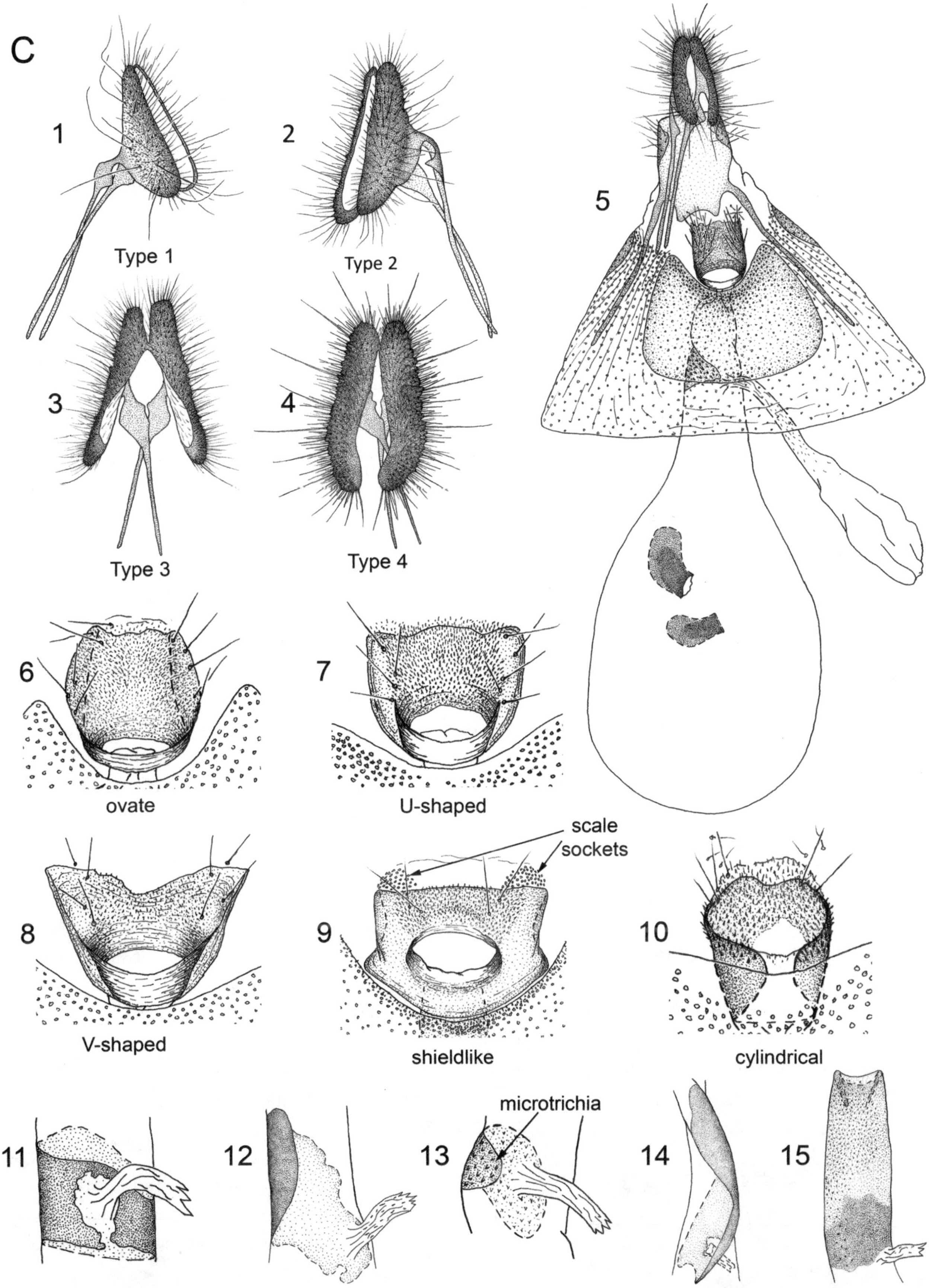


Figure C. Morphological terms used to describe the different types of female genitalia (explained in the text).

We recognize two principal forms for the tegumen together with its appendages:

Type 1. (Figs. B1, B2). Uncus moderately developed and clearly differentiated from the dorsolateral shoulders of the tegumen; lateral extremities of the gnathos arising on the tegumen and well-separated from the socii.

Type 2. (Figs. B3, B4). Differentiation of the uncus from the dorsolateral shoulders of the tegumen absent to barely discernable; lateral extremities of the gnathos arising partially on the ventral margins of the socii and partially on the adjacent portions of the tegumen.

Both Types occur in Nearctic *Epiblema*: about 77% Type 1 (Figs. B1, B2) and 23% Type 2 (Fig. B3). *Sonia* species are Type 2 (Fig. B4); *Notocelia* species are Type 1; and *Suleima* species have a mixture of Type 1 and Type 2 characteristics. In all cases, the dorsal surface of the uncus is scaled, and the ventral surface is scaleless. Some of the genitalia illustrations include a line that indicates the common boundary between the scaled and scaleless regions (e.g., Plate 18, Figs. 43b–e).

In nearly all *Epiblema*, *Sonia*, and *Notocelia*, the medial surface of the valva has a prominent projection (the **clasper**) located near the distal margin of the basal excavation. Some species of *Notocelia* also have a distally directed tablike projection (the **auxiliary clasper**) located near the ventrodistal corner of the basal excavation. *Suleima* species do not have a clasper. We recognize five primary clasper Types (Figs. B5–B9):

Type 1. Tetrahedron (Fig. B5). This tentlike structure has a triangular base (subsumed in the medial surface of the valva) and three triangular lateral surfaces, one facing distally, one facing dorsally, and one facing ventrally. The latter two faces meet in a common ridgeline that rises from or near the distal margin of the basal excavation and is obliquely inclined relative to the medial surface of the valva. The distal face is perpendicular to the medial surface of the valva and often seems to be absent (as in an open tent door), but one can usually detect (under sufficiently high magnification) a thin membrane that spans that face. Projecting from the dorsal and ventral faces are numerous, stiff, moderately short, setae.

Type 2. Wedge (Fig. B6). This form has the shape of a wedge or ramp whose flat surface rises from or near the distal margin of the basal excavation, is inclined obliquely to the medial surface of the valva, and attenuates to a rounded or angular distal extremity. Its narrow lateral surface is perpendicular to the medial surface of the valva. Numerous stiff moderately short setae project from the flat surface.

Type 3. Bulge (Fig. B7). This lumplike protrusion on the medial surface of the valva is located at the distal margin of the basal excavation and supports numerous stiff moderately short setae.

Type 4. Displaced Peak (Fig. B8). This form features a small tetrahedron located mid-way between the distal margin of the basal excavation and the neck. Numerous stiff moderately short, setae project from the medial surface of the valva in the area between the peak and the distal margin of the basal excavation.

Type 5. Clasper + Auxiliary Clasper (Fig. B9). This form is comprised of a Type 1 clasper and an auxiliary clasper located, respectively, near the dorsodistal and ventrodistal corners of the basal excavation.

Of the 40 Nearctic *Epiblema* with tegumen + appendages Type 1, 38 have clasper Type 1, one (*E. tandana*) has clasper Type 3, and one (*E. fulminana*) is of indeterminant clasper type. Of the 12 *Epiblema* species with tegumen + appendages Type 2, eight have clasper Type 2, and the rest have clasper Type 4. All *Sonia* species have clasper Type 4. In *Notocelia*, four species have clasper Type 5, and two have clasper Type 1.

In a little over one-third of Nearctic *Epiblema* species (about 36%), the distal margin of the basal excavation has a basally directed tablike projection that we call the **basal process** (Figs. B1, B11). This character is either absent or greatly diminished in *Sonia*, *Notocelia*, and most *Suleima*.

In typical slide-mounted male genitalia preparations, many of the species treated here (particularly *Sonia* and *Suleima*) show a series of spiniform setae distributed along the distal margin of the cucullus from the anal angle to about three-fourths the distance to the apex (e.g., Figs. B1, B11–B14). The sockets of the two or three setae nearest the anal angle usually are embedded in the edge of the cucullus, and those setae naturally project outward from the distal margin. The remaining setae in the series have their sockets on the medial surface of the cucullus and are often forced by the cover slip into an outwardly projecting orientation relative to the distal margin. The valva outlines in the genitalia plates (illustrating variation in valva shape) show these setae as they appear under the microscope.

In the female genitalia, we identify four primary forms of the papillae anales (Figs. C1–C4). All four Types occur in *Epiblema*. *Sonia* has Type 3, *Suleima* has Type 2, and *Notocelia* has Type 4. The Types are defined as follows:

Type 1. (Fig. C1). Pads substantially flat and facing laterally.

Type 2. (Fig. C2). Pads convexly curved and facing laterally.

Type 3. (Fig. C3). Pads with posterior lobes separated and facing ventrally, anterior lobes facing laterally.

Type 4. (Fig. C4). Pads substantially flat and facing ventrally.

In almost all *Epiblema*, the shape of the sterigma is ovate (Fig. C6), U-shaped (Fig. C7), or V-shaped (Fig. C8); in *Sonia*, it is always U-shaped (Fig. C5); in *Notocelia*, it is shieldlike (Fig. C9) or moderately V-shaped (Plate 37:84f, h, i and 85f, g, h); in *Suleima* it is cylindrical (Fig. C10) or U-shaped. In some instances (especially the ovate and shieldlike forms), the lateral margins of the sterigma fold under the lamella postvaginalis to an extent indicated in the illustrations by dashed lines (Figs. C6 and C9). Most of the species treated here exhibit some sclerotization of the ductus bursae between the constriction anterior to the ostium and the juncture with the ductus seminalis. Frequently encountered forms are illustrated in Figs. C11–C15. In all species of *Sonia* except *S. hodgesi*, the sclerotized portion is patchlike (Fig. C13) and has dentate microtrichia on its inner surface. All Nearctic *Epiblema* except *E. insidiosana* and *E. exacerbatricana* lack such microtrichia. In *Notocelia*, the ductus bursae is weakly sclerotized from the ostium to the ductus seminalis (often with a more strongly sclerotized ringlike or patchlike section near the ductus seminalis), and much of its inner surface is covered with microtrichia (Fig. C15). Females of *Suleima* have an elongate semi-helical sclerotized band extending from the ductus seminalis nearly to the constriction anterior to the ostium (Fig. C14).

### Immature Stages

Not much is known about the morphology of the immatures of Nearctic Eucosmini. MacKay (1959) provided the most comprehensive larval descriptions for North American taxa, including 17 species treated here. She arranged species into groups based on similarities in larval morphology. Although these groups do not correspond exactly with current generic concepts, four of them approximate the four genera covered in this volume: *Epiblema* with Type 1 male genitalia are represented by "*Epiblema* Hübner, Group 2;" *Epiblema* with Type 2 male genitalia and *Sonia* are represented by "*Sonia* Heinrich, *Epiblema* Hübner, Group 1;" *Suleima* are represented in "*Suleima* Heinrich, *Thiodia* Hübner, Group 2;" and *Notocelia* are represented by "*Epiblema* Hübner, Group 3." In most cases, MacKay (1959) examined few individuals of each species, so her conclusions regarding morphological characters and potential intraspecific variation are preliminary at best.

Swatschek (1958) provided descriptions of the larvae of Palearctic species of *Epiblema* and *Notocelia*, treating them as separate genera. He used characters similar to those employed by MacKay (e.g., crotchet arrangement) to distinguish between the two genera. *Sonia* and *Suleima*, being strictly Nearctic taxa, were not treated in his monograph.

Larval characters listed under the generic accounts in this volume are based on MacKay's (1959) keys, descriptions for her specific groups, and descriptions of Eucosmini larvae by Brown (1987) and Horak (1998). Information on eggs of *E. strenuana* (representing Eucosmini) is summarized from Peterson (1965) and Passoa (2008). Information on pupae of *E. strenuana* and *E. scudderiana* (representing Eucosmini) is summarized from Moscher (1916) and Horak (1998).

### Examination and Illustration Techniques

Adults were examined with a Leica MZ95 stereomicroscope equipped with an ocular micrometer; genitalia were examined with a Leica DME compound microscope. Adult photographs were taken with a Canon EOS digital SLR and edited using Adobe Photoshop. Lighting was provided by an Aristo DA-10 light box with a T55 (5500K) "daylight" bulb to maintain color correctness or with a Canon 430EX II or MR-14EX flash, in which case color balance was manually adjusted in Adobe Photoshop. Some photos are combinations of several images stacked with Zerene Stacker. Several photographs are provided for most species to illustrate different forewing phenotypes. Genitalia drawings were based on pencil tracings of projected images produced with a Ken-A-Vision X1000-1 microprojector. They were refined during examination of the genitalia with a compound microscope and then inked with various types of pens, including KOH-I-NOOR Rapidograph (4×0 and 3×0 nibs), Micron (003–02 tips), and COPIC (0.03 tip). Genitalia drawings were scanned with an Epson Perfection 2400 Photo scanner and compiled into plates using Adobe Photoshop. Multiple drawings are provided for most species in order to illustrate intraspecific variation in features such as valva, uncus, and sterigma shape.

The adult images were cropped and sized in a uniform manner to fit in the rectangular 5×3 arrays that comprise the adult plates. They do not illustrate size differences between species. Size information for a particular species is presented in the caption heading on the page facing the associated plate, and the data for all the species are tabulated in Appendix A. The genitalia drawings were sized to fit allotted spaces in the genitalia plates and do not conform to any standard scale. They are, therefore, not comparable as to size, either within a given species or between different species. Many illustrations were flipped horizontally in Adobe Photoshop, so what appears on the page to be a right forewing or right valva may in fact be the left such item on the insect. These adjustments were applied to some of the type specimens illustrated here, but in most cases we did not alter photos of primary types.

## Miscellaneous

Plant names follow the most recent taxonomy listed in the United States Department of Agriculture (USDA) Natural Resources Conservation Service (NRCS) *PLANTS Database* (plants.usda.gov). This database was last accessed in April, 2023.

Many North American species of *Epiblema*, *Sonia*, *Suleima*, and *Notocelia* have not been sufficiently sampled to yield accurate determinations of their geographic distributions. The comments on the range of a particular species are based on selected specimens from institutional collections, specimens from various private collections, and our own field experience. The result is, at best, a crude approximation to the actual range of the insect.

Many of the species discussed here were described by authors such as Clemens (1860, 1865), Walsingham (1879, 1884, 1885), and Kearfott (1905, 1907, 1908, 1910) prior to the practice of designating a unique name-bearing specimen (the holotype) for each new taxon. Those early descriptions were based on series of one or more specimens (syntypes), and most of those specimens eventually were deposited in various institutional collections. The task of stabilizing those names by selecting a name-bearing syntype (the lectotype) for each species remains a work in progress. Heinrich (1923) provided some guidance in this regard, but his comments about the "types" often were not sufficiently detailed to identify unique specimens. He did not have access to the Walsingham material (in the BMNH), so his remarks about those species were based on Walsingham's published descriptions, which usually mention the number of syntypes, a date, and the collection locality. In the 1960s, Obraztsov examined the Walsingham syntypes and selected specimens to serve as lectotypes, but he did not live to publish those designations. His photographs (35 mm slides) of the lectotypes and their genitalia were deposited in the AMNH and are currently on long-term loan to the MEM. Heinrich (1923) did examine Kearfott's material, most of which is in the AMNH, and for each of those species he selected a syntype (which he called the "type") and reported its repository and collection site. There is often more than one syntype in the AMNH satisfying Heinrich's specifications, and in those cases, the identity of the "type" remained ambiguous. Frequently, Heinrich illustrated the genitalia of the "type," and the data on the corresponding slide together with a green slide label attached to the specimen served to identify a unique name-bearing specimen. Klots (1942) published a catalogue of the microlepidoptera type material in the AMNH in which he reported lectotypes for numerous Kearfott species, crediting Heinrich (1923) with the designations. Most of those specimens bear a bluish-green "LECTOTYPE" label that we assume was attached by Klots. In the cases in which Heinrich's (1923) remarks are ambiguous, that label presumably represents Klots' interpretation of Heinrich's intended "type." Type fixation for the Clemens species proved to be especially challenging due to difficulty in locating the syntypes. The efforts of various authors to find and designate appropriate lectotypes are recounted in Darlington (1947) and Miller (1973). In the species accounts below we designate lectotypes for 10 taxa whose names have not yet been stabilized.

The following abbreviations are used for the various types reported here: HTP = holotype, ALTP = allotype, PTP = paratype, LTP = lectotype, PLTP = paralectotype, CoTP = cotype.

# Species Accounts

Each species is assigned a number (1, 2, 3, etc.) in the order of its appearance in the text. The species number also serves to identify the adult images and genitalia illustrations associated with that taxon. Intraspecific variation in forewing appearance and genitalia are illustrated with multiple images and multiple drawings, respectively. Multiple images for a particular species are indicated by appending lower case letters to the species number (1a, 1b, 1c, etc.). Two or more genitalia drawings based on a single specimen are assigned a single letter; different letters indicate different specimens. We utilize 87 numbers, of which 83 are associated with validly recognized names. The other four refer to specimens of uncertain taxonomic status or complexes thereof.

The species accounts are arranged in sections according to genus: *Epiblema* (species 1–52), *Sonia* (species 53–69), *Suleima* (species 70–81), and *Notocelia* (species 82–87). *Epiblema* is further subdivided into two subsections corresponding to two distinctive forms of the Tegumen + Appendices (Type 1: species 1–40; Type 2: species 41–52). In each genus/section, most species are assigned to groups based on similarity in forewing appearance and/or genitalia, with the name of the earliest described group member providing the group title.

Preceding each species account is a list of literature references selected to document synonymies, previous diagnostic treatments, historical nomenclature usage, and changes in taxonomic status. It is not intended to be a complete bibliography.

## *Epiblema* Hübner

*Epiblema* Hübner, [1825] 1816, Verz. bekannter Schmett. 375. Type species: *Phalaena* (*Tinea*) *foenella* Linnaeus, 1758 [subsequent designation by Fernald 1908].
*Cacochroea* Lederer, 1859, Wien. ent. Monatschr. 3: 331. Type species: *Paedisca grandaevana* [Lienig and] Zeller, 1846.
*Epiblemma* Hübner, [1825] 1816 Verz. bekannter Schmett. 63 [misspelling of *Epiblema*].
*Euryptychia* Clemens, 1865, Proc. Ent. Soc. Philad. 5: 140. Type species: *Euryptychia saligneana* Clemens, 1865.
*Eurytychia* Heinrich, 1923, Bull. U.S. natn. Mus. 123: 137 [misspelling of *Euryptychia*].
*Monosphragis* Clemens, 1860, Proc. Acad. Nat. Sci. Philad. 12: 354. Type species: *Monosphragis otiosana* Clemens, 1860.

**Diagnosis.** The North American eucosmine genera that historically have been confused with *Epiblema* are *Eucosma* Hübner, *Pelochrista* Lederer, *Suleima* Heinrich, *Sonia* Heinrich, and *Notocelia* Hübner. *Epiblema* differs from *Eucosma*, *Pelochrista*, and *Suleima* in the presence of a rudimentary clasper on the medial surface of the male valva but agrees in that regard with *Sonia* and *Notocelia*. All but two Nearctic *Epiblema* have a prominent Type 1 or Type 2 clasper, the exceptions being *E. tandana* (with a bulgelike Type 3 clasper) and *E. fulminana* (with a weakly expressed and sometimes obsolete Type 4 clasper). In *Sonia*, the clasper is always Type 4. In *Notocelia*, it is Type 1 and often is accompanied by a small auxiliary clasper (Type 5). *Epiblema* differs from *Sonia* and *Suleima* in forewing venation ($R_5$ and $R_4$ are separated vs. united, Fig. A1a vs. A1b, c) but agrees with *Eucosma*, *Pelochrista*, and *Notocelia* in that respect. *Epiblema* differs from *Suleima* in the presence vs. absence of a costal fold at the base of the male forewing, and it differs from *Notocelia* in the absence vs. presence of two non-deciduous cornuti in the vesica.

**Description.** *Forewing* (Fig. A4). Mean FWL: 5.1–11.9 mm, average = 7.2 mm; AR: 2.32–3.18, average = 2.74; CFR: 0.25–0.65, average = 0.39; male with costal fold (except *E. alba*); costa weakly arched; termen straight to weakly concave; venation as in Fig. A1a; forewing pattern usually semifasciate (Fig. A4); interfascial spot usually discernable, varying from white and conspicuous to concolorous with ground, often with weakly contrasting transverse reticulations or a few dark spots on the inner margin; fasciate markings often poorly defined, sometimes weakly represented by slightly darker shades of ground color; subbasal fascia usually subsumed in basal patch; median fascia often fragmented; postmedian fascia narrow, arching over apical margin of ocellus, extending to distal margin, sometimes expanding into conspicuous spot anterior to ocellus; ocellus obscure to conspicuous, with lustrous gray or pinkish bars on lateral margins and 2–4 black dots or dashes on central field; costal strigulae obscure to conspicuous, varying from white to concolorous with ground, delimited by dark marks or dashes, often accompanied by thin striae extending toward termen.

*Hindwing*. Venation as in Fig. A1b; color uniform in shades of gray or brown.

*Male genitalia*. Tegumen + appendages Type 1 (Figs. B1, B2) or Type 2 (Fig. B3); socii fingerlike; phallus stout, gradually attenuating distally, with base closely surrounded by anellus; vesica with cluster of deciduous cornuti (average number per species: 2–43); valva with costal margin concave, saccular corner angulate, SA: 106°–166°, neck defined by shallow to moderate ventral emargination, NR: 0.35–0.84; medial surface of valva with conspicuous clasper at or near distal margin of basal excavation (Figs. B1, B11, B12); cucullus with dorsal lobe moderately to strongly developed, apex rounded, distal margin convexly curved to nearly straight, ventral lobe weakly developed, anal angle rounded to triangular; setae on medial surface of cucullus coarse along ventral three-fourths of distal margin, usually grading to hairlike near costal margin and apex.

*Female genitalia*. Papillae anales in four primary forms (Figs. C1–C4); pads of papillae anales long and narrow, usually densely microtrichiate and moderately setose; sterigma ovate (Fig. C6), U-shaped (Fig. C7), or V-shaped (Fig. C8) and nearly always microtrichiate; lamella postvaginalis with shallow central trough flanked laterally by a few hairlike setae; lamella antevaginalis ringlike; posterior margin of sternum 7 concavely emarginated to one-fourth to one-half length of sterigma; sclerotization of ductus bursae ringlike (Fig. C11), patchlike (Fig. C12), or absent, rarely with microtrichia on inner surface (present only in *E. insidiosana* and *E. exacerbatricana*); corpus bursae usually with two signa, occasionally with none.

**Immature stages.** The immature stages of most species are unknown or poorly documented. In general, eggs and pupae of *Epiblema* are typical Eucosmini, and the larvae are typical of other internal feeders in the tribe that lack an anal comb, such as *Eucosma* and *Rhyacionia* (Brown 1987; Horak 1998). MacKay (1959) separated larvae of *Epiblema* and *Sonia* from those of *Suleima* by the structure of the spinneret (rounded or somewhat broadened versus blunt or indented).

*Egg*. Oval, scalelike ("flat"), watery white, and variably translucent; chorion bright, with a rough surface possessing irregular, prominent, rounded elevations and depressions; deposited in small clusters of 2–5 eggs.

*Larva*. Stemmata most often spaced far apart (III and IV sometimes closer together than others), usually flat and irregular in shape; head in dorsal view with stemmatal area strongly rounded; vertical angle more often obtuse or a right angle than acute; spinneret with rounded tip; labial palpus with a distinctly stout basal segment; L-group on T1 trisetose, never enlarged or extending under the spiracle; setal pinacula most often large; pinacula of D2s on A8 always less than their own diameter apart; SD1 on A8 anterodorsal, anterior, or slightly anteroventral to spiracle; SD2 often included on the same pinaculum with SD1 on A1–8; L1 and L2 setae more or less directly ventral to spiracle on A2–8; SV-group setal counts on A1, 2, 7, 8, 9 are 2(3):2(3):2:2:2(1); crotchets uniordinal or unevenly biordinal but never distinctly biordinal; anal comb absent.

*Pupa*. A2–7 with two lateral rows of spines; A8–10 usually with one row of spines; maxillary palpus extending into the proximolateral angles of the maxilla; maxilla about 0.4 × FWL; labial palpus usually 0.5 × length of the maxilla; cauda with 4–8 flat hooked setae; anal rise with 1–2 or rarely 3 setae on each side; cremaster absent.

**Biology.** The life histories of several North American *Epiblema* have been described in detail, likely as a result of the conspicuous gall-inducing habits of those species. The summary provided here is from Thompson and Guelph (1926), Bottimer (1926), Decker (1932), Putnam (1942), Miller (1976), and McClay (1987). Larvae are stem or root borers in Asteraceae, including *Ambrosia* L. (ragweed), *Anaphalis* DC. (pearly everlasting), *Artemisia* L. (sagebrush), *Baccharis* L., *Bidens* L. (beggarticks), *Euthamia* Nutt. ex Cass. (goldentop), *Rudbeckia* L. (coneflower), *Parthenium* L. (feverfew), *Solidago* L. (goldenrod), *Xanthium* L. (cocklebur), and other genera. Females lay eggs on the blossoms or upper leaves, and early instar larvae mine leaves or feed in the blossoms, buds, or seed heads. Third instar larvae bore into the stem or drop to the base of the plant and bore into the stem and/or roots. Feeding within the stem can cause girdling and death of the upper part of the plant; concentrated feeding in a short area inside the stem causes the stem to swell, forming a gall. Gall-inducing species hollow out the stem below the gall, partition this space off from the main gall, and move into the lower chamber to overwinter. Root-boring species overwinter in a hollow cavity in the roots. Pupation occurs in the overwintering chamber in the spring. Most species complete one or two generations per year.

## **Section 1.** Type 1 Tegumen + Appendages

### The *strenuana* group (species 1–6)

The *strenuana* group is composed of five described species and an unnamed morphospecies. Much of its taxonomic history has revolved around the question of whether its two most widespread members, *E. strenuana* (Walker, 1863) and *E. minutana* (Kearfott, 1905), are truly distinct species. Kearfott (1905) distinguished *E. minutana* from *E. strenuana* by size (smaller), color (grayer), forewing geometry (narrower), and unspecified differences in the ocellus and the costal strigulae. Heinrich (1923) noted variation in forewing appearance and valva shape amongst specimens referred to these two taxa but found no criteria that he thought would unambiguously partition that material into two or more distinct species, so he treated *minutana* as a synonym of *strenuana*. In an unpublished Master's thesis, R. L. Brown (1973) concluded that the two taxa are distinct based on forewing color, markings, socius size, sterigma shape, and signa shape. Shortly thereafter, Blanchard (1979) resurrected *E. minutana*, agreeing with Kearfott's assessment of forewing geometry and independently pointing out differences in sterigma shape that had been observed by Brown (1973). He also considered the base diameter of the signa in the corpus bursae to be larger in *E. strenuana* than in *E. minutana*. Miller and Pogue (1984) tested these and other putative differences with an allometric analysis. They concluded that the two species could not be separated based on what previously had been regarded as diagnostic character states, but they proposed no taxonomic changes. Shortly before this last study, and possibly unknown to its authors at the time they submitted their results, Powell (1983), without explanation, listed *minutana* as a junior synonym of *strenuana*. Some subsequent authors (Brown 2005; Powell and Opler 2009; Gilligan et al. 2018) adopted Powell's position; others (Miller 1987; Gilligan et al. 2008) viewed the specimens previously determined as *E. strenuana* or *E. minutana* as comprising the *E. strenuana* "complex."

In a recent study examining taxonomic issues in the *strenuana-minutana* complex as related to biocontrol, Gilligan et al. (2020b) analyzed COI DNA barcodes from 79 specimens and found that they segregate into two clusters that are supported by differences in several morphological characters. The specimens in the *strenuana* cluster differ from those in the *minutana* cluster in AR (noted by Kearfott and Blanchard), sterigma shape (noted by Blanchard and Brown), and the size and shape of the socius (noted by Brown). Prompted by these molecular and morphological differences, Gilligan et al. (2020b) elevated *E. minutana* to species status. These authors also noted a substantial amount of variation in the *E. minutana* cluster that is not present in the *E. strenuana* cluster, with some of the *minutana* individuals from Florida and California forming separate subgroups. This might be explained as population level differences, or it could indicate separate taxa. Lacking consistent morphological distinctions associated with the molecular subgroups, we tentatively refer all this material to *E. minutana*. Group characteristics include:

*Forewing*. Mean FWL: 5.3–7.1 mm; AR: 2.80–3.18; CFR: 0.36–0.43; predominant forewing color gray, brown, or orange brown; fasciate markings often weakly defined; basal patch concolorous with median fascia; interfascial spot often obscure, pale when clearly expressed, otherwise barely distinguishable from adjacent subbasal and postmedian areas; ocellus conspicuous (except in *E. ochraceana*), with grayish or pinkish bars on lateral margins and two black marks on a whitish central field; costal strigulae whitish.

*Male genitalia*. Uncus well developed, clearly differentiated from dorsolateral shoulders of tegumen, with apex rounded; socius fingerlike; vesica with 6–34 cornuti; valva with clasper Type 1, prominent basal process near ventrodistal corner of basal excavation, neck moderately wide, NR: 0.57–0.73, saccular corner angulate, SA: 140°–158°; cucullus ovate (length-to-width ratio: 1.91–2.13), with dorsal lobe strongly developed, ventral lobe weakly developed, anal angle rounded.

*Female genitalia*. Papillae anales Type 1; sterigma ovate or U-shaped, with lateral margins folded under lamella postvaginalis (as indicated by dashed lines in illustrations); lamella antevaginalis ringlike, with posterior extremities attenuating into lateral margins of lamella postvaginalis; posterior margin of sternum 7 concavely emarginated to approximately one-half length of sterigma; corpus bursae with two signa of nearly equal size located opposite one another in posterior one-half of bursa.

## 1. ***Epiblema strenuana*** (Walker, 1863)

(Plate A, 1a–o; Plate 1, 1a–h)

*Grapholita strenuana* Walker 1863:383.
*Grapholita exvagana* Walker 1863:383.
*Steganoptycha flavocellana* Clemens 1865:138; Miller 1973:214.
*Grapholita subversana* Zeller 1875:318.
*Paedisca strenuana*: Walsingham 1879:52; Fernald 1882:40; Walsingham 1884:140.
*Eucosma strenuana*: Fernald 1903:458; Barnes and McDunnough 1917:170.
*Epiblema strenuana*: Heinrich 1923:140, figs. 257, 258; McDunnough 1939:48; Darlington 1947:95; MacKay 1959: 31; Stegmaier 1971:260; Miller 1973:214; Powell 1983:35; Miller and Pogue 1984:227; McClay 1987:23; Miller 1987:56; Brown 2005:286; Gilligan et al. 2008:121; Horak 2008:310; Powell and Opler 2009:135; Gilligan et al. 2020b:352.

**Types**. *Grapholita strenuana*. Lectotype (Plate A:1a, designated by Miller and Pogue 1984). ♂, North America, Carter Collection, BMNH(E) 819923, slide 5737, BMNH. *Grapholita exvagana*. Lectotype (Plate A:1b, designated by Miller and Pogue 1984). ♂, North America, Carter Collection, BMNH(E) 819924, BMNH (hindwings and abdomen missing). *Steganoptycha flavocellana*. Lectotype (designated by Darlington 1947). ♂, Type No. 7214, ANSP (abdomen missing). *Grapholita subversana*. Syntypes?

Walker's descriptions of *G. strenuana* and *G. exvagana* appear on the same page of his 1863 list of Tortricites & Tineites in the BMNH. In neither case did he mention the number of specimens examined. The lectotypes (indicated above) were formally designated by Miller and Pogue (1984), who credited N. S. Obraztsov with their selection. Each of those specimens (Plate A:1a–b) has a moderately prominent white interfascial spot. The number of specimens supporting the description of *S. flavocellana* is unknown. Darlington (1947) reported the above indicated specimen as the type, attributing the designation to Heinrich (1923). Miller (1973: fig. 14) presented an image of that specimen, referred to it as the lectotype, and credited Darlington with the designation. The image shows the interfascial spot to be obscure. Zeller (1875) seems to have examined three syntypes for *Grapholita subversana*, two females collected by Burgess on 27 June and 3 July in Massachusetts and one male collected by Boll in Texas. He reported retaining a female in his collection and returning a male and a female to the MCZ, but Miller and Hodges (1990) did not mention this species in their list of microlepidoptera types in the MCZ. Brown (2005) reported a male holotype from Texas in the MCZ but then indicated that it is lost. All the above synonymies date back to Fernald (1882).

*Epiblema strenuana* is a brownish (Plate A:1a–b, n–o) to grayish (Plate A:1g, j–k) species of variable size (FWL: 4.0–9.0 mm, mean = 7.1) and medium forewing geometry (AR = 2.80). The interfascial spot extends from the inner margin to the radius and varies in color from whitish (Plate A:1a–b, m–o) to shades of bronze or gray (Plate A:1d, g, j), in the latter case being weakly distinguishable from the adjacent subbasal and median areas by its lack of white-tipped scales. Its proximal margin is sometimes indicated by a thin pale line (Plate A:1h). The ocellus is prominent, with its white central field flanked laterally by grayish transverse bands and bisected by a longitudinal slightly-curved blackish dash. Its proximal margin is edged with a fragmented black line. The costal strigulae on the distal one-half of the wing are white or gray, usually inconspicuous (except for strigula 9), with associated gray striae and orange-brown streaks extending toward the termen. The individuals depicted in Plate A:1n–o are representative of a few specimens from California with especially prominent interfascial spots and costal strigulae, the latter separated by brown rather than orange-brown streaks. In other respects, they resemble typical *E. strenuana*.

In males, the socius is long and narrow, with lateral margins nearly parallel; the vesica has 23–34 cornuti; NR = 0.57; and SA = 140°. In females, the sterigma is U-shaped and elongate (length-to-width ratio = 1.95), the posterior margin of sternum 7 is semicircular, and the ductus bursae has a twistlike sclerotized contortion near the juncture with the ductus seminalis.

Heinrich (1923) stated that the larva is a stem borer on *Ambrosia artemisiifolia* L. (annual ragweed) and *Ambrosia trifida* L. (great ragweed) but did not provide a source for that information. Stegmaier (1971) reported rearing specimens in Florida from larvae feeding in fusiform galls in the lateral branches of *A. artemisiifolia*. He also reported rearing a similar species of *Epiblema* (later determined as *E. minutana* by Miller and Pogue 1984) from larvae boring in the stems of *A. artemisiifolia*. Other larval hosts have been mentioned in the literature, including *Parthenium hysterophorus* L. (Santa Maria feverfew), *P. confertum* A. Gray (Gray's feverfew), and *P. argentatum* A. Gray (guayule) (McClay 1987); *Xanthium* L. (cocklebur) (MacKay 1959; Miller 1987; Powell and Opler 2009); and *Chenopodium* L. (goosefoot) (Brown 1973; Miller 1987). This species has been introduced to Australia and China as a biocontrol agent for *Parthenium* L. (feverfew), *Ambrosia* L. (ragweed), and other invasive weeds in those countries (Gilligan et al. 2020b).

*Epiblema strenuana* is broadly distributed in eastern United States. Its full range probably mimics that of *A. artemisiifolia*, which has been recorded in all the contiguous 48 states and in all the southern Canadian provinces. Literature records must be viewed with caution due to the long-standing confusion between *E. strenuana* and *E. minutana*. Most of the material in our study sample was collected in the region extending from the Great Plains east to the Atlantic Coast and from Minnesota, Ohio, and Maryland south to the Gulf Coast. We also examined several specimens from central and southern California. In the Midwest, *E. strenuana* has two primary flights, one in late June, the other from mid-August to mid-September.

## 2. ***Epiblema minutana*** (Kearfott, 1905)

(Plate B, 2a–o; Plate 1, 2a–i)

*Eucosma minutana* Kearfott 1905:356; Barnes and McDunnough 1917:170.
*Epiblema minutana*: Blanchard 1979:179; Miller and Pogue 1984:227; Gilligan et al. 2020b:354.
*Eucosma antaxia* Meyrick 1920:344, unnecessary replacement name for *minutana*.
*Epiblema strenuana* (not Walker): Heinrich 1923:140; McDunnough 1939:48; Powell 1983:35; Brown 2005:286; Powell and Opler 2006:192; Gilligan et al. 2008:121; Powell and Opler 2009:135.

**Lectotype** (designated by Blanchard 1979). ♂, New Jersey, Essex County, Montclair, W. D. Kearfott, July 1908, slide 24505, USNM.

Kearfott (1905) mentioned a series of about 40 specimens from Tryon, North Carolina; Cincinnati, Ohio; New Brighton, Pennsylvania; Plummers Island, Maryland; Belvidere, Illinois; Smith County, Tennessee; Anglesea, New Jersey; and Essex County Park, New Jersey. Klots (1942) credited Heinrich (1923) with the designation of a lectotype from Essex County Park, New Jersey but did not agree with Heinrich's statement that the AMNH was the repository. Blanchard (1979) settled the matter by formally designating as lectotype the specimen in the USNM indicated above.

Typical *E. minutana* is a dark gray species (Plate B:1a–h) whose forewing appearance differs from that of *E. strenuana* in its sharply defined, prominent, white costal strigulae and its lack of orange-brown scaling in the subcostal area near the apex. It varies in size (FWL: 4.3–7.9 mm) much like *E. strenuana* (FWL: 4.0–9.0 mm) but is, on average, smaller than the latter species (mean FWL = 6.0 vs. 7.1 mm). Previous authors have noted, and we concur, that its forewing is narrower than that of *E. strenuana* (AR = 3.12 vs. 2.80). The interfascial spot usually is a pale shade of gray (Plate B:2c–d) that weakly contrasts with the adjacent subbasal and median areas, but in some individuals, it is barely detectable (Plate B:2a). The ocellus resembles that of *E. strenuana* except that its lateral margins often have a faint pinkish tint.

Typical *E. minutana* resembles *E. strenuana* in male genitalia except the socius is shorter. In many instances it is triangular (Plate 1:2a–b), attenuating from a broad base to a narrowly rounded apex, but we examined some specimens in which the lateral margins are parallel (as in *E. strenuana*). In females, the sterigma is ovate (instead of U-shaped) and less elongate (length-to-width ratio = 1.55 vs. 1.95).

The typical phenotype (Plate B:2a–f) is broadly distributed in eastern United States, where the larval host is presumed to be *Ambrosia artemisiifolia* L. (annual ragweed). It also occurs in central California (Plate B:2g–h), where it has been reared by J. Powell in Contra Costa County from *Ambrosia chamissonis* (Less.) Greene (silver bur ragweed) and *A. psilostachya* DC. (Cuman ragweed). The former plant ranges along the Pacific coast from southern California to Alaska; the range of the latter is transcontinental.

In addition to the typical form, there is a paler phenotype (Plate B:2i–l) with *minutana*-like forewing pattern and genitalia that is common in southern California and occurs infrequently in the East. These specimens are somewhat larger than the typical phenotype (mean FWL = 6.8 vs. 6.0 mm) but agree with it in forewing geometry (AR = 3.18 vs. 3.12). The EME has several examples of this phenotype that were reared from *A. chamissonis* by Powell in Contra Costa and Santa Barbara counties, California. Similar looking specimens from San Diego County, California were represented in the phylogenetic tree generated by Gilligan et al. (2020b: fig. 1), where they cluster with typical *E. minutana*. Their sequences show some minor consistent differences that might indicate additional taxa, but we found no morphological differences that would clearly distinguish them from typical *E. minutana*. Lacking a more comprehensive DNA analysis, we tentatively refer these pale individuals to *E. minutana*, but we report their numerical data (in Appendix A) separately from those of the typical phenotype.

The images in Plate B:2m–n show the extremes in forewing appearance in a population of *E. minutana* on Block Island, Washington County, Rhode Island. They, along with various intermediate phenotypes, comprise a series of 32 specimens collected on 6 August 2020 by A. S. Hunt at a single location in salt marsh habitat. Exemplars are deposited in the CUIC and the USNM. The nearly white specimen illustrated in Plate B:2o was collected on 1 April 1978 at the town of Rice in San Bernardino County, California. It appears to be a representative of an especially pale race of *E. minutana* from the Mojave Desert.

As discussed in Gilligan et al. (2020b), *E. minutana* (identified as *E. strenuana*) was accidentally introduced to Israel in 2008, where it has been reported feeding on *Ambrosia confertiflora* DC. (weakleaf bur ragweed) and *Ambrosia tenuifolia* Spreng. (slimleaf bur ragweed).

## 3. ***Epiblema luctuosissima*** Blanchard, 1984
(Plate C, 3a–h; Plate 2, 3a–f)

*Epiblema luctuosissima* Blanchard 1984:245; Miller 1987:57; Brown 2005:284; Gilligan et al. 2008:120.
*Epiblema luctuosana* Blanchard 1979:184 [preoccupied].

**Holotype.** ♂, Texas, Nueces County, North Padre Island, 6 April 1978, slide 69757, USNM.

This species originally was described as *E. luctuosana*, a name subsequently found to be preoccupied by the European *Ephippiphora luctuosana* Duponchel 1835, a synonym of *Epiblema scutulanum* Denis and Schiffenmüller 1775 (Blanchard 1984). The type series includes 51 paratypes (29 ♂, 22 ♀) from various localities in Texas.

*Epiblema luctuosissima* is similar in size to typical *E. minutana* (mean FWL = 6.1 vs. 6.0 mm) but has a broader forewing (AR = 2.82 vs. 3.15). The predominant forewing color is gray, sometimes lightly tinted with brown (Plate C:3d), usually with a fine salt-and-pepper appearance produced by numerous white-tipped scales (Plate C:3b–c). In typical specimens (Plate C:3a–c), the interfascial spot is whitish, extends from the inner margin into the discal cell, and may be suffused with pale pink. These features are usually sufficient to separate specimens of *E. luctuosissima* from those of *E. minutana*. Infrequently, the interfascial spot is nearly concolorous with the basal patch and the median fascia (Plate C:3d, g–h), in which case an examination of the genitalia may be necessary to obtain an accurate determination. The image in Plate C:3h depicts a particularly melanic male from Louisiana that we tentatively refer to this taxon based on similarity in genitalia. In all phenotypes, the ocellus is conspicuous, with pale-pink transverse bars on the lateral margins. Its narrow white central field has two longitudinal black dashes, one located centrally, the other near the anterior extremity. The white costal strigulae are sharply delimited by blackish dashes and have associated gray striae that extend toward the termen.

In males, the socius is moderately short, nearly uniform in width, and has an evenly rounded apex. Compared to typical *E. minutana*, the vesica has fewer cornuti (9–16, mean = 11, n = 7 vs. 14–25, mean = 20, n = 11), the valval neck is broader (NR = 0.73 vs. 0.64), and the cucullus is slightly more elongate (length-to-width ratio = 2.16 vs. 1.99). The saccular corner of the valva tends to be weakly defined, with SA = 158°, and the basal process is only moderately developed. In females, the sterigma is substantially ovate, with the posterior margin weakly emarginated. Compared to typical *E. minutana*, it is slightly more elongate (length-to-width ratio = 1.66 vs. 1.55). The ductus bursae and corpus bursae are conspicuously long and narrow, the former lacking sclerotization, the latter with two small signa of nearly equal size. These last features distinguish *E. luctuosissima* from all other members of the group except possibly *E. ochraceana*, for which the female is unknown.

We examined 40 specimens (26 ♂, 14 ♀). Most were collected in southeastern Colorado, southwestern Kansas, or Texas, but we found three records from Putnam County, Illinois and one from Boone County, Kentucky. Miller (1987) reported *E. luctuosissima* from Michigan and Minnesota. We have not seen specimens that support those last records and, therefore, cannot confirm the determinations, but Miller's illustrations of the valva and sterigma conform to ours. Adult capture dates range from 6 April to 27 November, with most records from June–August.

## 4. ***Epiblema adusta*** Wright and Gilligan, new species

(Plate C, 4a–d; Plate 2, 4a–e)

**Diagnosis.** *Epiblema adusta* is the smallest member of the *strenuana* group (mean FWL = 5.3 vs. 5.5–7.1 mm). It is the only member of the group with a medium-brown forewing. It resembles *E. ochraceana* in size (mean FWL= 5.3 vs. 5.5 mm), in the expression of the costal strigulae (subdued), and in the appearance of the interfascial spot (weakly contrasting), but it lacks the distinctive orange tint of the latter species. The male genitalia resemble those of *E. strenuana* and *E. ochraceana* in that the socius is long and narrow, with lateral margins parallel. The distribution of hairlike setae on the socius is conspicuously sparse in *E. adusta* and *E. ochraceana* but moderately dense in *E. strenuana*. Of the three taxa, *E. ochraceana* is the only one in which the basoventral margin of the cucullus projects somewhat onto the medial surface of the neck. Females of *E. adjusta* differ from those of *E. strenuana* in the shape of the sterigma (length-to-width ratio = 1.49 vs. 1.95) and in the sclerotization of the ductus bursae (ringlike but lacking a twistlike contortion). The female of *E. ochraceana* is unknown.

**Description.** *Head.* Labial palpus with medial surface white, lateral surface pale tan; frons whitish; vertex tan; antenna with dorsal surface brown, ventral surface whitish. *Thorax.* Dorsal surface pale brown; fore- and mid-leg with anterior surfaces pale brown, posterior surfaces pale tan; hind-leg uniformly pale tan; tarsus with alternating tan and brown annulations. Forewing: ♂ FWL 4.4–5.6 mm (mean = 5.0, n = 24), AR = 2.91, CFR = 0.39; ♀ FWL 5.3–6.3 mm (mean = 5.8, n = 11), AR = 2.77; dorsal surface brown with extensive blackish-brown suffusion; cubitus and $CuA_2$ highlighted by blackish-brown line from base to tornus; interfascial spot tan, usually extending from inner margin to cubital line, sometimes only weakly discernable; subbasal fascia obscure to undetectable; median fascia represented by blackish-brown quadrate pretornal mark; postmedian fascia dark brown, thin, and often interrupted near costa; ocellus pinkish, usually with brown longitudinal central dash and black spot at anterior extremity; costal strigulae white, separated by thin brown lines extending toward termen; fringe pale brown. Hindwing: Brownish gray with pale fringe. *Abdomen.* Male genitalia (n = 6): Socius long, slender, and sparsely setose, with lateral margins parallel; vesica with 12–24 cornuti; NR = 0.64; SA = 157°; otherwise, typical of the group. Female genitalia (n = 4): Papillae anales Type 1, with pads moderately setose; apophyses posteriores slightly shorter than apophyses anteriores; sterigma U-shaped; sclerotization of ductus bursae ringlike; corpus bursae with two moderately small signa of equal size.

**Holotype** (Plate C:4a). ♀, Mississippi, Jackson County, Grand Bay Savanna, 30.3433° N, 88.4192° W, R. L. Brown, S. M. Lee, and E. L. Martinez, 30 April 2004, slide DJW 4012, USNM.

**Paratypes** (24 ♂, 10 ♀). ALABAMA. Baldwin County, Plash Island, end of Highway 6, R. L. and B. B. Brown, 16–19 June 1984 (1 ♂, slide DJW 4200; 3 ♀, slides DJW 4016, 4017). LOUISIANA. Cameron Parish, Hackberry Prairie, D. M. Pollock, 29 March 1993 (1 ♂, slide DJW 4199). MISSISSIPPI. Jackson County, Grand Bay Savanna, 30.3433° N, 88.4192° W, R. L. Brown, S. M. Lee and E. L. Martinez, 30 April 2004 (14 ♂, slides DJW 4011, 4013; 1 ♀); Jackson County, Belle Fontaine Point, R. L. Brown, 21 May 2008 (1 ♂); Jackson County, Gulf Island National Seashore, R. L. Brown and G. Baker, 3–6 June 1984 (1 ♂, 2 ♀). NORTH CAROLINA. Carteret County, Fort Macon State Park, J. B. Sullivan, 16 July 1996 (1 ♂, genitalia in vial), 19 July 1998 (3 ♂, slide DJW 3957), 3 August 1997 (1 ♀), 10 September 1999 (1 ♂); Carteret County, Sunset Point, N. of Beaufort, J. B. Sullivan, 17 July 1999 (1 ♀, slide DJW 3958). TEXAS. Jefferson County, Sabine Pass, E. C. Knudson 11 June 1994 (2 ♀); Nueces County, North Padre Island, A. and M. E. Blanchard, 24 August 1979 (1 ♂, slide DJW 3970). Paratype depositories: MEM, MGCL, USNM.

**Etymology.** The specific epithet is the nominative singular feminine form of the Latin adjective *adustus*, meaning tanned, brown, or swarthy.

**Distribution.** This species has been collected in coastal localities along the Gulf of Mexico from Nueces County, Texas to Baldwin County, Alabama and on the Atlantic Seaboard in Carteret County, North Carolina.

## 5. Unnamed morphospecies

(Plate C, 5; Plate 2, 5a–b)

The EME has five specimens (4 ♂, 1 ♀) collected by J. Powell and P. Rude in Baja California, Mexico that are similar to *E. adusta* in size (mean FWL = 5.6 vs. 5.3 mm) and maculation but differ somewhat in forewing geometry (AR = 3.14 vs. 2.86), forewing color (grayish vs. brown), and subtle aspects of the genitalia. Only one of the five is in good condition. We suspect that they represent a distinct species but prefer not to propose another name at this time. Instead, we include illustrations to facilitate the discovery of additional material (perhaps from southern California) that might provide stronger support for a future description.

## 6. *Epiblema ochraceana* Fernald, 1901

(Plate C, 6a–b; Plate 2, 6a–b)

*Epiblema ochraceana* Fernald 1901:51; Heinrich 1923:144; McDunnough 1939:48; Powell 1983:35; Brown 2005:284.
*Eucosma ochraceana*: Fernald 1903:456; Barnes and McDunnough 1917:169.

**Holotype** (Plate C:6a; Plate 2:6a). ♂, Florida, Palm Beach, slide DJW 3925, USNM.

*Epiblema ochraceana* was described from a single specimen that bears the following four labels: "Type No. 5415 U.S.N.M.," "Collection Dr H G Dyar," "*Epiblema ochraceana* Fern. Type," and "15648." The first and second are printed on red and white paper, respectively; the third and fourth are handwritten on white paper. The type locality was reported by Fernald (1901). Heinrich (1923: fig. 268) illustrated under this name the genitalia of a specimen in the USNM which bears his handwritten slide label, the museum's green slide label (#70744), and a white label with the number "15646." He stated that this specimen also comes from Palm Beach, Florida, and his notes on the slide indicate that it was collected by Dyar and Caudell.

The two aforementioned specimens are the only ones we encountered that we can confidently refer to this taxon. They are small (FWL = 5.5 mm) and uniformly brownish orange. In the holotype (Plate C:6a), the interfascial spot and the ocellus are represented by pale orangish shades that weakly contrast with the rest of the wing; the other specimen (Plate C:6b) is in poor condition, with most of its forewing markings obscured. In both specimens, the male genitalia conform to those of *E. strenuana* in all respects except the number of cornuti (6–8 vs. 23–34), the distribution of hairlike setae on the socius (sparse vs. moderately dense), and the extent to which the basoventral margin of the cucullus extends onto the medial surface of the valval neck (weakly vs. not at all).

## The *sosana* group (species 7–9)

*Epiblema sosana* was described from Utah a little over a century ago. In the meantime, collections in North America have acquired numerous specimens from southwestern United States that tentatively have been referred to that taxon. No study has been undertaken to determine whether all these specimens are conspecific, probably because they have similarities with *E. strenuana* and *E. minutana*, two species with their own long-standing circumscription challenges. Here we propose a division of the "*sosana*" material into three taxa that constitute the *sosana* group: *E. sosana*, from southwestern United States; *E. cimarronensis*, from the southern Great Plains, Arizona, and southern California; and *E. wielgusi*, from Maricopa County, Arizona. Group characteristics include:

*Forewing*. Mean FWL: 7.0–8.5 mm; AR: 3.01–3.08; CFR: 0.36–0.40; predominant forewing color gray; subbasal mark moderately well defined from inner margin to discal cell; interfascial spot variably expressed in shades of white, gray, brown, or pale pink; pretornal mark usually discernable, often moderately prominent; ocellus conspicuous, with pale pink or gray bands on lateral margins and two black marks on narrow white central field; costal strigulae white; termen with narrow salt-and-pepper colored band from tornus to apex.

*Male genitalia*. Uncus clearly differentiated from dorsolateral shoulders of tegumen, with apex rounded; socius moderately short and fingerlike; vesica with 13–31 cornuti; valva with costal margin concave, ventral emargination shallow, neck moderately wide, NR: 0.63–0.76, saccular corner weakly defined, SA: 154°–159°, clasper Type 1, basal process weakly developed; cucullus ovate, with dorsal lobe strongly developed, ventral lobe weakly developed, length-to-width ratio: 1.74–1.88, setae on medial surface moderately coarse near distal margin, grading to hairlike at costal margin.

*Female genitalia*. Papillae anales Type 1, microtrichiate, and moderately setose; sterigma U-shaped or ovate, with lateral margins folded under lamella postvaginalis; lamella antevaginalis ringlike, with posterior extremities merging into lateral margins of lamella postvaginalis; posterior margin of sternum 7 concavely emarginated to approximately one-third length of sterigma; sclerotization of ductus bursae patchlike or ringlike; corpus bursae with two signa of nearly equal size located opposite one another in posterior one-half of bursa.

## 7. ***Epiblema sosana*** (Kearfott, 1907)
(Plate D, 7a–f; Plate 3, 7a–f)

*Eucosma sosana* Kearfott 1907:32; Barnes and McDunnough 1917:170.
*Eucosma pelina* Meyrick 1912:35, unnecessary replacement name for *sosana*.
*Epiblema sosana*: Heinrich 1923:145; McDunnough 1939:48; Powell 1983:35; Brown et al. 1983:225; Brown 2005:285; Powell and Opler 2009:135.

**Lectotype** (designated by Heinrich 1923). ♂, Utah, Stockton, T. Spalding, 2 August, slide CH 14 December 1919, AMNH.

Kearfott (1907) mentioned eight syntypes collected from 23 May to 2 August by Tom Spalding at Stockton, Utah. Heinrich (1923: fig. 269) illustrated the genitalia of a specimen in the AMNH that he referred to as the type, and Klots (1942) reported that specimen as a lectotype, crediting Heinrich with the designation.

*Epiblema sosana* is variable in size (FWL: 5.5–9.1 mm, mean = 7.3) and forewing color. Typical specimens from Utah (Plate D:7a–b) are medium gray; those in Plate D:7c–d are representative of a particularly dark phenotype from San Bernardino County, California; and Plate D:7e–f depict members of a pale desert race from Imperial County, California and Yuma County, Arizona. The basal area of the forewing has transverse blackish-gray reticulations, and the interfascial spot is generously suffused with pale pink. The last feature distinguishes *E. sosana* from the other species in the group.

The male genitalia do not provide any diagnostic character states that clearly separate this species from other members of the group. Females differ from those of *E. cimarronensis* and *E. wielgusi* in the form of the sterigma (U-shaped vs. ovate), the width of the ringlike lamella antevaginalis (more than twice as wide), the degree of proximity of the posterior projections of sternum 7 to the lateral margins of the sterigma (laterally divergent vs. approximate), and the sclerotization of the ductus bursae (small obscure patch vs. ringlike).

The USNM has several specimens (reported by Brown et al. 1983) that were reared by R. Goeden and D. Ricker from *Ambrosia acanthicarpa* Hook (flat spine bur ragweed) in Los Angeles, Riverside, and San Bernardino counties, California. Capture dates for wild caught specimens range from early May to late September. In addition to the material from Utah and southern California, we examined six other specimens that we refer to this taxon: 1 ♂ (ASUHIC) from Yuma County, Arizona, 1 ♀ (INHS) from Colorado Springs, Colorado, 1 ♂ (EME) and 1 ♀ (CMNH) from Dona Ana and Roosevelt counties, respectively, New Mexico, and 1 ♂ (USNM) and 1 ♀ (MGCL) from Culberson and Presidio counties, respectively, Texas.

## 8. *Epiblema cimarronensis* Wright and Gilligan, new species

(Plate D, 8a–i; Plate 3, 8a–d)

**Diagnosis.** *Epiblema cimarronensis* is distinguished from *E. sosana* by the following combination of characteristics: the interfascial spot lacks pinkish suffusion, the cucullus is a little more elongate (length-to-width ratio = 1.88 vs. 1.74), the sterigma is ovate (rather than U-shaped) and slightly more elongate (length-to-width ratio = 1.39 vs. 1.21), the lamella antevaginalis is narrower by half, and the sclerotization of the ductus bursae is prominent and ringlike vs. obscure and patchlike. *Epiblema cimarronensis* is the only member of the group that is sexually dimorphic with respect to size: ♂ mean FWL = 6.4 mm, ♀ mean FWL = 7.4 mm. Differences between *E. cimarronensis* and *E. wielgusi* are discussed under the latter species.

**Description.** *Head.* Whitish, sometimes with pale brownish-gray suffusion; scales on antenna gray brown with white apices. *Thorax.* Dorsal surface brownish gray; fore- and mid-leg with anterior surfaces grayish brown, posterior surfaces tan to whitish; hind-leg uniformly whitish; tarsus with alternating tan and blackish-brown annulations. Forewing: ♂ FWL 5.2–7.5 mm (mean = 6.4, n = 10), AR = 3.03, CFR = 0.36; ♀ FWL: 6.6–8.4 mm (mean = 7.4, n = 14); AR = 2.99; dorsal surface grayish brown, with blackish-brown subbasal and pretornal marks that vary from prominent (Plate D:8a) to weakly expressed (Plate D:8b); postmedian fascia varying from conspicuous (Plate D:8e) to barely discernable (Plate D:8f); interfascial spot concolorous with basal area, often with whitish edging on proximal margin; ocellus typical of the group, with lateral bands pinkish and proximal margin thinly edged with black; costal strigulae on distal one-half of wing white and sharply defined, with associated gray striae extending toward termen; fringe pale from tornus to $M_2$, darker at apex. Hindwing: Brownish gray. *Abdomen.* Male genitalia (n = 9): Typical of the group, with 14–24 cornuti, NR = 0.76, SA = 159°, cucullus with length-to-width ratio = 1.88. Female genitalia (n = 10): Typical of the group, with sterigma length-to-width ratio = 1.39, lamella antevaginalis narrow, sclerotization of ductus bursae ringlike.

**Holotype** (Plate D:8a). ♀, Kansas, Morton County, Cimarron National Grassland, 7.5 mi N of Elkhart, 101.8950° W, 37.1200° N, D. J. Wright, 2 August 1999, slide DJW 578, USNM.

**Paratypes** (7 ♂, 13 ♀). CALIFORNIA. San Bernardino County, Boulevard, R. D. Goeden and D. W. Ricker, 24 June 1970 (1 ♂, slide DJW 4639); San Diego County, Banner, R. D. Goeden and D. W. Ricker, 10 August 1970 (1 ♂, slide DJW 4635); San Diego County, San Diego, 16–23 August, (1 ♀, slide DJW 4625); Riverside County, Banning, R. D. Goeden and D. W. Ricker, 24 September 1970 (1 ♂, slide DJW 4636). KANSAS. Morton County, Cimarron National Grassland, 7.5 mi N of Elkhart, D. J. Wright, 25 August 2000 (1 ♀, slide DJW 3763). TEXAS. Briscoe County, Caprock Canyon State Park, E. C. Knudson, 29–30 August 1994 (1 ♀), 2–3 September 1996 (1 ♀); Coleman County, Santa

Anna, E. C. Knudson, 19 May 1985 (2 ♂, slides ECK 1155, DJW 3863; 3 ♀, slides ECK 1162, 1164, 1175); Cottle County, Paducah, A. and M. E. Blanchard, 17 April 1968 (1 ♀, slide 69884); Matador Wildlife Management Area, E. C. Knudson, 30 July 1983 (1 ♂, slide ECK 1062; 1 ♀, slide ECK 1143); Culberson County, Guadalupe Mountains, E. C. Knudson, 27–28 June 1992 (1 ♂); Jeff Davis County, Davis Mountain State Park, E. C. Knudson, 28 August 1977 (1 ♀); Kimble County, Junction, A. and M. E. Blanchard, 26 August 1973 (1 ♀, slide 69885); Potter County, Lake Meredith National Recreation Area, E. C. Knudson, 23 June 1985 (2 ♀, slide DJW 3852). Paratype depositories: MGCL, USNM.

**Etymology.** The specific epithet refers to the Cimarron River, which drains the sand prairie grasslands at the type locality.

**Biology.** Three males in the USNM from southern California were reared by Goeden and Ricker from larvae feeding on *Ambrosia confertiflora* DC. (weakleaf bur ragweed).

## 9. ***Epiblema wielgusi*** Wright and Gilligan, new species

(Plate E, 9a–c; Plate 3, 9a–e)

**Diagnosis.** *Epiblema wielgusi* resembles *E. cimarronensis* but is larger (mean FWL = 8.5 vs. 7.0 mm) and has a duller forewing appearance. The interfascial areas are more finely reticulate, the subbasal and pretornal marks are obscure (weakly contrasting with adjacent areas), and the costal strigulae are less clearly defined. A lack of pinkish tint in the interfascial area separates *E. wielgusi* from *E. sosana*. Differences in male genitalia between the three species are subtle and probably not reliable for diagnostic purposes. Females of *E. wielgusi*, *E. sosana*, and *E. cimarronensis* differ in the shape of the sterigma (length-to-width ratio = 0.98, 1.21, and 1.39, respectively) and the width of the lamella antevaginalis (narrowest, broadest, and intermediate, respectively). The sclerotization of the ductus bursae is prominent and ringlike in *E. wielgusi* and *E. cimarronensis* but patchlike and barely detectable in *E. sosana*. *Epiblema wielgusi* is known only from the type locality, near Phoenix, Arizona.

**Description.** *Head.* Labial palpus with medial surface whitish, lateral surface grayish brown; frons whitish; vertex grayish brown, scales on antenna grayish brown with white apices. *Thorax.* Dorsal surface concolorous with vertex; fore- and mid-leg with anterior surfaces grayish brown, posterior surfaces tan to whitish; hind-leg uniformly whitish; tarsus with alternating tan and grayish-brown annulations. Forewing: ♂ FWL 7.5–9.2 mm (mean = 8.5, n = 6), AR = 3.07, CFR = 0.38; ♀ FWL: 8.0–8.7 mm (mean = 8.5, n = 3), AR = 3.09; dorsal surface grayish brown; basal area crossed transversely by inconspicuous dark reticulations; subbasal fascia barely distinguishable from basal area; median fascia obscure, represented by weakly-expressed quadrate pretornal mark; postmedian fascia obscure but usually detectable; interfascial spot whitish, suffused with grayish brown, weakly contrasting with subbasal and pretornal marks; ocellus pale, mostly gray, with thin black edging on proximal margin; costal strigulae inconspicuous, whitish, with obscure gray stria emanating from strigula 8. Hindwing: Brownish gray, slightly paler than forewing. *Abdomen.* Male genitalia (n = 3): Typical of the group, with 21–31 cornuti, NR = 0.63, SA = 154°, cucullus length-to-width ratio = 1.86. Female genitalia (n = 3): Typical of the group, with sterigma length-to-width ratio = 0.98, lamella antevaginalis narrow, sclerotization of ductus bursae ringlike.

**Holotype** (Plate D:9a). ♀, Arizona, Maricopa County, Thunderbird Regional Mountain Park, R. S. Wielgus, 6 March 1977, slide DJW 3862, USNM.

**Paratypes** (6 ♂, 2 ♀). ARIZONA. Maricopa County, Thunderbird Regional Mountain Park, R. S. Wielgus, 20 March 1976 (1 ♂, slide DJW 3861; 1 ♀, slide DJW 3874), R. and J. Wielgus, 20 April 1975 (5 ♂, slides DJW 3995, 3996; 1 ♀, slide DJW 3397). Paratype depositories: ASUHIC, USNM.

**Etymology.** The specific epithet honors the late Ron Wielgus, who, with his son J. Wielgus, collected the specimens in the type series.

## The *carolinana* group (species 10–11)

This group consists of two similar species, *E. carolinana* and *E. arizonana*, the first broadly distributed in eastern North America, the second known primarily from Arizona. Group characteristics include:

*Forewing*. Mean FWL: 9.2, 11.0 mm; AR: 2.61, 2.57; CFR: 0.65, 0.60; basal patch brownish gray, with white and black speckling; distal margin of basal patch scalloped and highlighted with dashed black line from inner margin to radius; interfascial spot large, extending from inner margin to costal fold in males and from inner margin to costa in females, with numerous gray transverse reticulations; pretornal mark mottled, brown and/or gray, with prominent black speckling; ocellus with transverse gray bars on lateral margins, prominent black spot at apical corner, and up to four black spots/dashes on narrow whitish central field; costal strigulae anterior to interfascial spot gray, those from median fascia to apex white, the latter separated by reddish-brown quadrate marks; termen with narrow reddish-brown band from $CuA_1$ to apex; fringe gray.

*Male genitalia*. Uncus weakly developed but distinguishable from dorsolateral shoulders of tegumen; socius long, fingerlike, nearly uniform in width, weakly arched medially; phallus with base closely surrounded by anellus; vesica with 17–32 cornuti; valva with clasper Type 1, neck well defined by moderate ventral emargination, NR: 0.59, 0.64, saccular corner rounded, SA: 130°, 136°; cucullus elongate (length-to-width ratio: 3.61, 2.47), with dorsal lobe uniform in width, ventral lobe weakly developed, anal angle acute; distal periphery of cucullus with line of short stout setae extending from anal angle to three-fourths distance to apex, tracking ventral one-half of distal margin before proceeding onto medial surface; setae on medial surface of cucullus course near said line, grading to hairlike near costal margin and apex.

*Female genitalia*. Papillae anales Type 2; sterigma U-shaped, with posterolateral corners flared laterally; lamella postvaginalis microtrichiate; lamella antevaginalis broad and bandlike, with posterior extremities attenuating onto surface of lamella postvaginalis; sternum 7 trapezoidal, with anterior width about twice posterior width, and with posterior margin concavely emarginated to about one-third length of sterigma; ductus bursae with large sclerotized patch opposite juncture with ductus seminalis; corpus bursae with two signa of unequal size, the larger located near mid-bursa, the smaller near juncture with ductus bursae.

### 10. *Epiblema carolinana* (Walsingham, 1895)
(Plate E, 10a–f; Plate 4, 10a–h)

*Paedisca* (?) *carolinana* Walsingham 1895:509.
*Eucosma carolinana*: Fernald 1903:460; Barnes and McDunnough 1917:171.
*Epiblema carolinana*: Heinrich 1923:149; Thompson and Guelph 1928:73; McDunnough 1939:48; Putman 1942:221; MacKay 1959:37; Powell 1983:35; Miller 1987:58; Brown 2005:280; Gilligan et al. 2008:123.

**Holotype.** ♀, North Carolina, Morrison, 1882, BMNH(E) 819930, slide 11544, BMNH.

*Epiblema carolinana* was described from a single specimen, hence the holotype designation. It is smaller than *E. arizonana* (mean FWL = 9.2 vs. 11.0 mm) but is otherwise indistinguishable from its western counterpart based on superficial characteristics. Powell (1975) noted that males of the two species differ in the breadth of the cucullus (narrower in *E. carolinana*). We confirmed that observation by calculating the length-to-width ratio of the cucullus for several specimens of each taxon, obtaining average values of 3.61 (n = 6) for *E. carolinana* and 2.47 (n = 3) for *E. arizonana*. We also found that the basal process is strongly developed in *E. carolinana* but reduced to absent in *E. arizonana*. In females, the sterigma is more elongate in *E. carolinana*, with average length-to-ostium diameter = 1.95 (n = 5) vs. 1.48 (n = 6). Moreover, we observed a few scales on the lateral margins of the sterigma in *E. carolinana* but found none in *E. arizonana*.

Heinrich (1923) and Miller (1987) listed *Rudbeckia* L. (coneflower) as the larval host for *E. carolinana*. The USNM has specimens that were reared by Murray Glenn in Illinois from *Rudbeckia laciniata* L. (cutleaf coneflower) and from *Ratibida pinnata* (Ventenat) Barnhart (grayhead coneflower). Putman (1942) and Thompson and Guelph (1928) reported rearing this species from larvae boring in the stems and roots of *R. laciniata*. The latter authors included a detailed life history of the insect. Eggs are laid in the blossoms; early instar larvae feed at the base of the blossom; the third instar larva drops to the ground via a silk thread, bores into the stem about four inches above ground level, and then tunnels down into the root. Further larval development occurs in a cavity (as much as six to seven inches long) excavated in the root; the mature larva overwinters in that cavity; and pupation occurs there in the spring.

This species is broadly distributed in eastern North America. Most of the 46 specimens (30 ♂, 16 ♀) we examined were collected in the region extending from Massachusetts to North Carolina west to Michigan, Iowa, and Mississippi. Our sample included two males in the USNM from Albany County, Wyoming that we tentatively refer to this taxon. They are somewhat darker than typical *E. carolinana* but agree with that taxon in genitalia (Plate 4:10e). Adult flight occurs from late May to the end of August.

### 11. ***Epiblema arizonana*** Powell, 1975
(Plate E, 11a–f; Plate 4, 11a–e)

*Epiblema arizonana* Powell 1975:101; Powell 1983:35; Brown 2005:280.

**Holotype.** ♂, Arizona, Coconino County, Oak Creek Cãnon, near Flagstaff, G. P. Englehardt, reared from *Rudbeckia laciniata*, emerged 28 June 1936, slide 72844, USNM.

*Epiblema arizonana* is poorly represented in collections. Powell (1975) described it from the holotype, an allotype, and 18 paratypes (4 ♂, 14 ♀), all collected as larvae at the type locality. Two of the paratypes are in the EME; the others reside in the USNM. We examined the holotype, the allotype, and 11 paratypes (2 ♂, 9 ♀), all reared from *Rudbeckia laciniata* (cutleaf coneflower), with emergence dates ranging from 24 June to 2 July. We also tentatively refer to this taxon one additional female in the EME (Plate E:11f) that was collected on 2 August in Fort Collins, Colorado.

*Epiblema arizonana* is larger than *E. carolinana* (mean FWL = 11.0 vs. 9.2 mm) but is otherwise superficially indistinguishable from the latter. Genitalia differences between the two are discussed under *E. carolinana*.

The inability to separate *E. arizonana* from *E. carolinana* based on forewing appearance, together with the fact that both species utilize *R. laciniata* as a larval host, led Powell (1975) to suggest that these taxa might prove to be two races of a single species. At that time, the range of *E. arizonana* appeared to be disjunct from that of *E. carolinana* (Arizona vs. eastern North America), and that distinction, together with subtle differences in genitalia, seemed sufficient to diagnose the new species. Later discoveries of *arizonana/carolinana*-like specimens from Colorado and Wyoming (mentioned above under *E. carolinana*) suggest that the ranges of the two species may overlap or that there may be intermediate specimens that link the eastern and southwestern phenotypes. Given the scarcity of records of either taxon from the Southwest, the eastern foothills of the Rocky Mountains, and the high Central Plains, we are inclined to retain both names and await future sampling that might clarify this situation.

## The *tripartitana* group (species 12–13)

The *tripartitana* group consists of *E. tripartitana* and *E. glenni*, two species that are sympatric in much of eastern United States. They appear to be closely related to members of the *carolinana* group based on similarities in genitalia, but they differ from the latter taxa in size (mean FWL = 6.3, 6.1 vs. 9.2, 11.0 mm) and forewing appearance. Group characteristics include:

*Forewing*. Mean FWL: 6.3, 6.1 mm; AR: 2.69, 2.61; CFR: 0.56, 0.61; basal patch prominent, dark brown with black transverse reticulations; interfascial spot large, white, often with pale pinkish suffusion, extending from inner margin to costal fold in males and from inner margin to costa in females; median fascia brown or gray with black speckling; ocellus with transverse gray bands on lateral margins and up to four short black dashes on narrow grayish central field; costal strigulae beyond median fascia whitish; distal margin with narrow reddish-brown band from $CuA_1$ to apex; fringe gray from tornus to $CuA_1$, dark brownish gray from $CuA_1$ to apex.

*Male genitalia*. Uncus moderately developed and clearly differentiated from dorsolateral shoulders of tegumen; socius long, fingerlike, nearly uniform in width, weakly arched medially; phallus with base closely surrounded by anellus; vesica with 16–42 cornuti; valva with clasper Type 1, basal process prominent, costal margin concave, neck well defined by moderate ventral emargination, NR: 0.66, 0.65, saccular corner rounded, SA: 138°, 149°; cucullus elongate (length-to-with ratio: 3.03, 2.93), with dorsal lobe uniform in width, ventral lobe weakly developed, anal angle acute; distal periphery of cucullus with line of short stout setae extending from anal angle to three-fourths distance to apex, tracking ventral one-half of distal margin before proceeding onto medial surface; setae on medial surface of cucullus course near said line, grading to hairlike near costal margin and apex.

*Female genitalia*. Papillae anales Type 2; sterigma U- or slightly V-shaped; lamella postvaginalis microtrichiate; lamella antevaginalis broad and bandlike; sternum 7 trapezoidal, anterior width about twice posterior width, with posterior margin concavely emarginated to about one-third length of sterigma; ductus bursae with large sclerotized patch opposite juncture with ductus seminalis; corpus bursae with two signa of unequal size, the larger located near mid-bursa, the smaller near juncture with ductus bursae.

### 12. ***Epiblema tripartitana*** (Zeller, 1875)
(Plate F, 12a–i; Plate 4, 12a–f)

*Paedisca tripartitana* Zeller 1875:308; Fernald 1882:41.
*Eucosma tripartitana*: Fernald 1903:459; Barnes and McDunnough 1917:171.
*Epiblema tripartitana*: Heinrich 1923:146 (fig. 270 = *E. scudderiana*); Bottimer 1926:815; McDunnough 1939:48; MacKay 1959:34; Powell 1983:35; Miller 1987:57; Wright 2002:277; Brown 2005:286; Gilligan et al. 2008:122.

**Holotype**. ♀, Texas, Dallas, Boll, Type No. 14337, MCZ.

Zeller (1875) described *P. tripartitana* from a single female. Heinrich (1923) confused this species with darkly marked specimens of *E. scudderiana* from Florida (Plate G:14h–i) and illustrated the male genitalia of one of the latter individuals under the name *tripartitana* (Wright 2002). Consequently, historical literature records for this species must be treated with caution. For example, some specimens reported by Kimball (1965) from Florida as *E. tripartitana* proved upon recent examination to be the dark form of *E. scudderiana*.

*Epiblema tripartitana* is variable in size (FWL: 3.7–9.8 mm, mean = 6.3). The most distinctive forewing feature is the immaculate interfascial spot, which extends as a broad band from the inner margin to the costa, narrowing slightly toward the latter extremity, with distal margin weakly convex. The basal patch is dark brown and sharply defined; the median fascia is brownish to grayish, sometimes paler than the basal patch, with several black spots in the pretornal position; and the costal strigulae beyond the median fascia are white and well defined. The ocellus is usually dark and obscure, but in specimens from the Central Plains and the Southwest (Plate F:12c–f), it appears as a vaguely defined whitish patch.

This species differs from the similar looking *E. glenni* in the color and shape of the interfascial spot. In *E. tripartitana* the spot is white, often with a pale pinkish tint in fresh specimens. In females it merges with whitish costal strigulae, forming a continuous band from the inner margin to the costa (Plate F:12a, d–f, i); in males it is intercepted by the costal fold, where it contrasts weakly with pale gray costal strigulae (Plate F:12b–c, g–h). In *E. glenni* the spot is uniformly pinkish and narrows abruptly at the cubitus. In females it extends from the inner margin to the costa (Plate F:13a, d); in males it terminates at the contrasting brownish-gray costal fold (Plate F:13b–c, e–f).

We found no means of separating the two species based on male genitalia. In females of *E. tripartitana*, the posterior extremities of the bandlike lamella antevaginalis attenuate onto the surface of the lamella postvaginalis; in *E. glenni* they merge into the lateral margins of the lamella postvaginalis (Plate 5:12f vs. 13f).

The specimens we examined (55 ♂, 32 ♀) suggest a geographic range that extends from Colorado, New Mexico, and the Gulf Coast north and east to Wisconsin, New York, Connecticut, and the Atlantic Coast. Adult capture dates range from early April to the end of August. Bottimer (1926) reported *Rudbeckia maxima* Nutt. (great coneflower) as a larval host in east Texas and provided a detailed account of the insect's life history. He observed early instars feeding in the central part of the flower head and excavating a cavity downward for about an inch into the stem, producing a slight enlargement of the stem immediately below the flower head. The mature larva exits the stem through a hole at the bottom of the cavity, reenters the stem at 6–12 inches above the soil, and excavates a burrow from ground level to 2–3 inches above the entrance hole. During the winter the larva cuts off the stem from the inside and plugs the resulting opening. In the spring it pupates in the burrow, and the adult exits through a previously prepared hole at the uppermost part of the burrow.

## 13. *Epiblema glenni* Wright, 2002
(Plate F, 13a–f; Plate 5, 13a–f)

*Epiblema glenni* Wright 2002:279; Brown 2005:282; Gilligan et al. 2008:122.

**Holotype**. ♂, Ohio, Adams County, 1 mi S.E. of Lynx, D. J. Wright, 12 August 1998, slide 94171, USNM.

*Epiblema glenni* resembles *E. tripartitana* in size (mean FWL = 6.1 vs. 6.3 mm) and general forewing appearance but differs from the latter species in the shape and coloration of the interfascial spot, as discussed under *E. tripartitana*. The male genitalia of the two species are indistinguishable; females differ in subtle aspects of the sterigma (discussed under *E. tripartitana*).

We examined specimens of *E. glenni* from Illinois, Kentucky, Michigan, Missouri, North Carolina, Ohio, and Tennessee, states in which it is sympatric with *E. tripartitana*. Adult capture dates range from the end of May to mid-September.

## The *scudderiana* group (species 14–15)

This group consists of *E. scudderiana* and *E. benignata*, two species that are similar in forewing appearance and genitalia. The first is broadly distributed east of the Rocky Mountains, the second occurs west of the Mississippi River. Group characteristics include:

*Forewing*. Mean FWL: 8.2, 7.6; AR: 2.62, 2.84; CFR: 0.45, 0.40; basal patch sharply defined, dark brown, with slightly darker obscure transverse reticulations; interfascial spot large, white, and often sparsely speckled with gray; pretornal mark moderately well defined, gray with white and/or brown mottling and black speckling; ocellus with gray or pale pink bands on lateral margins and up to four black spots on narrow white central field; costal strigulae white; termen with reddish-brown band from $CuA_1$ to apex, often with small white spots on $M_1$ and $R_5$; fringe black to dark reddish brown from $CuA_1$ to apex, gray to whitish near tornus.

*Male genitalia*. Uncus semicircular; socius moderately short and stubby; vesica with 14–41 cornuti; valva with neck well defined, NR: 0.56, 0.66, saccular corner broadly rounded, SA: 149°, 155°, clasper Type 1, basal process moderately developed; cucullus with dorsal lobe attenuating slightly toward rounded apex, distal margin convex of uniform curvature, ventral lobe weakly developed, anal angle rounded; setae on medial surface of cucullus moderately coarse near distal margin, grading to hairlike near costal margin and apex.

*Female genitalia*. Papillae anales Type 2 and microtrichiate; sterigma U-shaped; lamella postvaginalis microtrichiate, with width of posterior margin slightly greater than ostium diameter; posterior margin of sternum 7 concavely emarginated to about one-third length of sterigma and laterally diverging therefrom; corpus bursae with two signa of nearly equal size located opposite one another near mid-bursa.

### 14. *Epiblema scudderiana* (Clemens, 1860)
(Plate G, 14a–i; Plate 6, 14a–f)

*Hedya scudderiana* Clemens 1860:358.
*Paedisca scudderiana*: Fernald 1882:41; Kellicott 1882:161; Walsingham 1884:140.
*Eucosma scudderiana*: Fernald 1903:459; Kearfott 1905:355; Barnes and McDunnough 1917:171.
*Epiblema scudderiana*: Heinrich 1923:147 (fig. 271); Darlington 1947:95; MacKay 1959:32; Miller 1973:221; Miller 1976:50; Powell 1983:35; Brown et al. 1983:225; Miller 1987:57; Brown 2005:285; Gilligan et al. 2008:123.
*Euryptychia saligneana* Clemens 1865:141; Miller 1973:221.
*Paedisca affusana* Zeller 1875:307.

**Types**. *Hedya scudderiana*. Lectotype (designated by Darlington 1947). ♂, Massachusetts, Type No. 7215, ANSP. *Euryptychia saligneana*. Holotype. Rock Island, Illinois, BMNH? *Paedisca affusana*. Holotype. ♀, North America, September, slide 5738, BMNH.

Clemens (1860) seems to have based *H. scudderiana* on a single specimen from Massachusetts that he had received from S. H. Scudder (he made no mention of other syntypes). Heinrich (1923) stated that the type is in the ANSP and that Massachusetts is the type locality. Darlington (1947) reported a type (a term he used as equivalent to lectotype) in the ANSP, identified it by its ANSP type number (7215), and credited Heinrich with the designation. Miller (1973: fig. 40) provided a photograph of that specimen, referred to it as a lectotype, and attributed the designation to Darlington. Clemens (1865) described *E. saligneana* from a single specimen which, according to the collector, B. D. Walsh, was bred from a willow gall. Miller (1973) could not find that specimen at the ANSP but pointed out that there is a male in the BMNH (Walsingham No. 36215) labeled "homo-type" and suggested that Walsingham may have inadvertently removed it from the ANSP. Brown (2005) reported the holotype as lost but recently examined the "homo-type" in the BMNH and reported (personal communication) that it agrees with *E. scudderiana* "in every aspect of forewing pattern and size." Zeller's description of *P. affusana* mentions one female "in Dr. Rössler's collection." Miller (1976) referred to that specimen as the type and stated that it is missing its left forewing and the distal part of its right forewing. *Euryptychia saligneana* and *Paedisca affusana* have been recognized as synonyms of *E. scudderiana* since Fernald (1882).

This species was last reviewed by Wright (2002). Historically it has been confused with *E. tripartitana* (see the discussion under the latter taxon). Both species have a large white interfascial spot that extends from the inner margin to the costa, substantially dividing the forewing into three regions: basal (dark), median (white), and distal (dark). In *E. tripartitana*, the interfascial spot is bandlike; in *E. scudderiana* it is symmetric about its medial axis, and its semicircular anterior extremity is tangent to the costa. Adding to the confusion has been the fact that Heinrich (1923, fig. 270) illustrated under the name *tripartitana* the genitalia of an especially dark male of *E. scudderiana* from Florida. In fact, the male genitalia of these two species are quite different in valva shape and socius size (Plate 6:14a vs. Plate 5:12a).

*Epiblema scudderiana* is a moderately large member of the genus (mean FWL = 8.2 mm) with a relatively broad forewing (AR = 2.62). It exhibits a fair amount of variation in forewing appearance. In the typical phenotype (Plate G:14a–c), the median fascia is substantially paler than the basal patch, its mixture of white, gray, and black scales producing a mottled grayish effect. Specimens from the Great Plains (Plate G:14d–e) often have pale pink suffusion in the interfascial spot and in the ocellus. A particularly dark phenotype in which the median fascia is concolorous with the basal patch (Plate G:14h–i) occurs in Florida and Georgia. The specimen in Plate G:14g is one of five males in the USNM collected by J. Vargo at Black Kettle National Grassland in Oklahoma. They have an ocellus that is unusually bright, an interfascial spot that falls short of reaching the costal margin, and a median fascia that is uniformly gray. They may represent an undescribed species, but we tentatively refer them to *E. scudderiana* until additional information (including females) becomes available. All the phenotypes have a prominent black dot centrally located on the proximal margin of the ocellus.

In males, the valva neck is moderately long and well defined, NR = 0.56, the saccular corner is weakly angulate, and SA = 149°. In females, the posterior extremities of the lamella antevaginalis attenuate into the lateral margins of the lamella postvaginalis, and the sclerotization of the ductus bursae is either absent or barely detectable as a small patch opposite the juncture with the ductus seminalis.

Miller (1976) reported larvae feeding in elongate galls on six different species of *Solidago* L. (goldenrod): *S. altissima* L. (tall goldenrod), *S. canadensis* L. (Canada goldenrod), *S. gigantea* Ait. (giant goldenrod), *S. juncea* Ait. (early goldenrod), *S. ulmifolia* Muhl. (elmleaf goldenrod), and *S. nemoralis* Ait. (gray goldenrod). He observed tip mining in plants with incipient galls, which he interpreted as the feeding behavior of early instar larvae prior to boring into the stem. After moving to the stem, concentrated feeding by the larva in a small area causes the stem to swell and a gall to develop. Pupation and overwintering occur in the gall. These elongate galls are much harder to detect in the field than the distinctive spherical galls induced by the goldenrod gall fly (*Eurosta solidaginis*; Diptera) on *Solidago*. The USNM has a female of the dark phenotype that was collected at Savannah, Georgia and bears a label with the inscription "reared from golden rod stem." Brown et al. (1983) reported three specimens in the FSCA that had been reared at Gainesville, Florida by D. H. Habeck on *Heterotheca subaxillaris* (Lam.) Britt. & Rusby (camphor weed).

We examined specimens from the region extending from southern Canada south to the Gulf Coast and from the Atlantic Coast west to North Dakota, western Kansas, and southern Texas. Adult capture dates range from mid-February to mid-August, with the highest concentration of records in May and June.

This species has been the subject of many years of research into cold hardiness and freeze avoidance led by Kenneth Storey at Carleton University. In the winter, larvae can survive to nearly –40°C by converting the glycogen in plants into ultrahigh concentrations of glycerol that account for 18–20% of their total body mass (e.g., Muise and Storey 2001; Williamson et al. 2021).

## 15. *Epiblema benignata* McDunnough, 1925
(Plate G, 15a–f; Plate 6, 15a–e).

*Epiblema benignata* McDunnough 1925a:21; McDunnough 1939:48; MacKay 1959:33; Brown et al. 1983:224.
*Epiblema benignatum* Powell 1983:35; Brown et al. 2005:280; Gilligan et al. 2008:123.

**Holotype** (Plate G:15a). ♂, Manitoba, Aweme, N. Criddle, 13 June 1923, slide 913, CNC.

McDunnough (1925a) designated a holotype and eight male paratypes, all with the same collection data. This species is remarkably similar to *E. scudderiana* in forewing appearance and genitalia but is slightly smaller (mean FWL = 7.6 vs. 8.2 mm) and has a somewhat narrower forewing (AR = 2.84 vs. 2.61). The specimens we examined document a range that includes the region extending westward from Manitoba to Washington State and south to Iowa, Kansas, New Mexico, and California, suggesting that *E. benignata* is a western sibling species to the predominantly midwestern and eastern *E. scudderiana*. Powell (1983) listed this species as "*benignatum*" in his checklist, and subsequent authors adopted that spelling. Here we revert to the original spelling by McDunnough (1925a).

*Epiblema benignata* differs from *E. scudderiana* in subtle aspects of forewing pattern and coloration. The interfascial spot is smaller, a little more triangular, and does not quite reach the costa, and the ocellus is white and prominent. In males, the neck of the valva is slightly broader (NR = 0.66 vs. 0.56). In females the sclerotization of the ductus bursae is more conspicuous (a distinct ring vs. a barely detectable patch), and the posterior extremities of the lamella antevaginalis attenuate onto the surface of vs. into the lateral margins of the lamella postvaginalis. We found no other genitalia features that separate the two species.

*Epiblema benignata* differs from *E. scudderiana* in larval host preference (*Artemisia* vs. *Solidago*) and in geographic distribution (western vs. eastern North America), but the latter distinction fails in the Great Plains, where the two species are sympatric. The USNM has several specimens that were reared by J. F. G. Clarke in Whitman County, Washington from larvae feeding in stem galls on two species of *Artemisia* L. (sagebrush): *A. dracunculus* L. (tarragon) (6 specimens, Plate G:15f), and *A. vulgaris* L. (common wormwood) (49 specimens, Plate G:15e). Brown et al. (1983) reported 15 specimens reared by Clarke from Wilma, Washington on *A. vulgaris* that did not form galls but do resemble the gall-feeding individuals in forewing appearance and genitalia.

## The *infelix* group (species 16–17)

The *infelix* group consists of *E. infelix* and *E. gibsoni*, two eastern species that are nearly identical in genitalia but readily separated by forewing appearance. Group characteristics include:

*Forewing*. Mean FWL: 7.9, 7.8 mm; AR: 2.71, 2.53; CFR: 0.51, 0.51; basal patch and median fascia concolorous; interfascial spot white; ocellus obscure, with gray transverse bands on lateral margins and 2–4 black spots on narrow brownish central field; costal strigulae from mid-wing to apex white, with associated gray striae extending toward termen.

*Male genitalia*. Uncus strongly developed, sharply differentiated from dorsolateral shoulders of tegumen, with weakly developed apical ridge; socius long and fingerlike, with lateral margins nearly parallel; vesica with 22–55 cornuti; valva with clasper Type 1, basal process absent to barely discernable, costal margin concave, neck moderately long, NR: 0.57, 0.58, saccular corner weakly angulate, SA: 148°, 154°; cucullus elongate (length-to-width ratio = 2.60, 2.94), with dorsal lobe nearly uniform in width, apex rounded, distal margin weakly convex, and ventral lobe barely developed.

*Female genitalia*. Papillae anales Type 4; ventral lobes of tergum 8 with mixture of scales and setae (Plate 7:16g); sterigma U-shaped and microtrichiate, with posterolateral corners of lamella postvaginalis acute;

lamella antevaginalis ringlike, with posterior extremities attenuating onto surface of lamella postvaginalis; posterior margin of sternum 7 concavely emarginated to about one-half length of sterigma, medially approximate to sterigma, laterally diverging therefrom; ductus bursae with weakly sclerotized patch near juncture with ductus seminalis; corpus bursae with two signa of equal size located opposite one another near mid-bursa.

## 16. ***Epiblema infelix*** Heinrich, 1923
(Plate H, 16a–e; Plate 7, 16a–g)

*Epiblema infelix* Heinrich 1923:151, fig. 276; McDunnough 1939:48; Powell 1983:35; Miller 1987:58 (in part, female sterigma); Wright 2002:282; Wright and Covell 2003:233; Brown 2005:283; Gilligan et al. 2008:126.

**Holotype.** ♂, North Carolina, Tryon, W. F. Fiske, 25 May 1904, slide 72880, USNM.

Heinrich (1923) described *E. infelix* from three specimens (2 ♂, 1 ♀) collected by W. F. Fiske at Tryon, North Carolina on 25 May, 5 July, and 4 July 1904, respectively. He identified the name-bearing specimen by illustrating its genitalia. Historically, this species has been confused with several other taxa, including *E. gibsoni*, *E. glenni*, and *E. walsinghami* (Wright 2002; Wright and Covell 2003). Miller's (1987) illustrations of the adult and male valva of *E. infelix* are, respectively, a photograph and a drawing of *E. glenni*, but his drawing of the sterigma is based on the female paratype of *E. infelix*.

*Epiblema infelix* agrees with *E. gibsoni* in size (mean FWL = 7.9 vs. 7.8 mm) but has a slightly more elongate forewing (AR = 2.71 vs. 2.53). The predominant forewing color is gray with generous amounts of brown suffusion. The basal and median areas are concolorous, the interfascial spot is triangular, and the obscure median fascia has some black speckling in the pretornal area. These features are sufficient to separate *E. infelix* from *E. gibsoni*. The latter species is predominantly brown and has a narrow, bright white, sharply-edged interfascial spot. In *E. infelix*, the strigulae on the distal one-half of the costa are separated by reddish-brown marks, and the termen has a narrow reddish-brown band from $CuA_1$ to the apex; in *E. gibsoni*, these markings lack the reddish tint. The genitalia of these two species are remarkably similar, and they exhibit enough subtle variation to render diagnosis on that basis unreliable.

*Epiblema walsinghami* resembles *E. infelix* but is smaller (mean FWL = 7.1 vs. 7.9 mm), differs in forewing geometry (AR = 2.32 vs. 2.71), and has more prominent forewing markings, most notably a black band on the inner margin of the basal patch (Plate L:28a). Males of *E. walsinghami* are easily distinguished from those of *E. infelix* by genitalia (Plate 12:28a vs. Plate 7:16a): the valval neck is wider (NR = 0.73 vs. 0.57), the cucullus is less elongate (length-to-width ratio = 1.66 vs. 2.60), and the basoventral margin of the cucullus extends ridgelike onto the medial surface of the neck. Females differ in papillae anales Type (2 vs. 4), the absence vs. presence of scales on tergum 8, the sclerotization of the ductus bursae (absent vs. patchlike), and the relative sizes of the signa (substantially different vs. equal).

*Epiblema glenni* differs from *E. infelix* in size (mean FWL = 6.1 vs. 7.9 mm), the shape and color of the interfascial spot (bandlike and pinkish vs. triangular and white), the presence of a basal process on the valva, the papillae anales Type (2 vs. 4), the size of the sclerotized patch on the ductus bursae (large vs. small), and the relative sizes of the signa (unequal vs. equal).

We examined 60 specimens (48 ♂, 12 ♀) from Alabama, Arkansas, Kentucky, Michigan, Mississippi, North Carolina, South Carolina, and Virginia. Collection dates ranged from 4 April to 5 July except for one North Carolina record of 22 October. In our experience in the Midwest, this species is most likely to be found in April or May in woodland openings or along woodland trails.

## 17. *Epiblema gibsoni* Wright and Covell, 2003
(Plate H, 17a–e; Plate 7, 17a–f)

*Epiblema gibsoni* Wright and Covell 2003:233; Brown 2005:282; Gilligan et al. 2008:126.
*Epiblema walsinghami* (not Kearfott 1907): Miller 1987:58

**Holotype.** ♂, Kentucky, Rowan County, Route 1274, 2 mi. W of Route 519, L. D. Gibson, 16 July 1994, USNM.

This species has been confused with *E. walsinghami* and *E. infelix* (Wright and Covell 2003). Kearfott (1907) described *E. walsinghami* from a mixed series of seven specimens, two of which (1 ♂, 1 ♀) were later identified by Heinrich (1923) as *E. infelix*. Heinrich considered the male to be "somewhat aberrant" and declined to include it as a paratype of *E. infelix*; it is, in fact, *E. gibsoni* (Wright and Covell 2003). Moreover, the photograph and male valva drawing in Miller (1987) for *E. walsinghami* are illustrations of *E. gibsoni*. A discussion of the diagnostic differences between these three species is included in the above account for *E. infelix*.

*Epiblema gibsoni* is distinguished from *E. infelix* and *E. walsinghami* by forewing color (rich medium brown vs. medium gray to blackish gray). Particularly fresh specimens often have a pale lavender tint. The interfascial spot is bright white, relatively narrow, sharply defined, and somewhat variable in shape. In one common phenotype (Plate G:17a) it widens slightly from the inner margin to $A_{1+2}$ and then tapers to a point in the discal cell. In some specimens from the Chicago area, the interfascial spot is greatly reduced or obsolete (Plate G:17e). The ocellus is obscure, with gray bands on the lateral margins and a few black dashes/spots on a narrow brownish central field.

We examined specimens from Arkansas, Illinois, Indiana, Kentucky, Louisiana, Michigan, Missouri, North Carolina, Ohio, South Carolina, Tennessee, and Virginia. Capture dates range from early June to mid-August, with 70% of the records in July. This species can be common in remnant prairie and glade habitat.

## The *desertana* group (species 18–21)

The *desertana* group is composed of four species of similar forewing pattern and genitalia: *E. desertana* and *E. rudei*, which have eastern and southwestern distributions, respectively; *E. discretivana*, which occurs in the southern and eastern coastal states from Texas to Rhode Island; and *E. obfuscana*, which is broadly distributed in eastern North America. Group characteristics include:

*Forewing*. Mean FWL: 6.9–9.4 mm; AR: 2.57–3.04; CFR: 0.42–0.57; fasciate markings pale gray to blackish gray; interfascial spot whitish or pinkish, quadrate, extending from inner margin to radius, often with gray transverse reticulations; ocellus obscure but usually detectable, with narrow white or gray central field flanked laterally by pinkish or grayish transverse bands; termen with conspicuous blackish band.

*Male genitalia*. Uncus strongly developed and, except for *E. obfuscana*, tapering to an angulate or pointed apex; socius moderately short, sometimes stubby; vesica with 0–24 cornuti; valva with clasper Type 1, basal process conspicuous, neck broad, NR: 0.67–0.82, saccular corner weakly angulate, SA: 150°–159°, cucullus with dorsal lobe strongly developed, apex rounded, distal margin convex, ventral lobe weakly developed, anal angle rounded.

*Female genitalia*. Papillae anales Type 4, with pads moderately narrow; sterigma U-shaped or V-shaped, in the latter case with length of posterior margin approximately twice ostium diameter; posterior margin of sternum 7 concavely emarginated to about one-half length of sterigma and diverging laterally from lamella postvaginalis; sclerotization of ductus bursae absent (*E. rudei*, *E. discretivana*), patchlike (*E. desertana*), or ringlike (*E. obfuscana*); corpus bursae lacking signa or with two signa of unequal size located opposite one another near mid-bursa.

## 18. ***Epiblema desertana*** (Zeller, 1875)

(Plate H, 18a–e; Plate 8, 18a–d)

*Paedisca desertana* Zeller 1875:306; Fernald 1882:41.
*Eucosma desertana*: Fernald 1903:459; Barnes and McDunnough 1917:171.
*Epiblema desertana*: Heinrich 1923:148; McDunnough 1939:48; MacKay 1959:38; Miller 1976:50; Powell 1983:35; Miller 1987:58; Brown 2005:281; Gilligan et al. 2008:124.

**Holotype.** ♂, Texas, Dallas, Boll, Type 14338, MCZ.

Zeller's (1875) description of *E. desertana* mentions only one specimen, hence the above holotype interpretation. Without explanation, Miller (1976) designated that specimen as the lectotype.

*Epiblema desertana* is medium in size (mean FWL = 8.0 mm) and average in forewing geometry (AR = 2.80). The fasciate markings include a prominent blackish-gray basal patch with transverse black reticulations, a variably expressed mid-costal bar, and a gray pretornal mark with black speckling. Some specimens show a thin postmedian fascia (Plate H:18a, c). The interfascial spot often has extensive pink suffusion, a feature that is usually sufficient to distinguish this species from other members of the group. The ocellus has a narrow white central field that is flanked laterally by pink transverse bands. The whitish to grayish costal strigulae are inconspicuous, and the termen has a prominent black band (including the fringe) that extends from $CuA_2$ to the apex.

The most distinctive feature of the male genitalia is the sharply pointed uncus, which has weakly concave lateral margins that are not clearly differentiated from the dorsolateral shoulders of the tegumen. The socius narrows rather abruptly from a relatively broad base to a papillate apex, and the vesica has 17–24 cornuti. The only other member of the group with a "pointed" uncus is *E. rudei*, but in that species the apex is not as sharp, the socius is broader, and the vesica has only 3–10 cornuti. The female genitalia have three distinctive characteristics, any one of which is sufficient to separate this species from other members of the group: sterigma shape (U vs. V), sclerotization of the ductus bursae (patchlike vs. absent or ringlike), and signa size (small vs. absent or moderately large).

*Epiblema desertana* is broadly distributed in eastern United States, from Minnesota, Kansas, and Texas east to the Atlantic Coast. Adult capture dates range from early March to the end of June. Miller (1976) reported rearing this species from larvae feeding in stem galls on *Euthamia graminifolia* (L.) Nutt. (flat-top goldenrod) and described larval development similar to that of *E. scudderiana*.

## 19. ***Epiblema rudei*** Powell, 1975

(Plate I, 19a–e; Plate 8, 19a–e)

*Epiblema rudei* Powell 1975:105; Powell 1983:35; Brown 2005:285; Powell and Opler 2009:136; Looney and Murray 2022:152.

**Holotype.** ♂, California, Fresno-San Benito County line, Big Panoche Creek, J. A. Powell, reared from *Gutierrezia californica*, emerged 17–19 March 1968, CAS.

*Epiblema rudei* is a southwestern species. It is larger than *E. desertana* (mean FWL = 9.4 vs. 8.0 mm) and has a slightly more elongate forewing (AR = 3.04 vs. 2.80). The two species are substantially similar in forewing pattern but differ in coloration. In *E. rudei*, the basal patch has a whitish ground with numerous gray transverse reticulations, producing an overall gray effect; the median fascia is concolorous with the basal patch; and the white interfascial spot lacks the pinkish tint characteristic of *E. desertana*. The ocellus is obscure and whitish, usually with distinguishable faint pink bands on its lateral margins. The black band on the termen is slightly longer in this species than in *E. desertana* (tornus to apex vs. $CuA_2$ to apex).

The male genitalia of *E. rudei* are much like those of *E. desertana*, but the uncus is not as sharply pointed, the socius is broader, the vesica has fewer cornuti (3–10 vs. 17–24), and the apex of the cucullus is more broadly rounded. In females, the ventral lobes of tergum 8 have a few scales mixed with hairlike setae (Plate 8:19e), a feature that is unique within the group to this species; the sterigma is V-shaped; the ductus bursae lacks sclerotization; and the corpus bursae has two signa, one of which is much larger than the other.

Powell (1975) reported rearing most of the type series, which includes an allotype and 15 paratypes (10 ♂, 5 ♀), from stem galls on *Gutierrezia* Lag. (snakeweed), specifically mentioning *G. californica* (DC.), Torr. and A. Gray (San Joaquin snakeweed) as the host for the holotype. Galls were collected from 2 February to 21 April, and adults emerged between 23 February and 24 April. Powell and Opler (2009) stated that the range of this insect includes "the dry western side of the Central Valley and interior southern California." We examined 6 ♂ and 4 ♀ from California that had been deposited in the EME or the USNM, 3 ♂ in the USNM collected by A. & M. E. Blanchard in Cameron and Hemphill counties, Texas, and 2 ♂ in the MGCL collected by E. C. Knudson in Brewster and Culberson counties, Texas. Capture dates for the wild caught specimens ranged from 28 February to 1 April.

Looney and Murray (2022) reported collecting specimens of *E. rudei* near Provo, Utah in traps baited with *Grapholita molesta* (Busck) pheromone lures. They also reared several specimens (5 ♂, 2 ♀) from samples of *Gutierrezia sarothrae* (Pursh) Britton & Rusby (broom snakeweed) that had been gathered in October near the collection site and placed in rearing chambers. Adults emerged in the second half of June. These specimens agree with *E. rudei* morphologically, but DNA sequence data is ambiguous and suggests that the Utah population could represent a cryptic species. More specimens and sequence data are needed to draw meaningful conclusions.

## 20. ***Epiblema discretivana*** Heinrich, 1921
(Plate I, 20a–e; Plate 9, 20a–e)

*Eucosma discretivana* Heinrich 1921:823, plate 99, fig. B.
*Epiblema discretivana*: Heinrich 1923:147, fig. 273; Bottimer 1926:816; McDunnough 1939:48; MacKay 1959:32; Miller 1976:51; Powell 1983:35; Brown 2005:281.

**Holotype.** ♂, Texas, Harris County, Sheldon, A. C. Johnson, larva collected 14 March 1919, adult emerged 10 April 1919, slide 72831, USNM.

Heinrich (1921) based this species on nine specimens (4 ♂, 5 ♀) reared by A. C. Johnson from larvae collected on 14 March 1919 at the type locality in stem galls on "wild myrtle." Heinrich (1921, 1923) illustrated the genitalia of a male that he called the type, thereby identifying a unique name-bearing specimen. Bottimer (1926) reported specimens collected in August, 1924 at the type locality that were reared by A. C. Johnson from larvae in stem galls on *Baccharis halimifolia* L. (eastern baccharis). It is likely that the larval host for the type series (reported as "wild myrtle") was incorrectly identified.

*Epiblema discretivana* is the smallest member of the group (mean FWL = 6.9 vs. 8.0–9.4 mm). It has a moderately broad forewing (AR = 2.58) with an overall gray appearance. The basal patch is dark gray, the interfascial spot is grayish white with gray transverse reticulations, the median fascia is represented by a vaguely defined blackish-gray pretornal mark, the ocellus has gray lateral bands, and the dark gray band on the termen extends from the tornus to the apex.

In males, the uncus is triangular and clearly differentiated from the dorsolateral shoulders of the tegumen, with lateral margins weakly convex; the socius is short and uniform in width; and the vesica has 0–5 cornuti. The neck of the valva is moderately broad, NR = 0.67, the saccular corner is weakly angulate, and SA = 150°. The dorsal lobe of the cucullus is somewhat tapered, producing a more angular apex than in other members of the group. In females, the sterigma is V-shaped, the ductus bursae lacks sclerotization, and the corpus bursae lacks signa, the last feature being unique within the group to this species.

Miller (1976) reared *E. discretivana* from three species of *Baccharis* L.: *B. halimifolia*, *B. angustifolia* Michx. (saltwater false willow), and *B. glomeruliflora* Pers. (silverling); and he illustrated the gall on *B. halimifolia*. The specimens we examined document a range extending from Lee, Manatee, and Putnam counties, Florida west along the Gulf Coast to the southern tip of Texas and north along the Atlantic Coast to Rhode Island. Adult capture dates range from mid-March to the end of the July, with one record from south Texas dated 3 October.

## 21. ***Epiblema obfuscana*** (Dyar, 1903)

(Plate I, 21a–e; Plate 9, 21a–d)

*Paedisca obfuscata* Riley 1888:33, *nomen nudum*.
*Eucosma obfuscana* Dyar, in Fernald 1903:459.
*Eucosma obfuscata*: Barnes and McDunnough 1917:171.
*Epiblema obfuscana*: Heinrich 1923:148; McDunnough 1939:48; MacKay 1959:36; Powell 1983:35; Miller 1987:57; Brown and Lewis 2000:1050; Brown 2005:284; Gilligan et al. 2008:124.

**Holotype.** ♂, USA, r.f. *Solidago*, 2 June 1884, USNM.

The name *Paedisca obfuscata* first appears in remarks by C. V. Riley in the report of the 3 December 1885 meeting of the Entomological Society of Washington. Riley commented on the larval behavior but did not include a description of the adult insect; hence the name is considered a *nomen nudum*. In Fernald's checklist (1903:459), under the heading "5140. 1. *obfuscana* Riley," Harrison Dyar inserted a footnote with a single sentence distinguishing Riley's type specimen from *E. desertana*. Heinrich (1923) stated that Riley's remarks could not be considered a description and credited the authorship of the species to Dyar with the spelling used by Fernald. In a catalogue of tortricid types in the USNM, Brown and Lewis (2000) reported a specimen of *obfuscana* labeled: "Type" ♂ USA, r.f. *Solidago*, 2 June 1884, and Brown (2005) reported this specimen as a holotype. Lacking any indication that Riley was referring to more than one specimen, we follow Brown (2005) in interpreting that male in the USNM as a holotype.

*Epiblema obfuscana* most closely resembles *E. discretivana* but is a little larger (mean FWL = 8.0 vs. 6.9 mm) and has brighter forewing coloration. The interfascial spot is larger and generously reticulated with gray, and the entire ocelloid region is more whitish than gray, rendering the ocellus barely detectable. This species differs from other members of the group in the length of the dark band on the termen ($CuA_1$ to apex vs. tornus or $CuA_2$ to apex). The hindwing is black, with white fringe on the posterior margin.

In males, the uncus is sharply differentiated from the dorsolateral shoulders of the tegumen, and its apical margin has a shallow medial indentation, producing a weakly bilobed appearance. The valva is essentially identical to that of *E. discretivana* in numerical characteristics, but the cucullus is slightly more elongate, and its basoventral margin overlaps the ventral one-fourth to one-third of the medial surface of the neck. The last feature distinguishes *E. obfuscana* from other members of the group. In females, the sterigma is V-shaped, the sclerotization of the ductus bursae is substantially ringlike, and the corpus bursae has two moderately large signa of slightly different size.

*Solidago* L. (goldenrod) has been reported by various authors as the larval host for *E. obfuscana*. Putnam (1942) gave a detailed description of the larval behavior on *S. altissima* L. (Canada goldenrod). He reported that the young larva enters the stem a few inches below the tip and girdles the stem, causing the death of the upper part of the plant. It then excavates a tunnel to the base of the plant, develops therein during later instars, and overwinters below ground. The specimens we examined indicate a range in eastern North America from Nova Scotia to Minnesota south to North Carolina, Mississippi, and Louisiana. The USNM has several specimens from Whitman County, Washington (e.g., Plate I:21b) that may represent a disjunct population or possibly an unrecognized species. They were reared on *Solidago* by J. F. G. Clarke and agree morphologically with typical eastern specimens. Most wild-caught specimens were captured in May or June.

## The *radicana* group (species 22–27)

The *radicana* group consists of six species that are similar in genitalia and forewing pattern. All but *E. periculosana* are poorly represented in collections. Three are known from a few locations in California and Oregon, one was described from Alaska, one is based on four specimens collected in Quebec Province, and one is broadly distributed in the Rocky Mountain Region from Alberta south to Wyoming. Group characteristics include:

*Forewing*. Mean FWL: 6.2–8.7 mm; AR: 2.54–2.75; CFR: 0.33–0.39; subbasal and median fasciae concolorous, varying from yellowish gray to dark brown, often mottled with black, sometimes vaguely defined; interfascial spot white and conspicuous; ocellus obscure, bordered laterally by gray transverse bands, usually with three or four prominent black dashes on a brownish central field; costal strigulae from mid-wing to apex whitish and usually well defined; fringe brown to dark gray.

*Male genitalia*. Uncus moderately developed, clearly differentiated from dorsolateral shoulders of tegumen, with apex rounded or triangular; socius of medium length and uniform width; vesica with 6–25 cornuti; valva with clasper Type 1, basal process absent, costal margin concave, neck short and broad, NR: 0.72–0.84, saccular corner weakly angulate, SA: 134°–160°; cucullus with dorsal lobe strongly developed, apex semicircular, distal margin evenly convex, ventral lobe weakly to moderately developed, anal angle rounded or triangular.

*Female genitalia* (unknown in *E. hirsutana*). Papillae anales Type 2 or Type 3; sterigma U-shaped or ovate; lamella antevaginalis ringlike; posterior margin of sternum 7 concavely emarginated to about one-third length of sterigma; sclerotization of ductus bursae absent or ringlike; corpus bursae with two signa of nearly equal size.

### 22. *Epiblema radicana* (Walsingham, 1879)
(Plate J, 22a–f; Plate 10: 22a–d)

*Paedisca radicana* Walsingham 1879:53; Fernald 1882:39; Obraztsov 1964:144.
*Eucosma radicana*: Fernald 1903:457; Barnes and McDunnough 1917:170.
*Griselda radicana*: Heinrich 1923:186; McDunnough 1939:50; Powell 1964:86.
*Epiblema radicana*: Obraztsov 1965b:3; Powell 1975:109; Powell 1983:35; Brown 2005:284.
*Eucosma vomonana* Kearfott 1907:90; Heinrich 1923:139; synonymy by Obraztsov (1965b).
*Epiblema vomonana*: McDunnough 1939:48; Brown 2005:284.
*Eucosma serangias* Meyrick 1912:35, unnecessary replacement name for *vomonana*.
*Epiblema serangias*: Heinrich 1923:139; McDunnough 1939:48.

**Types**. *Paedisca radicana*. Lectotype (Plate J:22a, designated by Obraztsov 1965b). ♂, Oregon, Jackson County, "to Rogue River," 4–6 May 1872, Walsingham No. 91909, slide 5735, BMNH. *Eucosma vomonana*. Lectotype (designated by Heinrich 1923). ♂, California, Placer County, Cisco, A. H. Vachell, 1 June 1905, slide 702-Obr. prepared by C. Heinrich 14 Dec 1919, AMNH.

Walsingham (1879) did not specify the number of syntypes for *P. radicana*, but he did mention wing-span values for both sexes. Obraztsov (1965b) designated the lectotype, which bears a circular red-bordered label with the inscription "Type H.T." He also reported two paralectotypes in the BMNH with the same collection data as the lectotype (1 ♂, Walsingham No. 91911, abdomen damaged; 1 ♀, Walsingham No. 91910, slide 5736). Walsingham (1879) reported the collection year for the lectotype as 1871, but the data on the pin label reads 1872. Kearfott (1907) described *E. vomonana* from six syntypes collected by A. H. Vachell in Placer County, California. Obraztsov (1965b) accounted for these specimens as follows: 2 ♂ from the unincorporated community of Cisco, 3 ♂ and 1 ♀ from the small town of Colfax. The two males from Cisco have the same collection data; one is in the AMNH, the other in the USNM. Heinrich (1923) treated this taxon under Meyrick's (1912) replacement name *E. serangias*; he considered *vomonana* to be preoccupied by *womonana* Kearfott, despite the difference in spelling. He referred to the Cisco syntype in the AMNH as the "type" (page 140) and illustrated its genitalia (fig. 256), thus identifying a name-bearing specimen. The repositories for the four paralectotypes from Colfax are: (1 ♂) in the USNM; (2 ♂, slide 700-Obr.; 1 ♀, slide 701-Obr.) in the AMNH.

Heinrich (1923) did not examine Walsingham's types of *P. radicana* and was unaware of their connection with *E. vomonana* (= *E. serangias*). He mistakenly interpreted Walsingham's description and illustration of *P. radicana* as applying to an aggregation of specimens he had chosen as exemplars of his newly proposed genus *Griselda*, and he designated *P. radicana* as the type species for *Griselda*. The resulting taxonomic confusion was addressed by Obraztsov (1965b), who proposed the current resolution of this situation and illustrated the adults and male genitalia of all the relevant types. Heinrich (1923) placed three species in *Griselda*: *radicana*, *pennsylvaniana*, and *gerulae*. They have since been reassigned, the first to *Epinotia* Hübner (R. L. Brown 1983; J. W. Brown 2005), the last two to *Chimoptesis* Powell (Powell 1964), and *Griselda* is now considered a synonym of *Epinotia*.

*Epiblema gratuitana* Heinrich is another taxon that has been associated with *E. radicana*. Heinrich (1923:268) described *E. gratuitana* from a male holotype and a female paratype in the USNM that were collected in Washington State at Eastsound and Whidbey Island, respectively. McDunnough (1939) treated this name as a junior synonym of *Epiblema purpurissatana* Heinrich (now assigned to *Notocelia* Hübner), but Powell (1983) and Brown (2005) placed *E. gratuitana* in the synonymy under *Epiblema radicana*. Heinrich's illustration of the male genitalia of *E. gratuitana* (1923: fig. 431) clearly shows two non-deciduous cornuti in the vesica, indicating that this taxon belongs in *Notocelia*. We examined both specimens and concluded they are *Notocelia purpurissatana*.

*Epiblema radicana* differs from other members of the group in having a pale grayish-white band extending continuously from the costa to the tornus along the distal margin of the median fascia. The basal and subbasal fasciae are distinguishable as a narrow line and a broad quadrate bar, respectively, both blackish in color and extending from the inner margin to the discal cell. The white interfascial spot is moderate in size and usually has some gray or black transverse reticulations. In males (Plate I:22a–b, d) it extends from the inner margin to the discal cell, in females (Plate I:22c, e) it merges with whitish or grayish costal strigulae, producing a pale band from the inner margin to the costa. The median fascia is a grayish-brown band from mid-costa to the inner margin that is overlaid with various small black marks. The subapical area contains a large brown triangular patch which has one side contiguous with the pale postmedian band, one vertex at the apex, and one vertex on the termen at approximately $CuA_1$. The distal side of the patch is edged with a narrow white line from $M_3$ to the apex. The termen has a thin black line from the tornus to the apex, and the fringe is mostly gray. This species does not have a clearly defined ocellus.

In males, the ventral surface of the uncus has a peaklike apical ridge, the neck of the valva is broad, NR = 0.83, the saccular angle is sometimes barely detectable, SA = 160°, and the basoventral margin of the cucullus extends ridgelike onto the ventral one-third of the medial surface of the neck. In females, the papillae anales are Type 2, the sterigma is U-shaped, the posterior margin of sternum 7 diverges laterally from the sterigma, and the ductus bursae lacks sclerotization. The signa in the corpus bursae are rather small, with one located at mid-bursa and the other near the juncture with the ductus bursae. In other members of the group for which the female is known, the signa are located opposite one another.

The types of *E. radicana* were collected in southwest Oregon in early May. Powell (1975) reported two colonies near San Francisco, California, one in San Mateo County, the other in Stanislaus County, with adults flying from mid-March to the end of April. The types of *E. vomonana* (e.g., Plate I:22f) were collected in east central California (Placer County) in June and July. Obraztsov (1965b) synonymized *E. vomonana* with *E. radicana*. He also determined as *E. radicana* two males in the USNM from Colorado, one collected in Jefferson County by Oslar (1965b: figs. 4, 9), the other at Bear Creek by Morrison (1965b: fig. 5). These two specimens and the types of *E. vomonana* exhibit subtle forewing features that raise some concern as to whether they are conspecific with *E. radicana*, but a thorough assessment of this situation will require a great deal more material from the Rocky Mountain Region.

## 23. ***Epiblema hirsutana*** (Walsingham, 1879)
(Plate J, 23a–f; Plate 10, 23a–c)

*Paedisca hirsutana* Walsingham 1879:50; Fernald 1882:40.
*Eucosma hirsutana*: Fernald 1903:458; Barnes and McDunnough 1917:170.
*Epiblema hirsutana*: Heinrich 1923:149; McDunnough 1939:48; Powell 1975:110; Powell 1983:35; Brown 2005:283.

**Lectotype** (Plate J:23a, here designated). ♂, California, Sonoma County, near Pluton Creek, 22–23 May 1871, BMNH(E) 819916, slide 11569, BMNH.

Walsingham (1879) mentioned four males that he collected on 23 May 1871 in Sonoma County, California and on 7 May 1872 at Rogue River, Oregon. The lectotype was selected by Obraztsov and bears a circular red-bordered label with the inscription "Type H.T."

*Epiblema hirsutana* is one of the smallest members of the group (mean FWL = 6.4 mm), surpassing only *E. deverrae* (mean FWL = 6.2 mm) in average size. Except for a relatively small white interfascial spot, the forewing is largely uniform in color (a mixture of browns and grays). The lectotype (Plate I:23a) is dark brown (possibly due to age and storage conditions), but specimens collected in the 1960s (Plate I:23d–f) are paler, with yellowish-gray tints. *Epiblema hirsutana* differs from *E. radicana* in the following respects: it lacks the pale postmedian band in the latter species; it has only a hint of a subapical patch; the subbasal and median fasciae are less conspicuous; the interfascial spot is smaller; the ocellus is well defined but somewhat obscure, with four prominent black dashes on the central field; and the white line on the termen from $M_3$ to the apex is less pronounced. In overall appearance, this species most closely resembles *E. lyallana*, but the latter is somewhat larger (mean FWL = 7.0 vs. 6.4 mm), has a larger interfascial spot, and has a more pronounced postmedian fascia that expands into a conspicuous black mark anterior to the ocellus.

The male genitalia resemble those of *E. radicana*, but the saccular angle is smaller (134° vs. 160°), the neck is slightly narrower (NR = 0.75 vs. 0.83), and the setae along the distal margin of the cucullus are a little larger. The female is unknown.

There are very few records for this species. The types were collected in southwest Oregon and Sonoma County, California. Powell (1975) reported a colony northeast of Oakland in Alameda County, California, and the USNM has a specimen obtained from the Fernald Collection (Plate J:23c) that lacks data but appears to be an exemplar given to Fernald by Walsingham. Known capture dates indicate that *E. hirsutana* flies in April and May.

## 24. ***Epiblema lyallana*** McDunnough, 1935

(Plate J, 24a–c; Plate 10, 24a–d)

*Epiblema lyallana* McDunnough 1935:141; McDunnough 1939:48; Powell 1983:35; Brown 2005:284.

**Holotype** (Plate J:24a). ♂, Quebec, Gaspe County, Mt. Lyall, W. J. Brown, 1,500 feet, 25 July 1933, slide TOR 4250, CNC.

*Epiblema lyallana* was described from the holotype, an allotype in the CNC collected by E. B. Watson on 27 July in Gaspe County, Quebec, and two paratypes (1 ♂, 1 ♀) collected by W. J. Brown on 2 August and 4 August at the type locality. The male paratype is in the USNM. We did not locate the female paratype.

*Epiblema lyallana* is a medium sized member of the group (mean FWL = 7.0 mm) with a moderately broad forewing (AR = 2.54). The predominant forewing color is brownish gray. The subbasal fascia is represented by a brown/gray bar from the inner margin to the cell that includes a prominent black spot on the inner margin and a few smaller black marks along the proximal edge of the interfascial spot. In the allotype (Plate I:24c), these spots coalesce into a moderately well-defined blackish subbasal mark. The interfascial spot is white and unmarked except for a couple of black dots on the inner margin; the median fascia is composed of a weakly discernable mid-costal bar and a black spotted pretornal mark; and the postmedian fascia is interrupted near the costa but expands into a conspicuous black band along the apical margin of the ocellus. In the allotype (Plate J:24c), the last feature is nearly as well developed as the triangular subapical mark in *E. radicana.* The ocellus is flanked laterally by gray bands and has three or four bold black dashes on a brownish central field. The distal one-half of the costa has four paired whitish strigulae delimited by black quadrate/wedge shaped marks and grayish striae, and the termen has a thin black line from tornus to apex that is bordered proximately by a series of small whitish dots that form a dashed line from tornus to apex.

In males, the valva has a short broad neck, with NR = 0.74, and a weakly angulate saccular corner, with SA = 143°. The cucullus has a strongly developed dorsal lobe that tapers slightly to a semicircular apex, an evenly curved convex distal margin, a weakly developed and narrowly rounded anal angle, and a basoventral margin that extends ridgelike onto the ventral one-fourth of the medial surface of the neck. In females, the papillae anales are Type 2, the lateral margins of the U-shaped sterigma are folded under the lamella postvaginalis, the ductus bursae has a well-developed sclerotized ring at the juncture with the ductus seminalis, and the signa are located opposite one another in the posterior one-half of the bursa.

The type series was collected in the Province of Quebec between 25 July and 2 August. We have seen only one other specimen that has been determined as *E. lyallana*, a male in the CNC collected on 23 August 1988 by G. S. Walley at Lake Louis, Alberta. That specimen lacks the bold dashes in the ocellus and the conspicuous postmedian fascia possessed by the types. We suspect it is not conspecific with *E. lyallana* but are unsure as to which, if any, of the described species it might belong. This is yet another example of the need for additional material in order to confidently circumscribe some of the members of this group.

## 25. ***Epiblema deverrae*** Brown, 1991
(Plate K, 25a–d; Plate 11, 25a–d)

*Epiblema deverrae* Brown and Powell 1991:110; Brown 2005:281.

**Holotype.** ♂, California, Monterey County, Landels-Hill Big Creek Reserve, 5 mi N. Lucia, J. Powell, 4–6 June 1982, UCB.

*Epiblema deverrae* is the smallest member of the group (mean FWL = 6.2 mm vs. 6.4–8.7). Aside from a small white interfascial spot that is sometimes tinted with pale orange, the forewing coloration is a mottled combination of black and brown. The fasciate markings are black and include a variably expressed subbasal mark, a prominent quadrate mid-costal bar, a weakly defined pretornal mark, and a postmedian fascia that is strongly constricted near the costa but expands into a conspicuous black patch anterior to the ocellus. The ocellus is obscure but discernable, with four black bars on the central field and dark gray bands on the lateral margins, the proximal band often connecting with a gray stria emanating from costal strigula 6. The costal strigulae are whitish and separated by black wedge-shaped marks. The termen has a thin black line from tornus to apex that is edged proximally by a narrow reddish-brown band, the last containing several small whitish spots.

The male genitalia resemble those of *E. lyallana* except that the basoventral margin of the cucullus does not extend onto the medial surface of the neck, a feature that also distinguishes *E. deverrae* from *E. radicana* and *E. hirsutana*. In females, the papillae anales are Type 2, the sterigma is ovate, the posterior projections of sternum 7 are approximate to the sterigma, the ductus bursae has a sclerotized ring near the juncture with the ductus seminalis, and the signa are located opposite one another at mid-bursa.

To our knowledge, this species is known only from the type locality (described in Brown and Powell, 1991). Those authors reported that the moth flies diurnally and appears to be associated with *Madia madioides* (Nutt.) Greene (woodland madia). They found no galls on the plant stems and suggested that the larva may be a root-borer. The capture dates for the 13 specimens we examined ranged from 26 May to 16 June.

## 26. ***Epiblema periculosana*** Heinrich, 1923
(Plate K, 26a–f; Plate 11, 26a–c)

*Epiblema periculosana* Heinrich 1923:268; McDunnough 1939:48; Powell 1983:35; Brown 2005:284.

**Holotype.** ♂, British Columbia, Mount McLean, A. W. Hanham, 5,500 feet, August, genitalia slide 72816, wing slide 72817, USNM.

Heinrich (1923) described *E. periculosana* from three males collected at the type locality. He identified the holotype by illustrating its genitalia.

*Epiblema periculosana* is the second largest member of the group (mean FWL = 8.0 mm), exceeded in size only by *E. arctica* (mean FWL = 8.7 mm). The forewing has a large white interfascial spot and white suffusion in the subcostal area beyond the median fascia. The rest of the forewing coloration is a mottled mixture of brown, gray, and black. The subbasal and pretornal marks are moderately well defined and speckled with black. The ocellus is weakly expressed but discernable, with transverse gray bands on the lateral margins and four black dashes/dots in the central field. The costal strigulae are white, and the termen has a thin black line from the tornus to the apex that is edged proximally by a reddish-brown band containing several whitish spots.

The male genitalia resemble those of *E. lyallana* in that the dorsal lobe of the cucullus is particularly well developed and tapers evenly to a rounded apex, but the valva neck is a little broader (NR = 0.80 vs. 0.74), the saccular angle is slightly larger (SA = 152° vs. 143°), and the basoventral margin of the cucullus does not extend onto the medial surface of the neck. In females, the papillae anales are Type 2, the sterigma is U-shaped, the posterior extremities of the bandlike lamella antevaginalis attenuate onto the lamella postvaginalis, and the ringlike sclerotization of the ductus bursae has a substantial lobelike extension that projects nearly halfway from the ductus seminalis to the ostium.

Current records indicate that *E. periculosana* is the most widely distributed member of the group. We examined specimens from Alberta, British Columbia, California, Colorado, Idaho, Oregon, Washington, and Wyoming. Most of the specimens were collected at elevations of 5,000–12,200 feet, but a few were taken at 2,500 feet in Bonner County, Idaho. Capture dates range from mid-June to late August. It has been our experience that males are readily collected in black-light bucket traps, but we have seen only one female (in the CNC) out of hundreds of museum specimens.

## 27. ***Epiblema arctica*** Miller, 1985
(Plate K, 27a–e; Plate 11, 27a–d)

*Epiblema arctica* Miller 1985:36; Gilligan et al. 2020a:36.
*Epiblema simplonianum* (not Duponchel): Miller 1995:251; Brown 2005:285.

**Holotype.** ♂, Alaska, Anaktuvuk Pass, K. W. Philip, 11 July 1970, slide JCL 1027839, AMNH.

Miller (1985) described *E. arctica* from the holotype and 28 paratypes (23 ♂; 5 ♀), all collected in Alaska. Ten years later, amidst of a discussion of Holarctic Beringian Tortricidae, Miller (1995) decided that *E. arctica* is conspecific with the Palearctic *E. simplonianum* (Duponchel), and Brown (2005) then listed the name in the synonymy under the later species. After analyzing DNA data, Gilligan et al. (2020a) elevated *E. arctica* to species status, concluding that *E. simplonianum* is not Holarctic and that the two taxa are distinct. We examined the male genitalia of both species, comparing the drawings for *E. simplonianum* in Kuznetsov (1987) and Razowski (2003) with four of our preparations from paratypes of *E. arctica*, and found differences in the valva that support this conclusion. Both drawings referred to above show the ventral two-thirds of the medial surface of the valva neck to be substantially “scooped out.” That is not the case in the paratypes of *E. arctica* that we examined.

*Epiblema arctica* is the largest member of the group (mean FWL = 8.7 mm, vs. 6.2–8.0). Its overall forewing appearance is blackish brown, except for a prominent whitish interfascial spot that extends from the inner margin to the discal cell. The maculation includes a blackish-brown basal patch with inconspicuous black reticulations/speckling and a blackish-brown, bandlike median fascia that is speckled with black, especially in the pretornal area. In males, the two marks merge anterior to the radius, but in females (Plate K:27d) they are separated by a white spot on the costa. The ocellus is somewhat obscure, with its narrow whitish central field edged laterally by grayish bands and bisected by a transverse line of black dots. The white costal strigulae are conspicuous and sharply defined by black quadrate or wedge-shaped marks, and the termen features a variably expressed black line from tornus to apex and a vaguely defined whitish bar from $M_2$ to $R_5$.

In males, the uncus is triangular, the vesica has 15–25 cornuti (n = 5), the valval neck is short and broad, NR = 0.84, the saccular corner is weakly developed, and SA = 155°. The cucullus is large, with a strongly developed, triangular, ventral lobe whose basoventral margin does not project onto the medial surface of the neck. In females, the papillae anales are Type 3, the U-shaped sterigma has acute posterolateral corners, sternum 7 is rather long and narrow (length about 1.5 times width) with lateral margins nearly parallel, the sclerotization of the ductus bursae is ringlike, and the signa are moderately large and located opposite one another.

We examined 24 specimens (20 ♂, 4 ♀), including 15 paratypes. They were collected between 26 June and 24 July at six different locations in Alaska.

## The *walsinghami* group (species 28–30)

The *walsinghami* group is composed of *E. walsinghami*, *E. iowana*, and *E. coracinana*, three species that are similar in genitalia but easily separated by forewing appearance. Group characteristics include:

*Male genitalia.* Uncus weakly to moderately developed and clearly differentiated from dorsolateral shoulders of tegumen; socius short and fingerlike; vesica with 7–30 cornuti; valva with clasper Type 1, basal process absent, costal margin concave, neck moderately broad, NR: 0.64–0.76, saccular corner angulate, SA: 149°–158°; ventral one-half of medial surface of neck densely setose; cucullus with dorsal lobe strongly developed, apex rounded, distal margin weakly convex, ventral lobe weakly to moderately developed, basoventral margin overlapping ventral one-third to one-half of medial surface of neck.

*Female genitalia.* Papillae anales Type 2; sterigma U-shaped or V-shaped; lamella postvaginalis microtrichiate, with lateral margins curled inward; lamella antevaginalis ringlike, merging posteriorly into lateral margins of lamella postvaginalis; posterior margin of sternum 7 concavely emarginated to about one-half length of sterigma and laterally divergent from lamella postvaginalis; sclerotization of ductus bursae absent or ringlike; corpus bursae with two signa, one somewhat larger than the other, located slightly posterior to mid-bursa.

### 28. ***Epiblema walsinghami*** (Kearfott, 1907)
(Plate L, 28a–c; Plate 12, 28a–d)

*Enarmonia walsinghami* Kearfott 1907:57.
*Laspeyresia walsinghami*: Barnes and McDunnough 1917:174.
*Epiblema walsinghami*: Heinrich 1923:150; McDunnough 1939:48; Powell 1983:35; Wright and Covell 2003:231; Brown 2005:286; Gilligan et al. 2008:126.

**Lectotype** (designated by Heinrich 1923). ♀, New Jersey, Essex County, W. D. Kearfott, 30 April 1899, AMNH.

Kearfott (1907) mentioned seven syntypes (sex not specified). Wright and Covell (2003) accounted for six of those specimens. There is only one in the AMNH labeled Essex County, New Jersey, so Heinrich's (1923) statements about the type locality and the specimen repository constitute a lectotype designation. Two of the syntypes were collected by W. F. Fiske at Tryon, North Carolina and are not conspecific with the lectotype (Wright and Covell 2003); one (in the USNM) is a female of *E. infelix*, the other (in the AMNH) is a male of *E. gibsoni*. The three remaining paralectotypes were collected in New Jersey, one at Great Notch in the Watchung Mountains (1 ♂, AMNH) and two at Hemlock Falls (1 ♂, USNM; 1 ♀, AMNH). Miller (1987) confused this species with *E. gibsoni*, presenting under the name *walsinghami* an adult photograph and a male valva drawing of *E. gibsoni*.

*Epiblema walsinghami* is a medium sized species (mean FWL = 7.1 mm) with one of the broadest forewings in the genus (AR = 2.32). Aside from a white triangular interfascial spot that extends into the discal cell, the forewing is dark brownish gray, with various black spots/patches. The last include a broad band on the inner margin from the base to the interfascial spot, a quadrate mid-costal bar, a mottled pretornal mark, and a prominent postmedian fascia that expands into a substantial black spot anterior to the ocellus. The ocellus is obscure but recognizable, with gray bands on the lateral margins and three black spots on a brown central field. The costal strigulae on the distal one-half of the wing are pale gray and separated by black wedges, and the termen has a thin black line from tornus to apex.

In males, the vesica has 15–21 cornuti, NR = 0.73, and SA = 154°. In females, the sterigma is U-shaped, with sharply-pointed posterior projections at the posterolateral corners, and the ductus bursae lacks sclerotization.

*Epiblema walsinghami* has been confused with *E. infelix* and *E. gibsoni* (see the account for *E. infelix*), but it differs from those two species in size (mean FWL = 7.1 vs. 7.9 and 7.8 mm), forewing geometry (AR = 2.32 vs. 2.71 and 2.53), valva shape (Plate 12:28a–c vs. Plate 7:16a–e and 17a–d), and the relative sizes of the signa in the corpus bursae (unequal vs. equal).

This species is poorly represented in collections. We examined 14 specimens (4 ♂, 10 ♀) collected in Illinois, Kentucky, New Jersey, Ohio, Pennsylvania, and Ontario (Ottawa). Capture dates range from 21 April to 11 June. One female from Montgomery County, Ohio was collected diurnally, and one female from Bullitt County, Kentucky was found in a malaise trap. We do not know of any specimens that have been collected at black light.

## 29. *Epiblema iowana* McDunnough, 1935

(Plate L, 29a–i; Plate 12, 29a–e)

*Epiblema iowana* McDunnough 1935:142; McDunnough 1939:48; Powell 1983:35; Miller 1985:33; Miller 1987:58; Brown 2005:283; Gilligan et al. 2008:153.
*Epiblema naoma* Clarke 1953:226; synonymy by Miller (1985).

**Types.** *Epiblema iowana.* Holotype (Plate L:29a). ♂, Iowa, Renwick, G. S. Walley, 19 May 1928, slide TOR 4057, CNC. *Epiblema naoma.* Holotype. ♂, Illinois, Putnam County, M. O. Glenn, 25 June 1950, USNM.

The description of *E. iowana* was based on the holotype and an allotype, both from Humboldt County in north central Iowa. Clarke (1953) described *E. naoma* from 18 specimens (9 ♂, 9 ♀) collected in Putnam County, Illinois, some having been reared from larvae boring in roots of *Ratibida pinnata* (Vent.) (prairie coneflower). The synonymy was recognized by Miller (1985).

*Epiblema iowana* is similar in size to *E. walsinghami* (mean FWL = 7.3 vs. 7.1 mm) but has a slightly more elongate forewing (AR = 2.59 vs. 2.32). This species varies in predominant forewing color from tan to dull brownish gray and has dark brown or blackish fasciate markings. In males (Plate L:29a–b, e, g, i), the subbasal fascia is represented by a quadrate mark from the inner margin to the radius; in females (Plate K:29c–d, f, h), it extends bandlike to the costa. The interfascial spot varies from white to tan to a mottled mixture of white, brown, and gray, often extending to the costa as a transverse band. In both sexes, the median fascia is moderately well defined as a gray-brown band that is overlaid with a prominent blackish mid-costal bar and black blotches in the pretornal area. The preterminal fascia expands into a conspicuous dark mark anterior to the ocellus; the ocellus is obscure, with a transverse medial line of black dots that sometimes appears to be a branch of the preterminal fascia; and the termen usually has a narrow black band from $CuA_2$ to the apex.

In males, the vesica has 7–10 cornuti, NR = 0.76, and SA = 158°. In females, the sterigma is V-shaped, with the anterior extremity broadly rounded, and the ductus bursae lacks sclerotization.

This species is associated with the tall grass prairie region of the upper Midwest but also occurs in remnant prairie habitat as far south as Louisiana and Mississippi. Adults fly in April and May.

**30. *Epiblema coracinana*** Wright and Gilligan, new species
(Plate L, 30a–c; Plate 12, 30a–c)

**Diagnosis.** *Epiblema coracinana* is slightly smaller than *E. walsinghami* and *E. iowana* (mean FWL = 6.6 mm vs. 7.1 and 7.3), has a narrower forewing (AR = 2.66 vs. 2.32 and 2.59), and has a longer male costal fold (CFR = 0.57 vs. 0.47 and 0.45). The three group members are substantially similar in forewing pattern but *E. coracinana* is easily distinguished from the other two by the color of its inconspicuous interfascial spot (blue gray vs. white or tan). Of the three species, *E. coracinana* is the only one with blue-gray scaling in the interfascial spot and in the ocellus, but this difference can be difficult to detect in flight-worn specimens. In male genitalia, *E. coracinana* differs from *E. walsinghami* and *E. iowana* in the following respects: it has the narrowest valva neck (NR = 0.64 vs. 0.73 and 0.76), the smallest saccular angle (149° vs. 154° and 158°), the largest number of cornuti in the vesica (23–30 vs. 15–21 and 7–10), and the most elongate cucullus (length-to-width ratio = 2.22 vs. 1.66 and 1.78). Moreover, the distal one-third of the medial surface of the cucullus has a distinctive band of short discretely distributed setae from the anal angle to three-fourths the distance to apex (Plate 12:30a). This feature is not present in other members of the genus. Females of *E. walsinghami* and *E. iowana* lack sclerotization of the ductus bursae; those of *E. coracinana* have a narrow weakly-sclerotized ring at the juncture with the ductus seminalis.

**Description.** *Head.* Labial palpus with medial surface tan, lateral surface brown; lower frons pale brown, upper frons and vertex brown to blackish brown; antenna brown dorsally, tan ventrally. *Thorax.* Dorsal surface blackish brown; fore- and mid-leg with anterior surfaces brown, posterior surfaces tan; hind-leg mostly tan; tarsus with alternating tan and dark brown annulations. Forewing: ♂ FWL 6.0–6.3 (mean = 6.2, n = 2), AR = 2.77, CFR = 0.57; ♀ FWL 5.6–7.4 (mean = 6.7, n = 6), AR = 2.62; dorsal surface dark brown, with fasciate markings accentuated in black; basal fascia represented by weakly defined black line from inner margin to radius; subbasal mark extending from inner margin to radius, with black edging on lateral margins; interfascial spot dull blue gray, quadrate, extending from inner margin to cell; pretornal mark brown with black speckling and black lateral margins; postmedian fascia black; costal strigulae barely discernable, brown/gray, separated by obscure black marks; ocellus lustrous blue gray, with medial line of black dots; termen with dark orange-brown band from $M_3$ to apex, bordered distally by thin black line from tornus to apex; fringe dark gray. Hindwing: Dark grayish brown. *Abdomen.* Male genitalia (n = 2): Typical of the group; vesica with 23–30 cornuti; NR = 0.64; SA = 149°; cucullus moderately elongate (length-to-width ratio = 2.22), with band of short, stout, discretely distributed setae on distal one-third of medial surface from anal angle to three-fourths distance to apex. Female genitalia: Typical of the group; sterigma U-shaped; ductus bursae with narrow weakly sclerotized ring at ductus seminalis; corpus bursae with one signum distinctly larger than the other.

**Holotype** (Plate L:30a; Plate 12:30c). ♀, Pennsylvania, Allegheny County, Oak Station, F. Marloff, 16 June 1908, slide DJW 4304, CMNH.

**Paratypes** (2 ♂, 5 ♀). PENNSYLVANIA, Allegheny County, Oak Station, F. Marloff, 21 May 1915 (1 ♂, MS 13292), 14 June 1907 (1 ♀, slide DJW 4305), 5 July 1908 (1 ♂, slide DJW 4306), 6 July 1906 (1 ♀); Allegheny County, Pittsburg, W. Zanol, 22 July 1985 (1 ♀, slide DJW 4307); Allegheny County, Sample Station, H. Kahl, 5 July 1909 (1 ♀, genitalia in vial); Huntingdon County, Stone Valley Recreation Area, F. D. Fee, 26 May 2008 (1 ♀, genitalia in vial). Paratype depositories: CMNH, USNM.

**Etymology.** The specific epithet is derived from the Latin *coracinus*, meaning ravenlike or black as a crow.

**Comments.** Four of the types were collected by Fred Marloff between mid-June and early July, 1906–1908. Two of those specimens bear a handwritten label with the inscription "*Olethreutes azurana* Kearf." It appears that W. D. Kearfott examined these specimens but never published the name.

## The *otiosana* group (species 31–32)

The *otiosana* group is composed of *E. otiosana* and *E. dorsisuffusana*, two species with similar genitalia but diagnostic differences in the expression of the interfascial spot and the ocellus. Group characteristics include:

*Forewing*. Mean FWL: 7.6, 8.1 mm; AR: 2.61, 2.68; CFR: 0.40, 0.40; forewing predominantly blackish brown to grayish brown, with reddish-brown suffusion in the apical area and along the termen from $CuA_1$ to apex; interfascial spot white and conspicuous; basal patch and median fascia confluent anterior to interfascial spot; ocellus with gray bands on lateral margins and one to three black spots on whitish central field; costal strigulae white or gray, subdued except for strigula 9, separated by reddish-brown and/or blackish marks.

*Male genitalia*. Uncus broad (basal width about twice height), clearly differentiated from dorsolateral shoulders of tegumen, with apex rounded; socius short, broad, and essentially uniform in width; vesica with 21–39 cornuti; valva with clasper Type 1, neck narrow, NR: 0.35, 0.36, saccular corner rounded, SA: 110°, 106°; cucullus with dorsal lobe strongly developed and tapering slightly toward rounded apex, distal margin weakly convex (sometimes with shallow indentation on ventral one-half), ventral lobe moderately developed, anal angle rounded, series of spiniform setae along distal margin from anal angle to one-half or three-fourths distance to apex.

*Female genitalia*. Papillae anales Type 2; sterigma ovate, with posterior margin nearly straight and lateral margins slightly folded under lamella postvaginalis; lamella antevaginalis bandlike, attenuating posteriorly into lateral margins of lamella postvaginalis; posterior margin of sternum 7 concavely emarginated to about one-third length of sterigma and laterally divergent from lamella postvaginalis; ductus bursae with sclerotized ring at juncture with ductus seminalis; corpus bursae with two signa, one larger than the other, located opposite one another and slightly posterior to mid-bursa.

### 31. *Epiblema otiosana* (Clemens, 1860)
(Plate M, 31a–f; Plate 13, 31a–e)

*Monosphragis otiosana* Clemens 1860:354.
*Paedisca otiosana*: Fernald 1882:41.
*Eucosma otiosana*: Fernald 1903:459; Kearfott 1905:355; Barnes and McDunnough 1917:171.
*Epiblema otiosana*: Heinrich 1923:154; Decker 1932:503; McDunnough 1939:48; Darlington 1947:95; Powell 1983:35; MacKay 1959:35; Miller 1973:218; Miller 1987:59; Brown 2005:284; Gilligan et al. 2008:125.
*Paedisca inclinana* Zeller 1875:301; Fernald 1882:41 (synonym of *otiosana*).

**Types**. *Monosphragis otiosana*. Lectotype (designated by Darlington 1947). ♂, Type No. 7216, ANSP. *Paedisca inclinana*. Lectotype (here designated). ♂, New York, July 1868, Speyer, slide 11563, BMNH.

Clemens' (1860) description of *M. otiosana* provides no information about the number of syntypes or the type locality. Darlington (1947) attributed the designation of the lectotype to Heinrich (1923), but Heinrich's comments provide no identifying information about the specimen except that it resides in the ANSP and that the type locality is Pennsylvania. Miller (1973) published a photograph of the lectotype and credited the designation to Darlington. Zeller's (1875) description of *P. inclinana* mentions two specimens (1 ♂, 1 ♀) that he retained in his collection and a male that he returned to the MCZ. Heinrich (1923) stated that the type was in the MCZ, but it is not mentioned in Miller and Hodges (1990), and it was reported missing by Brown (2005). The two specimens retained by Zeller are now in the BMNH, having passed upon Zeller's death (in 1883) to the Walsingham Collection and from there to the BMNH (in 1910). The specimen formally designated above as the lectotype bears, in addition to the collection data and slide number, a white "Type" label (presumably added by someone at the BMNH), an accession label with inscription "Zeller Coll. Walsingham Collection 1910-427," a round "Paralectotype" label, and a small white label with the inscription "N. Obraztsov." Zeller's illustration of the forewing of *P. inclinana* is an excellent match to Miller's (1973) photograph of *M. otiosana*. The synonymy of the two names has been recognized since Fernald (1882).

The forewing is uniformly dark grayish brown except for a white sharply-defined interfascial spot, a whitish/grayish ocellus, and some reddish-brown scaling near the apex and along the termen. The interfascial spot extends from the inner margin to midway through the discal cell and varies in shape, its most common form having a lateral projection on its distal margin that points toward the ocellus and attenuates distally (Plate M:31a, c, e, f). Infrequently, the apex of the projection is joined to the ocellus by a thin white line (Plate M:31e). The ocellus is well defined, with gray bands along the lateral margins and three short black dashes on a whitish/grayish central field. The costa has one conspicuous white strigula (number 9) and several obscure but discernable gray strigulae on the distal two-thirds of the wing. *Epiblema otiosana* is slightly smaller than *E. dorsisuffusana* (mean FWL = 7.6 vs. 8.1 mm). The two species are similar in genitalia and have a distinctive valva shape that is unlike that of any of their congeners. Diagnostic differences between this species and *E. dorsisuffusana* are discussed in the latter account.

*Epiblema otiosana* is broadly distributed in eastern United States, from Rhode Island to South Dakota south to Florida and Texas. Adults have been collected from May to September. This species is known in the applied entomology literature as the "bidens borer." Decker (1932) described rearings from four species of *Bidens* L. (beggarticks): *B. frondosa* L. (devil's beggartick), *B. cernua* L. (nodding beggartick), *B. connata* Muhl. ex Willd. (purplestem beggartick), and *B. vulgata* Greene (big devil's beggartick). By his account, the eggs are deposited on leaves of the host, and the first instar larva mines leaves and buds. Second and third instar larvae feed in the buds, blossoms, and seed heads. Thereafter, the larva migrates to the main stem and burrows downward in a spiral fashion, girdling the stem and causing the upper part of the plant to wilt. The mature larva proceeds to the bottom of the burrow, partitions off the upper portion of the burrow with silk and frass, and bores an exit hole in the lower chamber. Pupation occurs in the lower chamber, as does hibernation of the overwintering larva. Decker (1932) reported up to three generations per year in Iowa. He collected overwintering larvae from seven different plant genera other than *Bidens* but found no evidence of larval feeding and suggested that the larvae had migrated from their host and sought refuge in any suitable medium. Heinrich (1923) mentioned *Ambrosia* L. (ragweed) and *Polygonum* L. (knotweed) as larval hosts but provided no documentation.

### 32. ***Epiblema dorsisuffusana*** (Kearfott, 1908)
(Plate M, 32a–d; Plate 13, 32a–e)

*Eucosma dorsisuffusana* Kearfott 1908:167; Barnes and McDunnough 1917:171.
*Epiblema dorsisuffusana*: Heinrich 1923:152; McDunnough 1939:48; Powell 1983:35; Miller 1987:60; Brown 2005:281; Gilligan et al. 2008:124.

**Lectotype** (designated by Heinrich 1923). ♂, Ohio, Cincinnati, A. Braun, 23 June 1906, slide CH 15 December 1919, AMNH.

Kearfott (1908) described *E. dorsisuffusana* from 10 specimens (sex not reported) collected 6–23 June by Annette Braun at Cincinnati, Ohio and 3 July by Henry Engle at Pittsburgh, Pennsylvania. Heinrich (1923) illustrated the genitalia of a male that he called the type, reported the type locality as Cincinnati, Ohio, and named the AMNH as the repository. We examined Heinrich's genitalia slide for that specimen (in the AMNH), which bears the data listed above for the lectotype as well as the inscription "TYPE," but we did not find the specimen. There is an *E. dorsisuffusana* male from Cincinnati in the AMNH type collection that carries a blue-green "LECTOTYPE" label, but that specimen has an intact abdomen and, therefore, is not the type intended by Heinrich. It seems the actual type specimen might be lost, and its genitalia slide may have to suffice as the lectotype.

This species is similar in size to *E. otiosana* (FWL: 6.9–9.0 mm vs. 5.4–9.2 mm) but is slightly larger on average (mean FWL = 8.1, n = 24 vs. 7.6, n = 53). The two taxa are essentially indistinguishable in forewing geometry (AR = 2.68 vs. 2.61). *Epiblema dorsisuffusana* differs from *E. otiosana* in forewing color (almost black vs. grayish brown), in the expression of the interfascial spot (broader, projecting only halfway across the wing, and connecting distally to the ocellus), and in the expression of the ocellus (bright white with a gray band on the proximal margin vs. grayish with a gray band on each lateral margin). In *E. dorsisuffusana*, the interfascial spot has two black marks on the inner margin and is weakly separated from the ocellus by remnants of a fragmented pretornal mark. The result is a conspicuous white band along the inner margin from the subbasal position to the tornus that is speckled with black or gray spots in the pretornal position.

The male genitalia are as described for the group, but they exhibit a fair amount of variation in the shape of the cucullus and the development of a basal process. The valva of the lectotype (Heinrich 1923: fig. 280) resembles that of the specimen illustrated on Plate 13:32e. It has a slight incurvation of the ventral one-half of the distal margin of the cucullus and a moderately developed basal process. Variation in these two features is illustrated in Plate 13:32d–f with specimens from Ohio, Maryland, and North Carolina, respectively. A second valva form (which is substantially identical to that of *E. otiosana*) is illustrated in Plate 13:32a–c with individuals from Iowa, Oklahoma, and North Carolina, respectively. In the latter specimens, the cucullus is less elongate than in typical *E. dorsisuffusana*, its distal margin is weakly convex throughout, and the basal process is absent. We found no female genitalia characteristics that would reliably separate *E. dorsisuffusana* from *E. otiosana*. The variation in male genitalia might be indicative of two distinct species, but we found no stable differences in forewing appearance or geographic distribution to support the diagnosis of a new cryptic species.

We examined 37 specimens (32 ♂, 5 ♀) from the region extending from Maine to Iowa south to North Carolina, Mississippi, Arkansas, and Oklahoma. Adult capture dates range from early May to the end of July.

## Species not assigned to a group

### 33. ***Epiblema boxcana*** (Kearfott, 1907)
(Plate M, 33a–e; Plate 14, 33a–e)

*Eucosma boxcana* Kearfott 1907:87; Barnes and McDunnough 1917:170.
*Eucosma aspista* Meyrick 1912:35, unnecessary replacement name for *boxcana*.
*Epiblema boxcana*: Heinrich 1923:139; McDunnough 1939:48; Powell 1983:35; Miller 1987:56; Brown 2005:280; Gilligan et al. 2008:120.

**Lectotype** (designated by Heinrich 1923). ♂, Ohio, Cincinnati, Annette Braun, 21 May 1904, AMNH.

Kearfott (1907) based *E. boxcana* on 10 syntypes (sex not reported) collected between 17 May and 4 June at the following locations: Caldwell, New Jersey; Plummers Island, Maryland (by A. Busck); Philadelphia, Pennsylvania (by F. Haimbach); Cincinnati, Ohio (by A. Braun); Kerrville, Texas (by Dr. Barnes); and Iowa (by C. P. Gillette). The AMNH has only one male syntype from Cincinnati, so Heinrich's (1923) comments on the type serve to designate a name-bearing specimen. That specimen was mentioned by Klots (1942) as a lectotype, and it bears a blue-green "LECTOTYPE" label that presumably was attached by Klots.

Typical *E. boxcana* (Plate M:33a–d) is a medium sized grayish-brown species (mean FWL = 7.0 mm) with brown to blackish-brown markings. The proximal margin of the interfascial spot is sharply defined by a dark brown weakly-sinuous line along the distal edge of the subbasal mark, the grayish-white interfascial spot extends to the cubitus, the quadrate pretornal mark is conspicuous, and the postmedian fascia is obscure near the costa but well defined along the distal margin of the ocellus. The ocellus is mostly gray, with one or two black dashes in the narrow pale central field. The costal strigulae are obscure, gray, and separated by brownish-gray striate markings. Some dark specimens from Texas (Plate M:33d) have an almost blackish appearance. Plate M:33e depicts one of two specimens in the USNM that J. F. G. Clarke collected in Whitman County, Washington. Clarke reported (on a pin label) that they were reared from *Xanthium spaciosum* L., a plant now known as *X. spinosum* L. (spiny cocklebur). This phenotype resembles typical *E. boxcana* in size, forewing pattern, and male genitalia (Plate 14:33d), but its predominant color is brown (instead of grayish brown), and it has an unusually large amount of white in the ocellus and the costal strigulae. These specimens might be representatives of a distinct western sibling species, but lacking sufficient material to make a proper comparison, we tentatively refer them to *E. boxcana*.

In males, the uncus is well developed and sharply differentiated from the dorsolateral shoulders of the tegumen. Its apical margin has a shallow medial indentation, producing a weakly bilobed effect. The socius is short and substantially uniform in width, and the vesica has 20–28 cornuti. The valva has a Type 1 clasper, a conspicuous basal process, a moderately broad neck, NR = 0.64, a weakly defined saccular corner, and SA = 157°. The cucullus has a strongly developed dorsal lobe that attenuates slightly to a semicircular apex, a convex distal margin, a weakly developed ventral lobe, and a rounded anal angle. In females, the papillae anales are Type 2, the ventral extremities of tergum 8 bear a few scales along with some hairlike setae (note scale sockets in Plate 14:33e), the sterigma is U-shaped, and the posterior extremities of the lamella antevaginalis attenuate onto the surface of the lamella postvaginalis. The posterior margin of sternum 7 is weakly concave and diverges laterally from the lamella postvaginalis; the ductus bursae has a small sclerotized patch near the juncture with the ductus seminalis, and the corpus bursae has two signa of nearly equal size located opposite one another near mid-bursa.

The specimens we examined document a range that includes the region from Nebraska to New Hampshire south to Texas, Louisiana, Mississippi, Alabama, and North Carolina. Miller (1987) reported the species from Michigan and Minnesota. Adult capture dates range from 18 March to 17 August, but most are in April or May. The only western records we encountered were the specimens mentioned above from Whitman County, Washington that were reared from *X. spinosum*. As far as we know, there have been no reports of larval hosts for this species in eastern North America.

## 34. *Epiblema resumptana* (Walker, 1863)

(Plate N, 34a–f; Plate 14, 34a–d)

*Penthina resumptana* Walker 1863:376; Heinrich 1923:261 [omitted species].
*Paedisca resumptana*: Walsingham 1879:44.
*Proteopteryx resumptana*: Fernald 1882:46; Fernald 1903:464; Barnes and McDunnough 1917:173.
*Eucosma resumptana*: McDunnough 1939:48.
*Epiblema resumptana*: McDunnough 1959:1; Powell 1983:35; Miller 1987:59; Brown 2005:284; Gilligan et al. 2008:122.
*Paedisca abbreviatana* Walsingham 1879:54.
*Eucosma abbreviatana*: Fernald 1882:40; Fernald 1903:458; Barnes and McDunnough 1917:170.
*Epiblema abbreviatana*: Heinrich 1923:155; McDunnough 1939:49; Powell 1983:35; Brown 2005:285; Gilligan et al. 2008:122.

**Types**. *Paedisca resumptana*. Holotype. ♂, Nova Scotia, Lieutenant Redman, slide 5459, BMNH. *Paedisca abbreviatana*. Lectotype (here designated). ♂, Washington, D.C., 1 May 1871, Walsingham, BMNH.

Walker (1863) appears to have described this species from a single male specimen. Walsingham (1879) placed it in *Paedisca*, provided a short description, and mentioned a male type with wing expanse of "14 millims" (i.e., FWL = 6.4 mm). We follow Brown (2005) in recognizing that specimen as a holotype. Walsingham's (1879) description of *E. abbreviatana* mentions two male specimens collected at Washington, D.C. The BMNH has two such specimens that we believe to be the syntypes, one with a round "Type B.M." label, the other with a cotype label. We chose the former specimen as the lectotype. There is some confusion about its collection date. One of its labels gives the date as 1 May 1871, but Walsingham (1879) reported "the beginning of August, 1872." It may be impossible to determine which is correct, but the specimens of *E. resumptana* that we examined were all collected in April, May, or June, so we included in the lectotype designation the date that occurs on the specimen label. This specimen also bears labels with the inscriptions "Brit. Mus. 1880-75" and "*Paedisca abbreviatana* Wlsm. Type." Neither syntype is dissected. McDunnough (1959) conveyed comments he received from J. D. Bradley (at the BMNH) indicating that, despite being in poor condition, the type of *P. resumptana* is superficially identical to the type of *P. abbreviatana* and that its genitalia are identical to those illustrated by Heinrich (1923: fig. 255) of a specimen of *P. abbreviatana* (in the USNM) from New Haven, Connecticut. The synonymy was published by Brown (2005).

*Epiblema resumptana* is a small grayish-brown species (mean FWL = 5.6 mm). The forewing has brown markings and pale brownish-gray interfascial areas, the latter usually with brown transverse reticulations. The fasciate markings include a thin weakly discernable basal fascia, a moderately prominent subbasal fascia that extends from the inner margin to the radius in males or to the costa in females, and a median fascia represented by a mid-costal bar and a pretornal mark that merge in females into a complete fascia (Plate N:34c, d). The ocellus is inconspicuous but discernable, with gray bands on the lateral margins and three black dashes or spots on a brownish central field. The costal strigulae beyond the median fascia are grayish and moderately well delimited by brown striate marks.

In males, the uncus is triangular and clearly differentiated from the dorsolateral shoulders of the tegumen, the socius is short and tapers slightly to a rounded apex, and the vesica has 8–15 cornuti. The valva has a Type 1 clasper, lacks a basal process, has a concave dorsal margin, a broad neck, NR = 0.78, an angulate saccular corner, and SA = 148°. The cucullus is large compared to the basal portion of the valva. Its dorsal lobe is strongly developed and tapers to a rounded apex, its distal margin is convex except for an occasional shallow indentation hear the apex (Plate 14:34a–b), and its basoventral margin extends slightly onto the medial surface of the neck. In females, the papillae anales are Type 3, the sterigma is U-shaped, the posterior margin of the lamella postvaginalis has a medial indentation, and the posterior extremities of the bandlike lamella antevaginalis attenuate onto the surface of the lamella postvaginalis. Sternum 7 is elongate, with length about 1.25 times average width and with posterior margin approximate to the sterigma. The ductus bursae has a sclerotized ring at the juncture with the ductus seminalis, and the corpus bursae has two signa, one a little larger than the other, located opposite one another near the juncture with the ductus bursae.

McDunnough (1959) observed adults of *E. resumptana* in association with *Anaphalis* DC. (pearly everlasting), and Miller (1987) reported the likely larval host as *Anaphalis margaritacea* (L.) Benth (western pearly everlasting). In the East, the range of this moth includes the region from Nova Scotia to Saskatchewan south to New Jersey, Kentucky, and Illinois. The USNM has several specimens that were collected by J. F. G. Clarke in Whitman County, Washington and Baker County, Oregon that resemble *E. resumptana* in forewing appearance and male genitalia (e.g., Plate N:34e–f, Plate 14:34c). Additional material, including females, would be necessary to properly evaluate these individuals, but for the present we refer them to *E. resumptana*. Adult capture dates range from mid-April to mid-June.

## 35. *Epiblema macneilli* Powell, 1975
(Plate N, 35a–c; Plate 14, 35a–d)

*Epiblema macneilli* Powell 1975:102; Powell 1983:35; Brown 2005:284; Powell and Opler 2009:135.

**Holotype**. ♂, California, Inyo County, near Mono Pass, 12,000 feet, C. D. MacNeill, 11 August 1958, CAS.

The type series includes the holotype, an allotype, and 26 paratypes (8 ♂, 18 ♀) collected in the vicinity of Mono Pass, Inyo County, California between 30 August and 3 September at elevations at or above 12,000 feet. The paratypes are distributed amongst the CAS, the EME, and the USNM.

*Epiblema macneilli* is the largest Nearctic *Epiblema* (mean FWL = 11.9 mm). The forewing has white to pale yellowish interfascial areas and blackish-brown markings. The maculation includes a bandlike subbasal fascia that extends from the inner margin to the costal fold in males (Plate L:35c) and from the inner margin to the costa in females (Plate N:35a, b); a median fascia consisting of two prominent quadrate marks, one at mid-costa, the other on the inner margin in the pretornal position; a large somewhat circular subapical patch; and three prominent wedge-shaped marks on the distal one-half of the costa that separate the whitish costal strigulae. The interfascial spot is large, capped anteriorly in males by the black mid-costal mark but fused with costal strigulae in females to form a continuous band from the inner margin to the costa. The ocellus is whitish, barely distinguishable, and sometimes invaded medially by a thin streak emanating from the subapical patch.

In males, the uncus is semicircular and sharply differentiated from the dorsolateral shoulders of the tegumen, the socius is short and somewhat wrinkled, and the vesica has 6–10 cornuti (n = 2). The valva has a Type 1 clasper, lacks a basal process, has a strongly concave dorsal margin, a broad neck, NR = 0.80, a weakly defined saccular corner, and SA = 162°. The dorsal lobe of the cucullus is strongly developed and tapers slightly to a blunt apex, the distal margin is convex except for an occasional shallow indentation near the apex, the ventral lobe is broadly rounded, and the basoventral margin extends ridgelike onto the medial surface of the neck. In females, the papillae anales are Type 2, the sterigma is U-shaped with lateral margins curled inward, the lamella antevaginalis is ringlike with quarter-round cross sections, the ductus bursae lacks sclerotization, and the corpus bursae has two signa of unequal size located opposite one another at mid-bursa.

This species is diurnal and has been collected between 11 August and 3 September at altitudes above 12,000 feet in Inyo County, California and at 13,000 feet on Mount Whitney, California. Powell (1975) reported observing it at the type locality in association with *Hulsea algida* A. Gray (alpine gold) but did not find any evidence of larval activity on the plant.

## 36. *Epiblema brightonana* (Kearfott, 1907)
(Plate N, 36a–f; Plate 15, 36a–g)

*Eucosma brightonana* Kearfott 1907:23; Barnes and McDunnough 1917:170.
*Epiblema brightonana*: Heinrich 1923:154; McDunnough 1939:48; Powell 1983:35; Miller 1987:59; Brown 2005:280; Gilligan et al. 2008:127.

**Lectotype** (here designated). ♂, Pennsylvania, Beaver County, New Brighton, F. A. Merrick, 4 August 1904, AMNH.

Kearfott (1907) mentioned seven syntypes (sex not reported), all collected by F. A. Merrick at New Brighton, Pennsylvania between 29 July and 10 August. Klots (1942) reported a male lectotype and two paralectotypes in the AMNH, attributing the lectotype designation to Heinrich (1923). The three syntypes in the AMNH are males, and two of them bear a blue "PARATYPE" label that probably was attached by Klots. We attached a red "LECTOTYPE" label to the third (which we presume is the specimen mentioned by Klots) and include here its formal designation as the name-bearing specimen.

*Epiblema brightonana* is a small species (mean FWL = 6.5 mm) with a broad forewing (AR = 2.32). The interfascial areas are brownish gray, and the markings are dark brown. Typical maculation (Plate N:36a–c) includes a broad triangular subbasal mark whose rounded apex reaches slightly beyond the cubitus, a median fascia represented by a prominent oblique band emanating from mid-costa together with a small triangular pretornal mark (the two usually at least weakly connected), and a postmedian fascia that is narrow (sometimes interrupted) near the costa and broad along the apical margin of the ocellus. The subbasal and pretornal marks are thinly edged with white, as are the distal margin of the median fascia and the proximal margin of the postmedian fascia. The ocellus is inconspicuous but clearly defined, with broad gray bands on the lateral margins and three or four small black dots on a thin brown central field. The costal strigulae beyond mid-wing are white or gray and separated by brown wedges.

Besides the typical individuals, we examined numerous specimens that exhibit varying developments of an oblique bar anterior to the ocellus that, when fully expressed, connects the median and postmedian fasciae (Plate N:36d–f). In this phenotype, the pretornal mark varies from small to absent. The genitalia of the two phenotypes are indistinguishable, so we interpret these differences in maculation as variation within a single species.

In males, the uncus is small, with an apical extension that varies from peaklike to bilobed; the socius is long, with sinuous lateral margins; and the vesica has 12–24 cornuti. The valva has a Type 1 clasper, a weakly discernable basal process, a strongly concave costal margin, a short broad neck, NR = 0.70, an angulate saccular corner, and SA = 122°. The medial surface of the neck has a distinctive patch of short stout setae. The cucullus has a strongly developed dorsal lobe that attenuates slightly toward a rounded apex, a weakly developed ventral lobe, and a convex distal margin that bears upwards of 12 medium sized spiniform setae evenly distributed from the anal angle to three-fourths the distance to the apex. The setae on the medial surface of the cucullus are coarse near the distal margin and grade to hairlike near the dorsal and apical margins. In females, the papillae anales are Type 3, the sterigma is U-shaped, the posterior margin of sternum 7 is weakly concave and diverges laterally from the sterigma, the ductus bursae has a broad sclerotized band at the ductus seminalis, and the corpus bursae has two large signa of nearly equal size, one near the juncture with the ductus bursae, the other in the anterior one-half of the bursa.

*Epiblema brightonana* ranges from Nova Scotia to South Dakota south to North Carolina, Alabama, Mississippi, and Louisiana. Adults have been collected from late June to mid-September.

## 37. ***Epiblema browni*** Wright and Gilligan, new species

(Plate O, 37a–e; Plate 15, 37a–d)

**Diagnosis.** *Epiblema browni* is unique among North American *Epiblema* in having a rather bright grayish-brown forewing and a tan bandlike interfascial spot, the latter narrowing slightly from the inner margin to the radius (in males) or to the costa (in females). Males are distinguished from other members of the genus by the shape of the cucullus: dorsal lobe long and narrow, with distal margin straight except for a shallow indentation near the apex; ventral lobe triangular, with an acute anal angle supporting a single short spine. Diagnostic features of the female genitalia include: papillae anales Type 1, with pads long and narrow; apophyses long and slender, with apophyses posteriores longer than apophyses anteriores; sterigma broad anteriorly, narrowing posteriorly, with posterior margin shorter than ostium diameter; and lamella antevaginalis bandlike and broad, with posterior extremities attenuating onto the surface of the lamella postvaginalis.

**Description.** *Head.* Labial palpus with medial surface white, lateral surface tan; frons whitish; vertex tan to grayish brown; antenna with dorsal surface brown, ventral surface whitish. *Thorax.* Dorsal surface brown; fore- and mid-leg with anterior surfaces brown, posterior surfaces whitish; hind-leg uniformly whitish; tibia with alternating brown and whitish annulations. Forewing: ♂ FWL 5.0–6.1 mm (mean = 5.7, n = 14), AR = 2.52, CFR = 0.53; ♀ FWL 5.9–7.0 mm (mean = 6.3, n = 6), AR = 2.59; basal patch and median area concolorous, with white-tipped blackish-brown scales producing uniform overall grayish brown appearance; interfascial spot bandlike, tan, tapering slightly from inner margin to radius in males (Plate O:37a, c) and from inner margin to costa in females (Plate O:37b, d, e); postmedian fascia represented by narrow brown line anterior to ocellus; ocellus with pinkish-gray bands on lateral margins, up to three black longitudinal dashes on whitish central field, and two or three black spots on proximal margin; costal strigulae whitish, inconspicuous (except for strigula 9), separated by orange-brown striae extending toward termen, with a particularly well expressed silver stria emanating from strigula 7; termen with narrow, salt-and-pepper colored band from tornus to apex; fringe tan near tornus, brownish gray from $CuA_1$ to apex. Hindwing: Uniformly brownish gray. *Abdomen.* Male genitalia (n = 7): Uncus triangular, sharply differentiated from dorsolateral shoulders of tegumen, with small peaklike apical extension; socius long, narrow basally, broad medially, tapering to narrowly round apex; phallus with base closely surrounded by anellus; vesica with 18–24 cornuti; valva with clasper Type 1, basal process absent, costal margin concave at neck but nearly straight along dorsal lobe of cucullus, neck narrow, NR = 0.44, saccular corner angulate, SA = 120°; cucullus long, narrow, and nearly uniform in width (length-to-width ratio = 3.65), with apex semicircular, ventral lobe triangular, and distal margin straight except for shallow incurvation near apex; anal angle acute, with short stout spine at vertex. Female genitalia (n = 2): Papillae anales Type 1, with pads narrow and sparsely setose; apophyses posteriores long, slender, and slightly longer than apophyses anteriores; sterigma broad anteriorly, tapering posteriorly, with lateral margins of lamella postvaginalis nearly straight and length of posterior margin about one-half ostium diameter; lamella antevaginalis bandlike, broad, with posterior extremities attenuating onto surface of lamella postvaginalis; sternum 7 trapezoidal, with lateral margins straight, width at anterior margin about twice that at posterior margin, posterior margin concavely emarginated to about one-third length of sterigma; ductus bursae with sclerotized ring at juncture with ductus seminalis; corpus bursae with two signa located near mid-bursa, one larger than the other.

**Holotype** (Plate O:37a, Plate 15:37a). ♂, Mississippi, Oktibbeha County, Osborn Prairie, 33.5114° N, 88.7356° W, R. L. Brown and L. Koehn, 30 August 1997, slide DJW 4019, USNM.

**Paratypes** (13 ♂, 6 ♀). ARKANSAS. Arkansas County, 3 mi E of Ethel, R. L. Brown, 9 September 1970 (1 ♀), 4 mi E of Ethel, R. L. Brown, 10 September 1970 (1 ♂, wing slide RLB 5, genitalia slide RLB 8). LOUISIANA. Bossier Parish, Bodcau Wildlife Management Area, 13 September 1996, R. L. Brown (2 ♂, slide DJW 4459; 1 ♀, slide DJW 4020). MISSISSIPPI. Chickasaw County, Tombigbee National Forest, R. Brown and J. Hill, 4 September 2005 (1 ♀); Oktibbeha County, Black Belt Prairie, R. L. Brown and D. M. Pollock, 24 August 1993 (1 ♂); T19N R14E Sec 16, D. M. Pollock, 28 August 1990 (1 ♂, slide DJW 4458); Osborn Prairie, R. L. Brown and L. Koehn, 30 August 1997 (5 ♂, slide DJW 4021; 2 ♀, slide DJW 4018); Osborn Prairie, R. Paterson, 6 September 2008 (1 ♂); 7 mi S of Starkville, D. M. Pollock, 31 August 1995 (1 ♂, slide DJW 327); T18N R14E Sec 23, R. Brown and B. Brown, 1 September 1987 (1 ♂). TEXAS. Harris County, J. Schneider, 18 September 2020, (1 ♀, slide DJW 4245). Paratype depositories: MEM, USNM.

**Etymology.** This species is named after Richard L. Brown, whose extensive collections in the states along the Gulf Coast have been especially useful in understanding the eucosmine fauna of southeastern United States.

**Comments.** In addition to the 20 specimens in the type series, we examined photographs (supplied by M. Sabourin) of a male collected by J. Rawlins and W. Zanol at WaKeeney, Kansas on 14 August 1991 (deposited in the CMNH). Moreover, we have seen images (posted on BugGuide and iNaturalist) of live moths from Arkansas, eastern Oklahoma, and Texas that appear to be *E. browni*. The combined records indicate that the range of this insect includes the region from Trego County, Kansas south to east Texas and southeast to Mississippi.

## 38. *Epiblema fulminana* (Walsingham, 1879)

(Plate O, 38a–d; Plate 16, 38a–f)

*Paedisca fulminana* Walsingham 1879:50; Fernald 1882:38.
*Eucosma fulminana*: Fernald 1903:457; Barnes and McDunnough 1917:170; Heinrich 1923:124; McDunnough 1939:47; Powell 1983:35; Miller 1987:53; Brown 2005:320; Gilligan et al. 2008:113.
*Sonia fulminana*: Gilligan and Wright 2013:336.
*Epiblema fulminana*: Wright and Gilligan 2021:132.

**Lectotype** (designated by Wright and Gilligan 2021). ♂, Texas, Bosque County, Belfrage, 15 May 1876, slide 11511, BMNH.

Walsingham (1879) described *P. fulminana* from two males collected by Gustaf W. Belfrage in Bosque County, Texas. The specimen designated above as lectotype was selected by Obraztsov. It bears a blue-bordered circular label with the inscription "LECTOTYPE," a red-bordered circular label with the inscription "Type H.T," and a white label that reads: "*fulminana Wlsm. Lectotype*, N. Obraztsov det. 1965." Brown (2005) interpreted this specimen as a holotype. The second syntype is dated 23 June 1876 and has been dissected (slide 6839, BMNH). Obraztsov photographed both syntypes and their genitalia and noted on a glassine envelope containing the 35 mm slides that the second specimen was "no doubt, a completely different species." We examined Obraztsov's images and are confident that the two specimens are conspecific.

Gilligan and Wright (2013) transferred *fulminana* from *Eucosma* (its home for more than a century) to *Sonia* based on the phylogenetic analysis in Gilligan et al. (2014) and an inaccurate count of the forewing radial veins. We recently proposed that this species be tentatively placed in *Epiblema* (Wright and Gilligan 2021), the hesitancy being due to inconsistency in the presence of a well-defined clasper on the male valva.

*Epiblema fulminana* is a large species (mean FWL = 11.3 mm) of average forewing geometry (AR = 2.62). The dorsal surface is uniformly yellowish brown to grayish brown with extensive whitish speckling. The fasciate markings are brown and barely distinguishable from the interfascial areas. They include a subbasal mark that is often interrupted on $A_{1+2}$, a weakly defined median fascia that is sometimes fragmented, and a well-defined but obscure postmedian fascia. The ocellus is gray and inconspicuous, the costal strigulae are white to gray and sharply defined, and the termen has a narrow salt-and-pepper colored band from the tornus to the apex.

In males, the weakly developed uncus is broad (basal width about four times height), has a shallow medial indentation in the apical margin, and is weakly differentiated from the dorsolateral shoulders of the tegumen. The socius is short and nearly uniform in width, and the vesica has numerous cornuti (average = 77, n = 3). The valva often (but not always) has a vestigial bulgelike clasper located midway between the distal margin of the basal excavation and the neck. The valval neck is broad, with NR = 0.70; the saccular corner is broadly rounded, with SA = 155°; and the distal margin of the basal excavation lacks a basal process. The cucullus has a well developed dorsal lobe that tapers to a weakly angular apex, a barely developed ventral lobe, and a convex distal margin with short stout setae uniformly distributed from the rounded anal angle nearly to the apex. In females, the papillae anales are Type 2, with narrow pads moderately setose and densely microtrichiate; the apophyses anteriores are about twice as long as the apophyses posteriores; and the sterigma narrows posteriorly, with the posterior margin about equal in length to the ostium diameter. The lamella postvaginalis has straight lateral margins that curl inward, and the posterior extremities of the ringlike lamella antevaginalis attenuate onto the surface of the lamella postvaginalis. The ductus bursae has a small sclerotized patch near the juncture with the ductus seminalis, and the corpus bursae has two moderately large signa, one located near the juncture with the ductus bursae, the other in the anterior one-half of the bursa.

We examined 44 specimens (37 ♂, 7 ♀) from Illinois, Iowa, Kansas, Kentucky, Ohio, and Texas. Miller (1987) reported a record from Michigan. Adult capture dates range from 15 May to 10 September. No larval host has been confirmed by rearing, but we have captured numerous adults in Adams County, Ohio in association with *Silphium terebinthinaceum* Jacq. (prairie dock) and a few in Riley County, Kansas near *Silphium laciniatum* L. (compass plant). The same two plant associations have been observed in Black Belt Prairie remnants in Mississippi (Richard Brown, personal communication).

## 39. *Epiblema alba* Wright and Gilligan, 2021

(Plate O, 39a–c; Plate 16, 39a–d)

*Epiblema alba* Wright and Gilligan 2021:130.

**Holotype** (Plate O:39a, Plate 16:39c). ♂, Florida, Putnam County, Palatka, H. D. Baggett, 23 August 1992, slide DJW 4052, USNM.

*Epiblema alba* is one of the smallest Nearctic members of the genus (mean FWL = 5.1 vs. 5.2–11.9 mm) and the only one that lacks a forewing costal fold. An explanation for the assignment of this species to *Epiblema* despite the lack of a costal fold is presented by Wright and Gilligan (2021).

The forewing is white with black markings, the latter including a conspicuous subbasal mark that extends from the inner margin to the cell, an obscure mid-costal bar, and a sometimes fragmented pretornal mark. The interfascial spot is broad, often with transverse black reticulations, and the postmedian fascia is narrow, moderately well defined, and usually interrupted near the costa. The ocellus has pinkish or grayish bands on the lateral margins and one or two black marks on a white central field, the costal strigulae are conspicuous, and the termen has a narrow salt-and-pepper colored band from tornus to apex. The fringe is white from the tornus to about $M_1$ and blackish at the apex.

In males, the uncus is weakly developed, variable in shape (Plate 16:39a–c), and sharply differentiated from the dorsolateral shoulders of the tegumen; the socius is short and weakly constricted near the base; the base of the phallus is closely surrounded by the anellus; and the vesica has 31–39 cornuti. The valva has a Type 1 clasper, lacks a basal process, has a weakly concave costal margin, a moderately broad neck, NR = 0.67, a weakly defined saccular corner, and SA = 149°. The cucullus has well developed dorsal and ventral lobes, a weakly convex distal margin, and a basoventral margin that overlaps the ventral one-half of the medial surface of the neck. In females, the pads of the Type 1 papillae anales are moderately narrow and sparsely setose, the apophyses posteriores and anteriores are long and about equal in length, the membrane posterior to tergum 8 has a pebbly texture, the sterigma is U-shaped, the lamella postvaginalis is rectangular and microtrichiate, and the posterior extremities of the narrow ringlike lamella antevaginalis attenuate into the lateral margins of the sterigma. The ductus bursae has a ringlike sclerotized contortion at the juncture with the ductus seminalis, and the corpus bursae has two signa located slightly posterior to mid-bursa, one somewhat larger than the other.

We examined 30 specimens (23 ♂, 7 ♀) from Baldwin County, Alabama; Hernando, Hillsborough, Marion, Martin, Putnam, and Santa Rosa counties, Florida; and Moore and Pender counties, North

Carolina. Adult capture dates range from 6 July to 18 September. *Chrysopis mariana* (L.) Elliott (Maryland golden aster) has been mentioned as a possible larval host (Wright and Gilligan 2021) based on its presence at collection sites in North Carolina (J. B. Sullivan, personal communication), but no rearings have been reported to confirm that conjecture.

## 40. ***Epiblema tandana*** (Kearfott, 1907)
(Plate O, 40a–c; Plate 16, 40a–d)

*Eucosma tandana* Kearfott 1907:23; Barnes and McDunnough 1917:171.
*Eucosma trapezitis* Meyrick 1912:34, unnecessary replacement name for *tandana*.
*Epiblema tandana*: Heinrich 1923:154; McDunnough 1939:48; Powell 1983:35; Brown 2005:286; Miller 1987:59; Gilligan et al. 2008:127.

**Lectotype** (designated by Heinrich 1923). ♂, Pennsylvania, New Brighton, F. A. Merrick, 18 July 1904, AMNH.

Kearfott (1907) mentioned four syntypes (sex not reported): one from Algonquin, Illinois collected by Dr. Nason on 8 July; one from New Brighton, Pennsylvania collected by F. A. Merrick on 18 July; one from Arlington Heights, Illinois collected by A. McElhose on 15 August; and one from Iowa collected by Gillette. There is only one syntype from New Brighton, Pennsylvania, so Heinrich's (1923) report of the type locality specifies a lectotype. It resides in the AMNH and bears a blue-green "LECTOTYPE" label that presumably was attached by Klots.

*Epiblema tandana* is a moderately large species (mean FWL = 8.7 mm) with a broad forewing (AR = 2.37). The dorsal surface is pale grayish brown with two dark brown marks: a triangular subbasal mark whose anterior vertex extends at least as far as the cubitus and a broad crescentlike mark that arises from the preapical portion of the costa, broadens in the subapical region, and attenuates toward the termen between $CuA_1$ and $CuA_2$. A faint indication of a partial median fascia is usually present in the form of a quadrate mid-costal bar that is slightly darker than the adjacent interfascial areas. This species lacks an ocellus and has weakly delimited costal strigulae that are concolorous with the interfascial areas.

In males, the uncus is weakly developed, with apex rounded; the socius is moderately long and broad, with lateral margins parallel; and the vesica has 23–30 cornuti. The valva has a well-defined neck, NR = 0.53, an angulate saccular corner, and SA = 115°. The distal margin of the basal excavation is weakly sinuous, with a Type 3 clasper located at the midpoint. The cucullus is large compared to the basal portion of the valva. Its dorsal lobe is strongly developed, with apex somewhat angular; its distal margin is straight near the apex and weakly convex otherwise; and its ventral lobe is weakly developed and broadly rounded. In females, the papillae anales are Type 4, with broad pads of nearly uniform width; the ventral lobes of tergum 8 have numerous scales and relatively few setae (Plate 16:40d); the sterigma is V-shaped; the posterolateral corners of the lamella postvaginalis are sharply pointed, each with a patch of scales on the posteriorly adjacent membrane; and the posterior extremities of the ringlike lamella antevaginalis attenuate onto the surface of the lamella postvaginalis. The posterior margin of sternum 7 is concavely emarginated to about one-half the length of the sterigma and diverges laterally from the lamella postvaginalis. The ductus bursae is sclerotized from the constriction anterior to the ostium to the ductus seminalis, and the corpus bursae has two signa of slightly different size, the larger anterior to mid-bursa, the smaller posterior thereto.

The specimens we examined document a range from New Jersey to Minnesota, south to Maryland, Kentucky, Illinois, and Iowa. Adult capture dates vary from the beginning of April to the end of August. Godfrey et al. (1987) reported a specimen that was reared by M. O. Glenn in Illinois from a larva boring in the roots of *Rudbeckia laciniata* L. (green headed coneflower).

# **Section 2.** Type 2 Tegumen + Appendages

## The *numerosana* group (species 41–44)

The *numerosana* group is composed of *E. numerosana*, *E. deflexana*, *E. abruptana*, and *E. mandana*, four species that are similar in size, color, and genitalia but are distinguished from one another by subtle aspects of forewing pattern, valva shape, and sterigma shape. Group characteristics include:

*Head.* Third segment of labial palpus black; second segment tan to whitish, with prominent black mark on lateral surface near dorsal margin; first segment whitish with small black mark on lateral surface; antenna with prominent black marks on pedicel and scape.

*Forewing.* Mean FWL: 6.3–6.9 mm; AR: 2.68–2.75; CFR: 0.25–0.28; dorsal surface brown to blackish brown; interfascial spot concolorous with basal area; fasciate markings usually discernible as slightly darker shades of brown or blackish brown; postmedian fascia thin, usually accentuated with black; cubitus overlaid with blackish line from base to mid-wing; ocellus white and conspicuous, with black dash in central field and black mark on anterior margin; costal strigulae white; termen with inconspicuous salt-and-pepper colored band from tornus to apex.

*Male genitalia.* Uncus triangular, barely differentiated from dorsolateral shoulders of tegumen, with apex rounded and basal width 2–3 times height; socius short, stubby, and tapering from moderately broad base to narrowly rounded apex; phallus with base closely surrounded by anellus; vesica with 22–56 cornuti; valva with clasper Type 2, basal process absent, costal margin weakly concave, ventral emargination long and shallow, neck moderately narrow, NR: 0.48–0.59, saccular corner broadly rounded, SA: 148°–150°; cucullus with dorsal lobe weakly to strongly developed, apex rounded, distal margin weakly convex, ventral lobe weakly developed; distal margin of cucullus with series of 4–8 long spiniform setae uniformly distributed from anal angle to about three-fourths distance to apex; setae on medial surface of cucullus coarse throughout.

*Female genitalia.* Papillae anales Type 2; sterigma ovate or U-shaped; lamella antevaginalis narrow and ringlike; posterior margin of sternum 7 concavely emarginated to about one-half length of sterigma, medially approximate to lamella antevaginalis, moderately divergent laterally from lamella postvaginalis; ductus bursae long and narrow, with small sclerotized patch opposite juncture with ductus seminalis; corpus bursae bulbous, with two signa of unequal size, the larger located near juncture with ductus bursae, the smaller at mid-bursa or in anterior portion of bursa.

## 41. ***Epiblema numerosana*** (Zeller, 1875)
(Plate P, 41a–f; Plate 17, 41a–f)

*Paedisca numerosana* Zeller 1875:317; Fernald 1882:38.
*Eucosma numerosana*: Fernald 1903:456; Barnes and McDunnough 1917:170.
*Epiblema numerosana*: Heinrich 1923:141; McDunnough 1939:48; Powell 1983:35; Miller 1985:34; Miller 1987:56; Brown 2005:284.

**Holotype** (Plate P:41a). ♂, Texas, Dallas, Boll, Type No. 14339, slide TMG 882, MCZ.

This species was described from a single male collected by Jacob Boll, probably in 1870 during a field trip to Texas sponsored by Louis Agassiz for the MCZ. The holotype is in poor condition. It is covered with what appears to be one or more forms of organic debris that likely were acquired in storage or in transit between Texas, the MCZ, and Europe (in the last case for examination by Zeller). Our illustration of the holotype (Plate P:41a) depicts the left fore- and hindwings, which appear to have been "cleaned" at some time in the past. The orangish speckles in the image are remnants of the debris mentioned above. This specimen bears a label with the inscription "*Grapholitha numerosana*, Z., LECTOTYPE, des. W. E. Miller." Miller may have believed this specimen to be part of a series of syntypes, or the label might have been added by someone else. The genitalia illustrated in Heinrich (1923: fig. 260) are those of a specimen of *E. chromata* (Miller 1985).

Heinrich (1923) considered *E. numerosana*, *E. deflexana*, *E. abruptana*, *E. praesumptiosa*, and *E. grossbecki* to be closely related taxa that would eventually prove to be conspecific, but no synonymies involving these names have been proposed. We examined the holotypes of *E. numerosana* and *E. deflexana*, together with 51 other specimens (35 ♂, 16 ♀) that have been associated with one or the other of those names. We found subtle differences in forewing maculation and male valva shape that partition this material into two cohorts, each containing one of the holotypes, so we retain the two names as referring to distinct species. *Epiblema praesumptiosa* and *E. grossbecki* are treated below as comprising the *praesumptiosa* group based on a shared distinctive forewing appearance and nearly identical genitalia. The name *mandana* was proposed by Kearfott (1907) for four specimens that he placed in *Eucosma*, but it has fallen out of use (for reasons discussed below under *E. mandana*). We revive it here as properly applying to a previously unrecognized species of *Epiblema* that often has been confused with *E. abruptana* in collections.

*Epiblema numerosana* is a moderately small species (mean FWL = 6.5 mm) with average forewing geometry (AR = 2.74). The forewing is blackish brown with extensive white suffusion, the latter consisting primarily of transverse reticulations in the basal and interfascial areas. The subbasal and median fasciae are inconspicuous but recognizable as shades of blackish brown. The subbasal fascia is represented by a dark patch on the cubitus and an associated mark on the inner margin, the two often separated by a pale line on $A_{1+2}$; the median fascia is bandlike and vaguely defined from mid-costa to the pretornal portion of the inner margin. The forewing of *E. numerosana* differs from that of *E. deflexana* in color (blackish brown vs. medium brown), the form of the white suffusion (reticulations vs. speckling), and the expression of the subbasal and median fasciae (vague but discernable vs. undetectable). It differs from that of *E. abruptana* in the lack of orange scaling in the interfascial spot and/or the subbasal fascia. *Epiblema mandana* is distinguished from other members of the group by the uniformity of its forewing coloration (medium brown without white or orange suffusion).

In males, the combination of a long narrow neck and a weakly developed cucullus contributes to a somewhat spatulate valva shape that distinguishes this species from other members of the group. In *E. deflexana*, the ventral lobe of the cucullus is more strongly developed, with a more conspicuous anal angle; in *E. abruptana*, the dorsal lobe of the cucullus is more strongly developed and tapers evenly toward the apex; and in *E. mandana*, the neck is noticeably broader (NR = 0.59 vs. 0.48). In females, the sterigma is ovate, with posterior margin medially indented, and the posterior extremities of the lamella antevaginalis attenuate into the lateral margins of the lamella postvaginalis. In these last two respects, *E. numerosana* resembles *E. deflexana* and *E. mandana*. Females of *E. abruptana* have a more elongate and somewhat U-shaped sterigma, with the posterior extremities of the lamella antevaginalis attenuating onto the surface of the lamella postvaginalis. In *E. numerosana* and *E. deflexana*, the inner surface of the ductus bursae usually has a patch of microtrichia near the juncture with the ductus seminalis. This feature does not occur in *E. mandana* and is rarely found in *E. abruptana*.

The meager available distributional data suggests that the range of this species extends through the central Midwest from North Dakota to Texas and east to central Illinois. In addition to the holotype, we examined 4 ♂ and 5 ♀ in the USNM collected by A. & M. E. Blanchard in Anderson County, Texas; 2 ♂ and 2 ♀ in the USNM collected by R. W. Hodges in Cherry County, Nebraska; 3 ♂ in the EME collected by R. Leuschner and J. Powell at Sand Ridge State Forest in Mason County, Illinois; 4 ♂ and 1 ♀ in the INHS collected by M. Glenn in Putnam County, Illinois; 1 ♀ in the EME collected by Chemsak and Powell in Medina County, Texas; 1 ♂ in the EME from Ransom County, North Dakota; 1 ♂ in the

MEM collected by W. E. Miller in Ramsey County, Minnesota; 1 ♂ in the USNM collected by C. N. Ainslie at Sioux City, Iowa; and 7 ♂ and 4 ♀ in the MGCL collected by E. C. Knudson in Anderson, Hemphill, Polk, Terrel, Tyler, and Washington counties, Texas. Adult capture dates range from 23 June to 28 September.

## 42. *Epiblema deflexana* Heinrich, 1923
(Plate P, 42a–c; Plate 17, 42a–f)

*Epiblema deflexana* Heinrich 1923:144; McDunnough 1939:48; Powell 1983:35; Brown 2005:321.

**Holotype** (Plate P:42a, Plate 17:42a). ♂, Texas, Brownsville, A. Busck, 27 May 1917, slide 72861, USNM.

Heinrich (1923) described *E. deflexana* from two males collected in Cameron County, Texas. He identified the name-bearing specimen by illustrating its genitalia.

In terms of the numerical statistics utilized here to quantify forewing and valva geometry, this species is essentially identical to *E. numerosana* (see Appendix A). *Epiblema deflexana* differs from the latter in forewing appearance (paler, more uniformly brown, with white speckling instead of white reticulations) and in male valva shape (cucullus larger, with ventral lobe more strongly developed). These differences are discussed above in the *E. numerosana* account. *Epiblema abruptana* is distinguished from *E. deflexana* and *E. numerosana* by its more mottled forewing appearance, which usually includes orangish accents in the interfascial spot and/or the fasciate markings.

Besides the holotype, we examined 5 ♂ and 2 ♀ in the MEM collected by D. M. Pollock in Bexar County, Texas and 4 ♂ and 2 ♀ in the EME collected by J. Powell and J. Chemsak in Nuevo Leon, Mexico. These specimens document a range that includes southern Texas (San Antonio to Brownsville) and extends west across the Rio Grande River into the Mexican state of Nuevo Leon. Current confirmed records suggest that the geographic distributions of *E. deflexana* and *E. numerosana* might be disjunct, lying south and north, respectively, of a line from Big Bend National Park to the southeastern corner of east Texas. The specimens from Texas were collected on 6 and 27 May, those from Mexico on 13 and 14 September.

## 43. *Epiblema abruptana* (Walsingham, 1879)
(Plate P, 43a–f; Plate 18, 43a–g)

*Paedisca abruptana* Walsingham 1879:53; Fernald 1882:40.
*Eucosma abruptana*: Fernald 1903:458; Kearfott 1905:354; Barnes and McDunnough 1917:170.
*Epiblema abruptana*: Heinrich 1923:141; McDunnough 1939:48; Powell 1983:35; Miller 1987:56; Brown 2005:555; Gilligan et al. 2008:121.

**Lectotype** (here designated). ♂, Texas, Bosque County, Belfrage, 10 October 1876, slide 11555, BMNH.

Walsingham (1879) described this species from three male syntypes with the same collection data. Obraztsov selected the lectotype, which bears a circular red-bordered label with the inscription "Type H.T." Brown (2005) reported that specimen as a holotype.

*Epiblema abruptana* is similar in size and forewing geometry to *E. numerosana* and *E. deflexana* (mean FWL = 6.4 vs. 6.5 and 6.3 mm; AR = 2.68 vs. 2.74 and 2.75). The forewing is grayish brown with variably expressed orangish-brown blotches and black markings. The fasciate markings include: a basal fascia in the form of a thin obscure line from the inner margin to the radius; an orangish-brown quadrate subbasal fascia that is variably accentuated with black, especially on the lateral margins; a fragmented median fascia comprised of a narrow mid-costal bar, a vaguely defined discal mark, and a weakly defined pretornal mark; and a prominent postmedian fascia. The interfascial spot is usually somewhat paler than the prevailing ground color but may be concolorous therewith (Plate P:43e) and is frequently suffused with orange.

The male genitalia differ from those of other members of the group in the shape of the dorsal lobe of the cucullus (more strongly developed and evenly tapering toward the apex). In females, the sterigma is U-shaped rather than ovate, and the posterior extremities of the lamella antevaginalis do not blend into the lateral margins of the lamella postvaginalis but terminate instead on its surface. The ductus bursae usually lacks microtrichia in the vicinity of the ductus seminalis.

Historically, *E. abruptana* has been confused with other members of the group, particularly *E. mandana* and *E. numerosana*, so literature records need to be treated with caution. We found this species to be broadly distributed in southcentral and southeastern United States, with numerous records from the region extending from Kansas to New Jersey, south to the Gulf coast from Texas to Florida. Adult capture dates range from mid-April to the end of October.

## 44. ***Epiblema mandana*** (Kearfott 1907), new combination
(Plate Q, 44a–e; Plate 18, 44a–g)

*Eucosma mandana* Kearfott 1907:17; Barnes and McDunnough 1917:169; Heinrich 1923:124; McDunnough 1939:47; Powell 1983:35; Miller 1985:240; Brown 2005: 317; Wright and Gilligan 2017:49, 51.
*Eucosma amanda* Meyrick 1912:34, unnecessary replacement name for *mandana*.

**Lectotype** (designated by Heinrich 1923). ♀, Washington, D.C., A. Busck, June, slide WEM 104783, AMNH.

Kearfott (1907) described *Eucosma mandana* from four specimens: one female collected in June by A. Busck in Washington, D.C.; one male collected on 1 July by W. D. Kearfott in Montclair, New Jersey; and two males dated 9 July 1901 and July 1903 collected by A. Busck on Plummers Island in Montgomery County, Maryland. When Heinrich (1923) reviewed this species, the female had been deposited in the AMNH (where it currently resides), and the three males were in the USNM (as they are today). Since there was only one syntype in the AMNH, Heinrich's statement "Type. – In American Museum" sufficed to identify the female as the name-bearing specimen. Miller (1985) illustrated the genitalia of the female (slide WEM 104783), incorrectly interpreted that specimen as being conspecific with *Eucosma comatulana* (Zeller 1875) (see Wright and Gilligan 2017: p. 49), and treated *E. mandana* as a synonym of *E. comatulana*. Moreover, Miller (1985) observed that the genitalia of the male syntype illustrated by Heinrich (1923: fig. 168, slide USNM 70455) as *E. mandana* agrees with that of *Eucosma vagana* McDunnough. Nearly three decades later, *E. comatulana* and *E. vagana* were transferred to *Pelochrista* Lederer (Gilligan and Wright 2013). In preparing our monograph on *Pelochrista* (Wright and Gilligan 2017), we examined all four syntypes of *E. mandana* and concluded that the males are indeed *P. vagana* and that the female is an *Epiblema*. Moreover, our research for the present volume uncovered a large group of specimens representing a previously unrecognized species of *Epiblema*, and the females in that group are conspecific with the lectotype of *E. mandana*. We therefore transfer *Eucosma mandana* to *Epiblema* and refer those *Epiblema* specimens to that taxon. Due to the long-standing confusion regarding the application of the name *mandana*, we include here a detailed description of *Epiblema mandana* and provide data for a substantial sample of specimens examined.

**Diagnosis.** This species is distinguished from other members of the *numerosana* group by its uniformly medium-brown forewing, which lacks both the whitish reticulations and speckling of *E. numerosana* and

*E. deflexana*, respectively, and the black fasciate markings of *E. abruptana*. *Epiblema mandana* is a little larger than the other three species (mean FWL = 6.9 vs. 6.3–6.5 mm) but not substantially different in other forewing statistics. The neck of the male valva is shorter and broader than in *E. numerosana* and *E. deflexana* (NR = 0.59 vs. 0.48 and 0.50), and the dorsal lobe of the cucullus is less strongly developed than in *E. abruptana*. The female genitalia are indistinguishable from those of *E. numerosana* and *E. deflexana* but differ from those of *E. abruptana* in that the sterigma is ovate rather than U-shaped and the posterior extremities of the lamella antevaginalis blend into the lateral margins of the lamella postvaginalis instead of attenuating onto its surface.

**Description.** *Head.* Labial palpus with third segment black, medial surface of second segment tan, lateral surface of second segment pale brown with diffuse black mark near dorsal margin, lateral surface of first segment with faint black dot; frons tan; vertex brown; antenna brown, with conspicuous black marks on pedicel and scape. *Thorax.* Dorsal surface brown; fore- and mid-leg with anterior surfaces brown, posterior surfaces tan; hind-leg uniformly tan; tarsi with pale alternating tan and brown annulations. Forewing: ♂ FWL 5.3–7.7 mm (mean = 6.8, n = 63), AR = 2.68, CFR = 0.28; ♀ FWL 6.2–7.8 mm (mean = 7.2, n = 15), AR = 2.70; costa weakly arched; apical angle slightly acute; termen straight; dorsal surface medium brown, often with grayish tints; fasciate markings absent to barely discernable, in latter case as slightly darker shades of brown; interfascial spot slightly paler than adjacent subbasal and median areas, sometimes with faint orange-brown tints, never sharply defined; ocellus whitish and conspicuous; costal strigulae white, separated by orange-brown lines extending from costa toward termen; strigula 7 with associated prominent gray stria extending to termen at approximately $R_5$; termen with narrow salt-and-pepper colored band from tornus to apex. Hindwing: Uniformly gray brown. *Abdomen.* Male genitalia (n = 12): Typical of the group: vesica with 24–45 cornuti; valva with clasper Type 2 (sometimes with weakly developed distal peak), neck moderately broad, NR = 0.59, saccular corner angulate, SA = 150°; cucullus with moderately developed dorsal lobe, weakly developed ventral lobe, uniformly curved distal margin, coarse setae on medial surface, and 5–8 spiniform setae distributed along distal margin from anal angle to three-fourths distance to apex. Female genitalia (n = 7): Sterigma somewhat ovate, with lateral margins weakly curled inward (indicated by dashed lines in Plate 18:44f); lamella postvaginalis uniformly microtrichiate; ductus bursae lacking microtrichia; other features typical of the group.

**Specimens examined** (42 ♂, 8 ♀). ALABAMA. Baldwin County, Bon Secour NWR, R. Brown and D. Pollock, 8–9 August 1994 (1 ♀, slide DJW 4040); Lawrence County, Prairie Grove Glade, S. M. Lee, 24 May 2004 (1 ♂). ARKANSAS. Conway County, Petit Jean State Park, S. M. Lee, 5 August 2008 (1 ♀). KANSAS. Riley County, Konza Prairie, J. MacGown, 12 June 1999 (1 ♂, slide DJW 4006), 13 June 1999 (1 ♂); L. D. Gibson, 23 July 1995 (2 ♂). KENTUCKY. Gallatin County, Markland Dam, L. D. Gibson, 21 June 1990 (1 ♂); Boone County, Big Bone Lick State Park, L. D. Gibson, 24 June 1981 (1 ♂). LOUISIANA. Bossier Parish, Barksdale Air Force Base, R. Brown and D. Pollock, 21 May 1996 (1 ♀, slide DJW 4183); R. L. Brown, 13 September 1996 (2 ♂); Iberia Parrish, Avery Island, 6 June 2007, R. Brown and R. Clary (1 ♂). MARYLAND. Dorchester County, Taylors Island, J. Glaser, 14 June 1998 (1 ♂, DJW 3973), 26 August 1999 (1 ♀, slide DJW 3974); Worcester County, Pocomoke State Forest, J. Glaser, 1 June 2000 (1 ♀, slide DJW 3821). MISSISSIPPI. Franklin County, Middleton Creek, T. Schiefer and R. Fontenot, 1 June 1992 (1 ♂); George County, Mixon Lakes, J. A. MacGown, 9–10 April 1999 (1 ♂); Hancock County, Stennis Space Center, R. Kergosien, 15 May 1993 (1 ♂); Lowndes County, T17N R16E Sec 34, D. M. Pollock, 19 June 1991 (1 ♀, slide DJW 4007); Black Belt Prairie, R. Brown and D. Pollock, 16 May 1993 (1 ♂); Panola County, John W. Kyle State Park, R. L. Brown, 1–3 June 1997 (1 ♂); Smith County, Bienville National Forest, R. L. Brown, 30 May 2000 (1 ♂); 1 mile N of Raleigh, R. Brown and G. Burrows, 14–15 June 1985 (1 ♂); Winston County, Tombigbee National Forest, R. L. Brown, 25 May 1999 (1 ♂), D. M. Pollock, 19 July 1999 (1 ♂); T18N R14E Sec 23, R. and B. Brown, 16 June 1988 (1 ♂); Oktibbeha County, 6 mi SW Starkville, R. L. Brown, 9 July 1984 (1 ♂); Warren County,

Vicksburg, R. Patterson, 25 July 2006 (1 ♀). OHIO. Adams County, 1 mi SE of Lynx, D. J. Wright, 27 June 1998 (1 ♂, slide DJW 3779), 2 July 2003 (1 ♂, slide DJW 3820), 5 July 1996 (1 ♂), 12 July 1998 (2 ♂), 25 July 1997 (3 ♂), 25 July 1998 (1 ♂), 1 August 1997 (1 ♂); Greene County, Wright-Patterson Air Force Base, E. H. Metzler, 1 July 1994 (1 ♂); Hamilton County, Miami Whitewater Forest, L. D. Gibson, 26 June 1986 (1 ♂, slide DJW 4041). SOUTH CAROLINA. Horry County, Galivants Ferry, J. M. Burns, 17 June 1962 (1 ♂), 4 August 1963 (1 ♂, slide JAP 1782). TENNESSEE. Davidson County, Couchville Glade, R. Brown and J. MacGown, 10 June 1997 (1 ♂); Wilson County, Lane Farm Glade, R. L. Brown, 12 June 1997 (1 ♀, slide DJW 4180). TEXAS. Hardin County, Big Thicket National Preserve, Teel Road, E. C. Knudson, 22 April 1995 (1 ♂, slide DJW 3849); Tyler County, Lake Steinhagen, E. C. Knudson, 24 August 1996 (2 ♂, slide DJW 3975); Kirby State Forest, E. C. Knudson, 15 April 1994 (1 ♂, slide DJW 4172). WEST VIRGINIA. Morgan County, Sleepy Creek Forest, J. Glaser, 14 June 2012 (1 ♂). Depositories: EME, LDG, MEM, MGCL, USNM.

## The *exacerbatricana* group (species 45–46)

This group is composed of *E. exacerbatricana* and *E. glaseri*, two species that are similar in forewing appearance but easily separated by features of the genitalia. Group characteristics include:

*Forewing*. Mean FWL: 6.1, 6.1 mm; AR: 2.74, 2.66; CFR: 0.25, 0.25; interfascial areas gray; fasciate markings various combinations of blackish brown and orange brown; subbasal mark quadrate, extending from inner margin nearly to radius; median fascia fragmented, including a mid-costal bar and a pretornal mark, the latter with a triangular component based on $CuA_2$; postmedian fascia conspicuous; ocellus prominent, with pink or gray bands on lateral margins and two black marks on central field; costal strigulae beyond mid-wing white and conspicuous; termen with narrow salt-and-pepper colored band from tornus to apex and white bar from $M_1$ to apex.

### 45. *Epiblema exacerbatricana* Heinrich, 1923
(Plate Q, 45a–e; Plate 19, 45a–g)

*Epiblema exacerbatricana* Heinrich 1923:146; McDunnough 1939:48; Powell 1983:35; Brown and Lewis 2000:1033; Brown 2005:281.

**Holotype.** ♂, North Carolina, Southern Pines, 16–23 August, slide 72862, USNM.

This species was described from the holotype and 18 paratypes (14 ♂, 4 ♀), all from the type locality, with capture dates of 16 August to 23 September. Heinrich designated the name-bearing specimen by illustrating its genitalia.

*Epiblema exacerbatricana* is a moderately small species (mean FWL = 6.1 mm) with average forewing geometry (AR = 2.75). In typical specimens (Plate Q:45a–b), the subbasal fascia is orange brown with distal margin edged in black. Individuals from the northern part of the range (Plate Q:45e) lack the orange-brown coloration, and those from the Gulf Coast (Plate Q:45c–d) tend to be paler and more brightly colored.

In males, the uncus is broad (basal width about four times height) and not differentiated from the dorsolateral shoulders of the tegumen, the short stubby socius tapers to a narrow apex, and the vesica has 20–39 cornuti. The valva has a Type 4 clasper, lacks a basal process, has a weakly concave costal margin, a long and shallow ventral emargination, NR = 0.72, a weakly defined saccular corner, and SA = 156°. The cucullus has a weakly developed dorsal lobe, usually with its apical margin somewhat flattened, a weakly produced ventral lobe, and several spiniform setae along the ventral one-half of the distal margin. In females, the papillae anales are Type 2, with 4–5 long hairlike setae along the mid-line of each pad and a distinctive band of discrete short stiff setae between the mid-line and the ventral margin (Plate 19:45f); the apophyses posteriores are especially stout; the sterigma is elongate (length about two times width); the lamella postvaginalis is microtrichiate and rectangular, with slightly flared posterolateral corners; and the lamella antevaginalis is narrow and ringlike, with its posterior extremities merging into the lateral margins of the lamella postvaginalis. The ductus bursae has a sclerotized ring near the ductus seminalis with inner surface microtrichiate, and the corpus bursae has two signa of nearly equal size, both located near mid-bursa.

The range of this species extends along the Atlantic coast from Massachusetts to Georgia, includes central Florida, and continues along the Gulf Coast from Sarasota County, Florida to Mississippi. Most of the specimens we examined had capture dates ranging from mid-August to mid-November, but we encountered one January record from Florida and one June record from Long Island, New York.

## 46. ***Epiblema glaseri*** Wright and Gilligan, new species

(Plate Q, 46a–e; Plate 19, 46a–f)

**Diagnosis.** *Epiblema glaseri* resembles *E. exacerbatricana* in size (each with mean FWL = 6.1 mm), forewing geometry (AR = 2.66 vs. 2.74), and forewing pattern. It differs from the latter in general forewing appearance (darker and more mottled) and head color (orange vs. gray). Males differ from those of *E. exacerbatricana* in the shape of the valva (neck shorter, apex of cucullus more angulate). Females differ from those of *E. exacerbatricana* in papillae anales Type (3 vs. 2), shape of the sterigma (U-shaped with length about equal to width vs. rectangular with length approximately twice width), sclerotization of the ductus bursae (absent vs. ringlike), and positioning of the signa in the corpus bursae (one near the juncture with the ductus bursae and one at the anterior extremity of the bursa vs. opposite one another at mid-bursa).

**Description.** *Head.* Third segment of labial palpus black; medial surface of second segment whitish; lateral surface of second segment tan, with elongate black mark near dorsal margin and long brown scales on ventral margin; lateral surface of first segment with black mark; lower frons white; upper frons and vertex orange brown; antenna brown with black marks on pedicel and scape. *Thorax.* Dorsal surface gray, with blackish transverse line on each thoracic segment; tegula pale gray with conspicuous black mark near base and obscure blackish mark near tip; fore- and mid-leg with anterior surfaces gray brown, posterior surfaces tan, black marks on tibia; hind-leg tan; tarsus with blackish annulations. Forewing: ♂ FWL 5.2–7.3 mm (mean = 6.0, n = 37), AR = 2.66, CFR = 0.25; ♀ FWL 5.4–7.9 mm (mean = 6.4, n = 8), AR = 2.64; costal margin weakly convex; apical angle slightly acute; termen straight; dorsal surface dark brownish gray with blackish-brown markings, sometimes with orangish tints in subbasal and pretornal marks; subbasal mark quadrate, extending from inner margin to cell, usually with black edging on lateral margins; median fascia vaguely defined as gray band from mid-costa to inner margin overlaid with various black marks, the latter including a mid-costal bar and a prominent triangular pretornal mark; postmedian fascia prominent, thin near costa, expanded anterior to ocellus; ocellus with gray bands on lateral margins and two black marks on central field; costal strigulae and termen typical of the group. *Abdomen.* Male genitalia (n = 20): Uncus barely differentiated from dorsolateral shoulders of tegumen, often with pair of weakly-developed ridgelike projections centrally located on apical margin; socius short, narrow, and sparsely setose, with setae clustered at apex; vesica with 28–60 cornuti; valva with clasper Type 4, basal process absent, dorsal margin weakly concave, ventral emargination shallow, neck broad, NR = 0.75, saccular corner weakly defined, SA = 166°; cucullus with dorsal lobe well developed and tapering to moderately angular apex, distal margin convex of uniform curvature, ventral lobe barely discernable, setae on medial surface coarse throughout. Female genitalia (n = 4): Papillae anales Type 3; sterigma U-shaped, with length approximately equal to width; lamella postvaginalis microtrichiate, with posterolateral corners rounded and lateral margins curled inward; lamella antevaginalis ringlike, with distal extremities merging into lateral margins of lamella postvaginalis; posterior margin of sternum 7 concavely emarginated to about one-half length of sterigma, weakly divergent laterally from sterigma; ductus bursae lacking sclerotization; corpus bursae with two signa of moderately different size, the larger near juncture with ductus bursae, the smaller near anterior extremity of bursa.

**Holotype** (Plate Q:46a). ♂, Georgia, Manuel County, Ohoopee Dunes, John Glaser, 16 September 2012, USNM.

**Paratypes** (32 ♂, 3 ♀). ALABAMA. Baldwin County, Bon Secour NWR, R. Brown and D. Pollock, 11 May 1994 (1 ♂), R. Brown and J MacGown, 5 September 1988, (1 ♂), 5–6 September 1988 (2 ♂, slides DJW 4057, 4121; 1 ♀, slide DJW 4122); 1 mi E of Oyster Bay, R. Brown and J. MacGown, 4 September 1988 (1 ♀, slide DJW 4079); Bibb County, Bibb County Glades Preserve, R. Brown and J. MacGown, 6 August 2003 (3 ♂, slide DJW 4058); Mobile County, Saraland, E. C. Knudson, 5 September 1982 (2 ♂, slide DJW 4198). FLORIDA. Highlands County, Archbold Biological Station, J. B. Heppner, 1 May 1975 (1 ♂, slide DJW 4078), 2 May 1975 (1 ♂); Levy County, Goethe State Forest, J. E. Hayden, 30 July 2011 (1 ♂), Hayden, McGowan, Paris, and Romero, 27 August 2011 (4 ♂, slide DJW 4197, wing slide DJW 6), J. Hayden and C. Romero, 3 September 2011 (1 ♂, slide DJW 4080), J. Hayden and F. Ponce, 12 October 2012 (1 ♂). GEORGIA. Emanuel County, Ohoopee Dunes Natural Area, R. Brown and S. Lee, 17 June 2002, (3 ♂, slide DJW 4056), R. Brown and J. MacGown, 19 June 2002 (1 ♀, slide DJW 4196); J. Glaser, 16 September 2012 (2 ♂, slide DJW 3807). KENTUCKY. Laurel County, ridgetop SW of Ned Branch, L. D. Gibson, 28 July 1992 (3 ♂, slides LDG 142, 157). LOUISIANA. Calcasieu Parish, 6 mi ESE of Buhler, R. Brown and D. Pollock, 15–16 September 1993 (1 ♂, slide DJW 4120). MISSISSIPPI. George County, 3 mi N of Lucedale, R. Kergosien, 19 August–17 September 1996 (1 ♂); Harrison County, Long Beach, R. Kergosien, 24 August 1997 (1 ♂); Tishomingo County, J. P. Coleman State Park, R. Kergosien, 9–19 July 1994 (1 ♂). NORTH CAROLINA. Moore County, Hoffman Road, Sandhills Gamelands, J. B. Sullivan and S. P. Hall, 10 September 2020 (1 ♂, slide DJW 4435). TEXAS. San Jacinto County, Double Lake Campground, 4 June 1989, E. C. Knudson (1 ♂, slide DJW 4174). Paratype depositories: LDG, MGCL, MEM, JBS, USNM.

**Etymology.** This species is named after John Glaser, whose collections in east central United States have been especially useful in developing many of the species accounts presented here.

**Comments.** In addition to the types, we examined six specimens (3 ♂, slides DJW 4543, MS 10048, 11159; 3 ♀, slides DJW 4544, 4550, MS 7236) from Collier and Highlands counties, Florida that were mostly in poor condition but which exhibited various amounts of reddish suffusion in the forewing (e.g., Plate Q:46e). Moreover, we were referred to several photographs on BugGuide of similar looking live moths from Florida whose red markings seem to be much more prominent than in the individuals to which we had access. The reddish specimens we examined are a little larger than typical *E. glaseri* (mean FWL = 6.8 vs. 5.8 mm) but are remarkably similar to that taxon in forewing pattern and genitalia. For the present, we interpret the reddish-marked specimens as color variants of *E. glaseri*, but we did not include them amongst the paratypes, and we did not include their numerical data with that of typical specimens.

## The *praesumptiosa* group (species 47–48)

Specimens traditionally identified as *E. praesumptiosa* are substantially uniform in genitalia but moderately variable in forewing appearance. They come from numerous coastal locations along the Gulf of Mexico (from south Texas to south Florida) and up the Atlantic seaboard as far as North Carolina. In our sample of 132 such specimens, we found two subgroups which, when viewed independently of the rest, are sufficiently well separated by forewing appearance and geographic distribution to warrant the recognition of two distinct species. One, consisting of 25 specimens, occurs primarily in southern Texas and includes the members of the type series for *E. praesumptiosa*. The other, with 60 specimens, is located primarily in coastal North Carolina. The remaining 47 specimens, which were collected along the Gulf Coast from east Texas to Florida and along the Atlantic Coast in Georgia and South Carolina, are intermediate in forewing appearance between the members of the two subgroups. This intergradation of morphological features and the overlap of the geographic ranges of the different forewing phenotypes, render reliable diagnoses of additional taxa problematic. Consequently, lacking evidence from molecular data or life history observations, we have adopted the traditional interpretation of the various forewing phenotypes as intraspecific variation in a single species. Nonetheless, we consider it likely that the resulting taxon is, in fact, a complex of two or more weakly differentiated species. In addition to its namesake, the *praesumptiosa* group includes *E. grossbecki,* a poorly known species from south Florida. Group characteristics include:

*Head.* Labial palpus white to tan, with black mark on third segment; second segment usually with some black scaling on lateral surface; first segment sometimes with small black dot on lateral surface; antenna unmarked or with black mark on pedicel.

*Forewing.* Mean FWL: 7.9, 8.3 mm; AR: 2.90, 2.72; CFR: 0.29, 0.31; dorsal surface predominantly grayish tan to yellowish tan; fasciate markings brown to black, substantially reduced to nearly absent; ocellus tan to white, usually with one or two dark spots on the central field and another on the anterior margin; costal strigulae small, sometimes inconspicuous, delimited from base to apex by fine blackish-brown dashes; termen with narrow salt-and-pepper colored band from tornus to apex.

*Male genitalia.* Uncus broad (basal width 2–3 times height), weakly differentiated from dorsolateral shoulders of tegumen, with apex evenly rounded; socius short; base of phallus closely surrounded by anellus; vesica with 26–60 cornuti; valva with clasper Type 2 (sometimes with weakly developed distal peak), basal process absent, costal margin weakly concave, neck moderately short, NR: 0.63, 0.56, saccular corner broadly rounded, SA: 149°, 150°; cucullus with dorsal lobe moderately developed and tapering to rounded apex, distal margin evenly curved, ventral lobe weakly developed, anal angle broadly rounded, 5–10 spiniform setae on distal margin from anal angle to about three-fourths distance to apex, setae on medial surface coarse.

*Female genitalia.* Papillae anales Type 2; sterigma U-shaped; lamella postvaginalis microtrichiate, with lateral margins curling inward; lamella antevaginalis ringlike, with the posterior extremities attenuating onto surface of lamella postvaginalis; posterior margin of sternum 7 concavely emarginated to about one-half length of sterigma and divergent from lateral margins of lamella postvaginalis; ductus bursae with small, sclerotized, medially-ridged patch near juncture with ductus seminalis; corpus bursae with moderately large signum near juncture with ductus bursae and smaller signum in anterior portion of bursa.

## 47. ***Epiblema praesumptiosa*** Heinrich, 1923
(Plate R, 47a–o; Plate 20, 47a–h)

*Epiblema praesumptiosa* Heinrich 1923:143; McDunnough 1939:48; Brown 1979:25; Powell 1983:35; Brown and Lewis 2000:1055; Brown 2005:284.

**Holotype.** ♂, Texas, Cameron County, Brownsville, A. Busck, 27 May 1917, slide 72818, USNM.

Heinrich (1923) proposed the name *praesumptiosa* for a series of six specimens (1 ♂, 5 ♀) collected by August Busck on 27 May 1917 at Brownsville, Texas. He illustrated the genitalia of the male and called it the type, thereby designating a unique name-bearing specimen.

We examined 25 specimens (15 ♂, 10 ♀) that we refer to as the typical phenotype (Plate R:47a–e). They agree in all respects with members of the type series, and they range through southeast Texas from Cameron County to Jefferson County. In these individuals, the third segment of the labial palpus has a conspicuous black mark on its dorsal surface, a whitish ventral surface, and a whitish tip; the second segment is white, with one or two small blackish spots or smudges on the lateral surface; and the first segment is white, often with a small black dot on the lateral surface. The antenna has a whitish scape and a black spot on the pedicel. The forewing is pale grayish tan, often with barely perceptible brownish transverse reticulations. The forewing markings include a small spot on the cubitus in the subbasal position, a small spot at the proximal margin of the ocellus, and an oblique blackish dash at the apex, the last accentuating the distal margin of strigula 9. The costal strigulae other than number 9 are short, whitish, and inconspicuous.

The specimens illustrated in Plate R:47f–j are representative of a more strongly marked phenotype that is the predominant form along the Gulf coast from Louisiana to Alabama and which also occurs along the Atlantic coast in South Carolina and Georgia. Compared to the typical phenotype, the forewing is slightly darker and has transverse reticulations that are a little more conspicuous, resulting in a slightly mottled forewing appearance. The subbasal mark is more strongly expressed, with the cubital spot larger than that of the typical phenotype and with an associated spot on the inner margin, the two sometimes connected. In these specimens, there often is a thin mid-costal bar extending into the cell. The head markings resemble those of the typical phenotype, but the dark mark on the third segment of the labial palpus varies from black to pale brown, and the lateral surface of the second segment of the labial palpus tends to be more tan than whitish.

The images in Plate R:47k–o illustrate the variation in forewing appearance found in most of the material from coastal North Carolina. These individuals have a subbasal mark (sometimes barely discernable) in the form of a faint brown shade extending from the inner margin to the cell, its distal margin sometimes accentuated by dark brown edging (Plate R:47k). The interfascial areas often have a mixture of whitish and pale brownish-gray patches that produce a blotchy effect (Plate R:47m–o). The head markings resemble those of the typical phenotype, except that the antenna lacks a black spot on the pedicel.

Specimens from Florida tend to be intermediate between the typical phenotype and the more strongly marked specimens from Louisiana, Mississippi, and Alabama. All the forms except the typical phenotype have at times been confused with *E. grossbecki*, but the latter taxon is more yellowish tan than grayish tan and has strongly expressed markings that include a subbasal mark, a discal mark, a pretornal mark, and a conspicuous postmedian fascia.

We found no genitalia characteristics amongst the various phenotypes that would support diagnosing more than one species. Males are remarkably uniform in the shape of the valva and in the structure of the tegumen and its appendages. In females, the lateral margins of the sterigma vary from slightly concave to slightly convex (Plate 20:47e, g–h).

We examined specimens from Cameron, Hidalgo, Jefferson, and Nueces counties, Texas; Lafourche Parish, Louisiana; Hancock and Jackson counties, Mississippi; Baldwin County, Alabama; Carteret and New Hanover counties, North Carolina; Charleston County, South Carolina, McIntosh County, Georgia; and Everglade, Gulf, and Monroe counties, Florida. Pin labels describe the collection sites in North Carolina as brackish marsh, maritime dry grassland, or coastal sandhills; those from Mississippi and Alabama as sand dunes or shell middens. Adult capture dates range from the end of February to late October.

### 48. ***Epiblema grossbecki*** Heinrich, 1923
(Plate S, 48a–d; Plate 20, 48a–e)

*Epiblema grossbecki* Heinrich 1923:142; McDunnough 1939:48; Powell 1983:35; Brown 2005:283.

**Holotype.** ♂, Florida, Everglades, 10 April 1912, slide CH 30 January 1920, #1, AMNH.

Heinrich (1923) described this species from 19 specimens from Everglades (10 ♂, 4 ♀) and Fort Myers (5 ♂), Florida. He singled out the holotype by illustrating its genitalia.

*Epiblema grossbecki* resembles *E. praesumptiosa* in many respects, but the forewing has a somewhat yellowish tint, and its blackish markings are more prominent. The subbasal fascia is represented by a large dark spot on the cubitus and, in males, an associated spot on the inner margin. The median fascia is composed of a thin mid-costal bar, a variably expressed discal mark, and a triangular pretornal mark that rests on $CuA_2$. This is the only phenotype in the group with a prominent postmedian fascia. This last feature is interrupted by yellowish-gray scaling near the costa but expands into a moderately broad band anterior to the ocellus that nearly connects with the apical dash. The labial palpus has black markings similar to those in *E. praesumptiosa*, but the base of the antenna is unmarked.

The male genitalia of *E. grossbecki* are similar to those of *E. praesumptiosa*, but the neck of the valva is a little narrower (NR = 0.56 vs 0.63) and the dorsal lobe of the cucullus is slightly more elongate. The female genitalia of the two taxa appear to be indistinguishable.

This species is poorly represented in collections. We examined 16 specimens (10 ♂, 6 ♀), all from south Florida. Most were captured in April or early May, but there was one record of 20 September from Monroe County and one of 19 November from Sarasota County. The presence of some intergradation in forewing markings between *E. grossbecki* and the Louisiana–Alabama phenotypes of *E. praesumptiosa* (e.g., Plate S:48b–c vs. Plate R:47i) suggest that the distinction between the two taxa is less than clear, but lacking additional evidence, we retain both names, applying *E. grossbecki* to especially well marked individuals from south Florida.

## Species not assigned to a group

### 49. ***Epiblema chromata*** Miller, 1985
(Plate S, 49a–e; Plate 21, 49a–e)

*Epiblema chromata* Miller 1985:35; Brown 2005:280.
*Epiblema numerosana* (not Zeller 1876:317): Heinrich 1923:141.

**Holotype.** ♀, Texas, Kerrville, H. Lacey, 8 July, slide CWT 156, AMNH.

*Epiblema chromata* was described from a holotype and five paratypes, all from southern Texas. Miller (1985) reported the paratype repositories as the USNM (2 ♂, 1 ♀) and the CNC (1 ♂, 1 ♀). He also pointed out that Heinrich (1923: fig. 260) mistakenly illustrated the genitalia of an *E. chromata* specimen as those of *E. numerosana*.

*Epiblema chromata* is a little larger than the average *Epiblema* (mean FWL = 8.3 mm) and has a moderately elongate forewing geometry (AR = 2.81). The third segment of the labial palpus is black;

the lateral surface of the second segment has a prominent black spot; and the lateral surface of the first segment sometimes has a black dot. The antenna has a black spot on the pedicel and another on the scape. The forewing exhibits two similar patterns that differ in the expression of the interfascial spot. In one (Plate S:49a–d), the interfascial spot is obscure, grayish brown, and sometimes weakly suffused with pale orange; in the other (Plate S:49e) it is conspicuous, somewhat better defined, and mostly pale orange. Both phenotypes have a blackish-brown band along the costa that is widest at mid-costa, narrows toward the base, and distally attenuates to a point at strigula 9. Other features include an obscure quadrate pretornal mark, a conspicuous tan streak from the distal end of the cell to the apex, and a whitish ocellus with a black dash on $M_2$.

The male genitalia are nearly identical to those of *E. praesumptiosa* and *E. grossbecki*, but the uncus is not quite as broad. Females differ from the latter two species in sterigma shape (ovate vs. U-shaped) and in the sclerotization of the ductus bursae (absent vs. patchlike).

We examined 41 specimens (28 ♂, 13 ♀) from Burnet, Cameron, Comal, Fort Bend, Gonzales, Kerr, San Benito, San Patricio, and Uvalde counties in central and southern Texas. Capture dates range from late February to late November, with at least one record in each of the intervening months.

## 50. *Epiblema insidiosana* Heinrich, 1923

(Plate S, 50a–f; Plate 21, 50a–d)

*Epiblema insidiosana* Heinrich 1923:145; McDunnough 1939:48; Powell 1983:35; Brown 2005:283.

**Holotype.** ♂, North Carolina, Southern Pines, 1–7 June, slide 72862, USNM.

Heinrich (1923) based this species on 13 syntypes (11 ♂, 2 ♀) with capture dates of 1–7 June and 16–23 July, all collected at Southern Pines, North Carolina. He illustrated the genitalia of a male he called the type, thus designating a name-bearing specimen.

*Epiblema insidiosana* is a small species (mean FWL = 5.5 mm) with a relatively broad forewing (AR = 2.50). The head is mostly white, but the lateral surface of the second segment of the labial palpus is tan, the vertex has a mixture of tan and white scales, and the dorsal surface of the scape is tan. The forewing is white with brown fasciate markings and brown transverse reticulations. The subbasal fascia is pale brown, moderately well defined from the inner margin to the radius, and overlaid with three or four dark brown reticulations; the interfascial spot merges with concolorous costal strigulae to form a vaguely defined whitish band from the inner margin to the costa; the median fascia is represented by a quadrate mid-costal bar and two triangular marks, the larger based on $CuA_2$, the other straddling the cubitus; and the postmedian fascia is thin but conspicuous. The ocellus has three pinkish to grayish components (a proximal band, an apical quadrant, and a tornal quadrant), the latter two separated by a blackish-brown longitudinal dash. The costal strigulae are white and bold, and the termen has a narrow salt-and-pepper colored band from the tornus to the apex.

In males, the uncus is broad (width about four times height) and barely differentiated from the tegumen, with a shallow medial indentation in the apical margin; the socius is short and stubby; and the vesica has 28–56 cornuti. The valva has a Type 4 clasper, lacks a basal process, has a concave costal margin, a shallow ventral emargination, NR = 0.59, a weakly defined saccular corner, and SA = 159°. The cucullus has a strongly developed dorsal lobe with a semicircular apex, a weakly developed ventral lobe, and a rounded anal angle. The female genitalia are unlike those of any other North American *Epiblema*. The sparsely setose pads of the papillae anales are narrow, attenuating to a point at each end; the apophyses posteriores are unusually stout; the membrane along the posterior margin of tergum 8 is microtrichiate; and the setae on tergum 8 are unusually long and stiff. The lamella postvaginalis is somewhat rectangular, has acute posterolateral corners, and narrows abruptly at the ostium; the lamella antevaginalis is narrow and ringlike, with its posterior extremities attenuating onto the surface of the lamella postvaginalis. The ductus bursae is encircled at the juncture with the ductus seminalis by a sclerotized band with inner surface microtrichiate, and the corpus bursae has two signa of nearly equal size located opposite one another at mid-bursa.

This species is poorly represented in collections. We examined 19 specimens (16 ♂, 3 ♀). Seven were collected at Southern Pines, South Carolina, the rest in Hernando, Levy, Marion, Putnam, and Suwannee counties, Florida and in Jackson County, Mississippi. Capture dates range from mid-May to early September.

## 51. *Epiblema separationis* Heinrich, 1923
(Plate T, 51a–l; Plate 22, 51a–h)

*Epiblema praesumptiosa separationis* Heinrich 1923:143; McDunnough 1939:48; MacKay 1959:27.
*Epiblema separationis* Brown 1979:25; Powell 1983:35; Brown 2005:285.

**Holotype.** ♂, Texas, Cameron County, Brownsville, A. Busck, 27 May 1917, slide 72819, USNM.

*Epiblema praesumptiosa separationis* was described from five specimens (3 ♂, 2 ♀) with the same collection data as the types for *E. praesumptiosa*. Heinrich (1923) referred to one of the males as the type and illustrated its genitalia, thereby designating a name-bearing specimen. He separated the subspecies from typical *E. praesumptiosa* by its smaller size, the shape of the cucullus, and the absence of brown spots on the forewing in the subbasal position and on the proximal margin of the ocellus. Brown (1979) elevated *separationis* to species status by quantifying the size difference, pointing out the intraspecific variation of forewing markings in both species, and reporting diagnostic differences in the female genitalia.

*Epiblema separationis* is one of the smallest of the North American *Epiblema* (FWL: 3.7–6.5 mm, mean = 5.1). The head is white, with no dark spots on the labial palpus; the antenna has a black mark on the pedicel that is conspicuous in specimens from south Texas but reduced to absent in individuals from North Carolina. The forewing is white with some grayish tan suffusion and is variably marked with one or more of the following arrays of blackish dots: 3–5 at the base of the wing, 1 near the cubitus in the subbasal position, 1 or 2 on the proximal margin of the ocellus, 2 in the ocellus, and 4–5 evenly distributed along the inner margin. The ocellus is whitish, often with pale pink tints; the white costal strigulae are delimited by numerous short blackish dashes; and the termen has a narrow salt-and-pepper colored band from tornus to apex.

In males, the uncus is strongly developed as a somewhat rectangular extension of the tegumen, usually with a shallow medial indentation in the apical margin. It varies in shape as illustrated in Plate 22:51a–f, with the base especially broad in specimens from North Carolina (Plate 22:51f). In all populations, the tegumen lacks differentiated dorsolateral shoulders, and the socius is short and narrow. The base of the phallus is closely surrounded by the anellus, and the vesica has 7–22 cornuti. The valva lacks a basal process and has a moderately developed clasper that varies in location (Type 1 or Type 4). The neck is wide, NR = 0.73, the saccular corner is vaguely defined, and SA = 162°. The cucullus has a moderately developed dorsal lobe that tapers to a somewhat angular apex, a strongly curved distal margin, a weakly developed ventral lobe, and a broadly rounded anal angle. The setae on the medial surface of the cucullus are especially coarse. In females, the papillae anales are Type 3, the pads attenuating from relatively broad posterior lobes to narrowly rounded anterior lobes; the sterigma is U-shaped, with posterolateral corners rounded and central trough microtrichiate; the lamella antevaginalis is ringlike; the posterior margin of sternum 7 is concavely emarginated to about one-half the length of the sterigma; the ductus bursae lacks sclerotization; and the corpus bursae lacks signa.

The 63 specimens we examined (45 ♂, 18 ♀) document a range that includes several disjunct areas on the Gulf and Atlantic Coasts, including: south Texas; Brazoria and Jefferson counties, Texas (south and east of Houston); Cameron Parish, Louisiana; Jackson County, Mississippi; Baldwin County, Alabama; south Florida; and Carteret and New Hanover counties, North Carolina. In Texas there appear to be two primary adult flights, one from late March to mid-June, and one from mid-September to late December. A few of the specimens from south Texas and south Florida bear labels with the inscriptions "ex. *Borrichia frutescens*" or "reared from *Borrichia frutescens* gall." MacKay (1959) described one such larva from a rearing at Brownsville, Texas. That specimen is in the USNM and does not have an associated pupal case. In the United States, *Borrichia frutescens* (L.) DC. (bushy seaside tansy) has a distribution that extends along the Gulf Coast from Texas to Florida and north along the Atlantic Coast as far as Maryland.

The specimens illustrated in Plate T:51k–l (from Mississippi and North Carolina) differ somewhat in forewing appearance from typical specimens, particularly in the expression of the postmedian fascia. They might represent a distinct species, but we tentatively refer them to *E. separationis* for lack of sufficient material to support a thorough diagnosis.

## 52. ***Epiblema colfaxana*** Wright and Gilligan, new species

(Plate T, 52a–c; Plate 22, 52a–c)

**Diagnosis.** Among the *Epiblema* species with Type 2 tegumen + appendices, there are two members of the *numerosana* group that might be confused with *E. colfaxana* based on forewing appearance, namely *E. deflexana* and *E. mandana*. *Epiblema colfaxana* resembles these species in size (mean FWL = 6.5 mm vs. 6.3 and 6.9, respectively) but has a more elongate forewing (AR = 3.21 vs. 2.75 and 2.68) and differs from each in subtle aspects of forewing coloration. The pale suffusion in the proximal one-half of the forewing is tan or whitish and reticulate as opposed to white and speckled in *E. deflexana*, and the subcostal area near the apex does not have the orange streaking typical of *E. mandana*. Moreover, the ocellus is less conspicuous (tan vs. bright white), and the base of the antenna has only one black mark (on the scape) vs. two in *E. deflexana* and *E. mandana* (one on the scape and one on the pedicel). In male genitalia, *E. colfaxana* most closely resembles *E. glaseri*, but these species are easily distinguished by forewing appearance and geographical distribution (New Mexico and Colorado vs. southeastern United States). Moreover, females of *E. colfaxana* differ from those of *E. glaseri* in papillae anales Type (2 vs. 3) and in sterigma shape (posterolateral corners of sterigma sharply acute vs. rounded.

**Description.** *Head.* Third segment of labial palpus blackish brown with whitish tip; second segment with medial surface whitish, lateral surface whitish with some brown scaling and with long brownish-gray scales on distal margin; first segment white; lower frons whitish; upper frons whitish with brown central spot; vertex with medial scales whitish, lateral scales brown; antenna tan with dark mark on dorsal surface of scape. *Thorax.* Dorsal surface tan with brown cross-band on each thoracic segment; tegula tan with brown basal, medial, and apical cross-bands; fore- and mid-leg with anterior surface of femur brown, posterior surface pale tan; tibia pale tan with three brown markings; hind-leg mostly pale tan; tarsus with alternating tan and brown annulations. Forewing: ♂ FWL 6.6 mm (n = 2), AR = 3.30, CFR = 0.30; ♀ FWL 6.2 mm (n = 1), AR = 3.02; costal margin nearly straight; apex acute; termen straight; dorsal surface brown, with tan transverse reticulations on basal one-half of wing; fasciate markings not clearly defined; ocellus tan, with black mark on costal margin and some black scales in central field; costal strigulae white; fringe scales pale brown with white apices and brown cross-bands. *Abdomen.* Male genitalia (n = 2): Uncus weakly developed, not clearly differentiated from dorsolateral shoulders of tegumen, with apical margin rounded and basal width about three times height; socius short, broad basally, tapering to narrowly rounded apex; vesica with 9–12 cornuti (n = 2); valva with clasper Type 2, basal process absent, costal margin weakly concave, ventral emargination shallow, NR = 0.67, saccular corner weakly defined, SA = 157°; cucullus with dorsal lobe moderately developed, apex rounded, ventral lobe barely discernable; distal margin of cucullus weakly convex, with series of about 12 short moderately-stout setae from anal angle to three-fourths distance to apex; setae on medial surface of cucullus moderately course near distal margin, grading to hairlike near costal and apical margins. Female genitalia (n = 1): Papillae anales Type 2; apophysis anteriores substantially longer than apophysis posteriores; sterigma moderately U-shaped, with posterolateral corners acute and length of posterior margin about 1.5 times ostium diameter; lamella postvaginalis microtrichiate; lamella antevaginalis ringlike, with posterior extremities attenuating onto surface of lamella postvaginalis; ductus bursae without sclerotization; corpus bursae with two signa, the larger located near juncture with ductus bursae, the smaller near anterior extremity of bursa.

**Holotype** (Plate T:52a, Plate 22:52a). ♂, New Mexico, Colfax County, Exit 419 on I-25, 36.4515° N, 104.5938° W, 6050 feet, T. M Gilligan, 5 August 2008, slide DJW 3371, USNM.

**Paratypes.** COLORADO. Weld County, Pawnee National Grassland, T8N R64W S3, D. J. Wright, 8 August 2004 (1 ♂, slide 153373), USNM; Cheyenne County, Wild Horse Post Office, T. M. Gilligan and C. E. Harp, 38.8267° N, 103.0100° W, 4453 feet, 28 September 2007 (1 ♀, slide DJW 2036), USNM.

**Etymology.** The specific epithet derives from the name of the county in which the holotype was collected.

**Remarks.** The male illustrated in Plate T:52c and Plate 22:52c (in the EME) is sufficiently similar to *E. colfaxana* in size, forewing appearance, and genitalia to be referred to that species, but it was collected in northern California (Lassen County), and we have not seen any specimens that bridge the geographical gap between there and New Mexico/Colorado. We did not designate it a paratype and did not include its data in the account for *E. colfaxana*, but we present the image to encourage the collection of additional material that might clarify this situation.

## *Sonia* Heinrich

*Sonia* Heinrich, 1923, Bull. U.S. natn. Mus. 123: 160. Type species: *Paedisca constrictana* Zeller, 1875.

Heinrich (1923) proposed the genus *Sonia* to accommodate three species that previously had been placed in *Eucosma* (*constrictana*, *filiana*, and *vovana*) but which share aspects of forewing venation and male genitalia that distinguish them from other members of that genus. During the next 67 years, four additional species of *Sonia* were described, one each by McDunnough (1925b), Clarke (1952), Blanchard (1979), and Miller (1990), bringing the species count to seven.

**Diagnosis.** In nearly all Nearctic eucosmine genera, there are five distinct branches of the radial vein ($R_{1-5}$) that extend to the margin of the forewing, and in those genera the relationship between $R_4$ and $R_5$ is variously described as separate (Fig. A1a), stalked, approximate, or connate. In *Sonia* and *Suleima* there are only four such branches; $R_4$ and $R_5$ are considered to be united (Figs. A2a and A3a). *Sonia* differs from *Suleima* in having hindwing veins $M_3$ and $CuA_1$ stalked (Fig. A2b) instead of united (Fig. A3b). Males of *Sonia* differ from those of *Suleima* in the presence of a costal fold at the base of the forewing, in the shape of the valva (Fig. B13 vs. B15), and in the presence of a clasper on the medial surface of the valva. In *Sonia* females, the sterigma is U-shaped, with a well developed lamella postvaginalis (Fig. C1); in *Suleima* it is either cylindrical (Fig. C10) or U-shaped. With the single exception of *S. hodgesi*, the sclerotized portion of the ductus bursae in *Sonia* is patchlike, with inner surface microtrichiate (Fig. C13); in *Suleima* it is elongate, semi-helical, and lacks microtrichia (Fig. C14).

**Description.** *Head.* Third segment of labial palpus usually conspicuously darker than second segment, medial surface of second segment white, lateral surface of second segment usually with dark spot near dorsal margin; antenna usually with black mark on pedicel and/or dark gray patch on ventral surface of scape.

*Forewing* (Fig. A6). Mean FWL: 5.7–11.1 mm, average = 7.5; AR: 2.47–2.91, average = 2.69; CFR: 0.25–0.31, average = 0.29; venation as in Fig. A2a, with $R_4$ and $R_5$ united; male with costal fold; costa weakly to moderately arched; apex slightly acute, sometimes weakly or moderately falcate; termen straight or slightly concave; dorsal surface gray, brown, orange brown, or blackish brown; semifasciate markings pale brown, reddish brown, or blackish brown, often with white edging; subbasal mark nearly always conspicuous, extending from inner margin to cell, infrequently reaching costa, often with proximal margin vaguely defined; median fascia usually fragmented into one or more of a narrow mid-costal bar, an irregularly shaped discal mark, and a prominent pretornal mark; interfascial spot sometimes obscure, usually paler than forewing ground color, frequently with whitish suffusion; postmedian fascia narrow, usually discernable; ocellus present but often inconspicuous; costal strigulae white to gray, separated by dark dashes, wedges, and/or quadrate marks; termen with narrow salt-and-pepper colored band from tornus to apex.

*Hindwing.* Venation as in Fig. A2b ($M_3$ and $CuA_1$ stalked); uniformly gray brown to orange brown.

*Male genitalia.* Tegumen + appendages Type 2 (Fig. B4); uncus weakly developed and triangular, with basal width about four times height, apex rounded or weakly indented medially, dorsal surface densely scaled, ventral surface unscaled; socius broad and moderately short, with apex rounded or bluntly truncated; phallus stout, with base closely surrounded by anellus; vesica with 7–88 deciduous cornuti (average = 41); valva (Fig. B13) with clasper Type 4 (Fig. B8), basal process absent, costal margin concave, ventral emargination long and shallow, NR: 0.48–0.69, saccular corner angulate, SA: 105°–166°; cucullus with dorsal lobe strongly developed, apex rounded, distal margin convex or straight, ventral lobe weakly developed to absent, setae along ventral three-fourths of distal margin long and spiniform, setae on medial surface coarse near distal margin and grading to hairlike near costal and apical margins.

*Female genitalia* (excluding *S. hodgesi*). Papillae anales Type 3 (Fig. C3); sterigma U-shaped (Fig. C5); lamella postvaginalis rectangular and microtrichiate, often with width at posterior margin slightly greater than ostium diameter; lamella antevaginalis ringlike, moderately narrow, and weakly sclerotized, with posterior extremities merging into lateral margins of sterigma; sternum 7 with posterior margin concavely emarginated to about one-half length of sterigma, diverging laterally from lamella postvaginalis; scales on sternum 7 densely spaced in band along posterior margin, otherwise uniformly and more sparsely distributed; sclerotization of ductus bursae patchlike, located opposite ductus seminalis, with inner surface microtrichiate (Fig. C13); corpus bursae with two finlike signa of slightly different size, one located at or near mid-bursa, the other closer to juncture with ductus bursae.

*Sonia hodgesi* differs from its congeners in its lack of microtrichia on the lamella postvaginalis, the number of signa in the corpus bursae (1 vs. 2), and in the form of the sclerotized portion of the ductus bursae (ringlike and lacking microtrichia vs. patchlike and microtrichiate). The other members of the genus are remarkably uniform in female genitalia, to the extent that those characteristics are of little use in discriminating among species. Consequently, in the *Sonia* species accounts that follow, the statement "typical of the genus," as applied to the female genitalia, refers to the comments in the preceding paragraph, *S. hodgesi* notwithstanding.

**Immature stages.** MacKay (1959) provided a larval description for only one species of *Sonia* (*S. canadana*), which she placed in one of her groups (based on similarity of larval characters) with *Epiblema praesumptiosa separationis* (now *E. separationis*). In our treatment, the latter species is included in the subgroup of *Epiblema* species with Type 2 male genitalia. This suggests that there may not be much in the way of larval differences between *Sonia* and Type 2 *Epiblema*. MacKay did separate the larvae of *Epiblema* and *Sonia* from those of *Suleima* by the structure of the spinneret (rounded or somewhat broadened versus blunt or indented).

*Egg*. As described for *Epiblema*.

*Larva*. Head with adfrontals not distinct, and not always extending to vertical angle; stemmatal area well rounded; vertical angle acute; spinneret with rounded tip; labial palpus with basal segment short and stout; L-group on T1 trisetose, never enlarged or extending under the spiracle; SD1 and SD2 on T1 close together; L1 and L2 on A1–8 with a strong tendency to be posterior to a vertical line through spiracle; SV-group setal counts on A1, 2, 7, 8, 9 few in number; L1 and L2 on segments A1–8 with a strong tendency to be posterior to a vertical line through spiracle; SD1 on segment A8 anterodorsal or dorsal to spiracle; crotchets uniordinal; anal shield rounded posteriorly; anal comb absent.

*Pupa*. As described for *Epiblema*.

**Biology.** Host plants have been recorded for four species of *Sonia*. In each case, the larva is a root-borer in Asteraceae. The most detailed accounts come from Powell and Opler (2006), who reported rearing *S. comstocki* and *S. filiana* from *Isocoma acradenia* (Greene) Greene (alkali goldenbush) and *S. vovana* from various species of *Gutierrezia* Lag. (snakeweed) and *Isocoma* Nutt. (goldenbush). The USNM has a specimen of *S. canadana* labeled "host: *Aster novae-angliae*" (New England aster). Ĉapek (1971) reported rearing a species of *Sonia* (determined as *S. canadana*) from *Solidago* L. (goldenrod), but he may have confused *S. canadana* with the similar looking *S. nigrana*.

## The *constrictana* group (species 53–55)

This group is composed of *S. constrictana*, *S. paraplesiana*, and *S. roseomaculana*, three species that often are confused with one another and with members of the *ferruginana* group (species 56–58). Blanchard (1979) distinguished *S. paraplesiana* from *S. constrictana* by the shape and markings of the apex of the forewing, the venation of the hindwing, subtle differences in male genitalia, and the shape of the sterigma. We found that subtle differences in wing pattern, wing shape, valva shape, and sterigma structure are reliable in separating the two species, but other distinctions reported by Blanchard (1979) (e.g., hindwing veins Rs and $M_1$ stalked in *S. paraplesiana* vs. separate in *S. constrictana*) proved to be less stable. Group characteristics include:

*Head.* Third segment of labial palpus black with pale tip; second segment with prominent dark bar on lateral surface and long gray scales along ventral margin; antenna with black scaling on pedicel and patch of grayish scales on ventral surface of scape.

*Forewing.* Mean FWL: 6.3–7.9 mm; AR: 2.47–2.81; CFR: 0.27–0.28; dorsal surface pale gray to brownish gray with dark brown to rosy-brown markings; subbasal mark prominent, extending from inner margin to radius, with proximal margin vaguely defined and distal margin edged with white; interfascial spot nearly concolorous with ground, often with diffuse whitish suffusion; mid-costal bar inconspicuous, sometimes barely discernable; discal and pretornal marks prominent, usually disjunct, with thin white edging on proximal margins; costal strigulae beyond mid-costa white; fringe gray.

*Male genitalia.* Typical of the genus (Fig. B13); neck long and narrow; NR: 0.48–0.62; saccular corner angulate; SA: 120°–146°.

*Female genitalia.* Typical of the genus (Fig. C1).

## 53. *Sonia constrictana* (Zeller, 1875)

(Plate U, 53a–f; Plate 23, 53a–c)

*Paedisca* (?) *constrictana* Zeller 1875:305, Plate IX, fig. 36; Fernald 1882:40; Walsingham 1884:140.
*Eucosma constrictana*: Fernald 1903:458; Kearfott 1905:359; Barnes and McDunnough 1917:170.
*Sonia constrictana*: Heinrich 1923:160; McDunnough 1939:49; Blanchard 1979:185–187; Powell 1983:35; Brown 2005:555.

**Holotype.** ♂, Texas, Dallas, Boll, Type 14335, slide V. Adams XII.4.69 (genitalia, right forewing, and right hindwing), MCZ.

Zeller (1875) mentioned only one specimen, a male in the MCZ that was collected by Jacob Boll at Dallas, Texas, so we concur with Brown (2005) in interpreting that male as the holotype. Nevertheless, it bears a red label with the hand-written inscription "LECTOTYPE, EPIBLEMA CONSTRICTANA (ZELLER), By A. BLANCHARD (1979)." It is possible that the label was not attached by Blanchard because he also refers to the specimen as a holotype in his 1979 paper.

Heinrich (1923) selected this taxon as the type species for *Sonia* without examining the genitalia of the holotype (that slide was prepared in 1969). He illustrated under this name (1923: fig. 291) the genitalia of a male in the USNM from Tryon, North Carolina that we believe to be *S. paraplesiana*. During the nearly 50 years that elapsed between the publication of Zeller's description and the appearance of Heinrich's monograph, a substantial amount of *constrictana*-like material accumulated in North American collections. Heinrich (1923) noted that those specimens exhibited considerable variation in forewing appearance, male genitalia, and hindwing venation, but finding no stable associations amongst the various character states, he interpreted this diversity as intraspecific variation in a single species. Fifty-six years later, Blanchard (1979) revisited this situation, prompted by differences in forewing pattern that he observed in the *constrictana*-like material he had collected in Texas. He examined the holotype of *S. constrictana* and found that it agreed with many of his specimens but consistently differed from others in subtle aspects of forewing appearance, genitalia, and hindwing venation. He proposed the name *S. paraplesiana* for the latter specimens and included images of the holotype of *P. constrictana*, its wing venation, and its genitalia (Blanchard 1979: figs. 30–32).

*Sonia constrictana* is a medium-sized member of the genus (mean FWL = 7.2 mm) with a moderately elongate forewing (AR = 2.81). The lateral surfaces of the first and second segments of the labial palpus are whitish, the first with a black dot, the second with a black longitudinal bar near the dorsal margin. The dorsal surface of the head is grayish. The forewing has a moderately falcate apex and a black apical spot (two features mentioned by Blanchard as distinguishing this taxon from *S. paraplesiana*). The interfascial spot is grayish, and the fasciate markings are typical of the group, but the holotype (which resembles the specimen in Plate A:1a) is unusual in that its discal and pretornal marks are confluent. The ocellus is gray and inconspicuous except for a prominent white spot in the apical quadrant.

The genitalia of *S. constrictana* are typical of the genus. Males usually can be distinguished from those of *S. paraplesiana* by the size of the SA (120° vs. 143°). Although these averages differ by more than 20°, there is some overlap in the ranges of the individual measurements (102°–136°, n = 9 vs. 122°–161°, n = 23). Females of *S. constrictana* differ from those of *S. paraplesiana* in the shape of the lamella postvaginalis (lateral margins parallel vs. weakly diverging posteriorly) and in the length-to-width ratio of the sterigma (average = 1.51, n = 5 vs. 1.09, n = 11).

We examined 44 specimens (31 ♂, 13 ♀), one from Cameron Parish, Louisiana, the rest from Brazoria, Cameron, Coleman, Cottle, Fort Bend, Hidalgo, Nueces, San Patricio, and Zapata counties, Texas. Adult capture dates range from 4 May to 14 November, with the greatest concentration of records in the period from June through September. Blanchard (1979) reported one specimen from Illinois (which we are unable to confirm) and others from Florida (which we suspect were misidentified). We examined numerous specimens from Florida that previously had been determined as *S. constrictana* and refer most of them to *S. paraplesiana*. *Sonia constrictana* appears to be broadly distributed in Texas, ranging from the base of the panhandle to the southern tip of the state, and southeast to the Louisiana coast.

## 54. ***Sonia paraplesiana*** Blanchard, 1979
(Plate U, 54a–f; Plate 23, 54a–e)

*Sonia paraplesiana* Blanchard 1979:184; Miller 1987:62; Brown 2005:555; Gilligan et al. 2008:131.

**Holotype.** ♂, Texas, [Harris County], Houston, A. Blanchard, 5 June 1968, slide 69605, USNM.

Blanchard (1979) described *S. paraplesiana* from 18 specimens (15 ♂, 3 ♀), all collected by André and May Blanchard in Anderson, Harris, San Patricio, and Tyler counties, Texas.

This species is a little smaller than *S. constrictana* (mean FWL = 6.3 vs. 7.2 mm) and has a slightly broader forewing (AR = 2.67 vs. 2.81). It differs from *S. constrictana* in the following respects: the lateral surface of the second segment of the labial palpus is more uniformly brown; the dorsal surface of the head is orange brown to dull brown rather than grayish; the forewing apex is not falcate and does not have a black circular spot; the forewing markings tend to be a brighter shade of brown; the interfascial spot is sometimes faintly suffused with pink; the mid-costal bar is more prominent; the discal mark usually merges with the postmedian fascia, forming a band that arches over the anterior margin of the ocellus; the postmedian fascia expands abruptly at the merger with the discal mark; and the ocellus is more clearly defined, with a pinkish bar on the proximal margin and a pinkish spot in each of the apical and tornal quadrants. Most specimens have a conspicuous, dark brown, V-shaped mark that arises on the costa between strigulae 8 and 9, bends around strigula 9, and terminates at the apex. A similar less prominent mark occurs in *S. constrictana*, but in that species the distal leg of the V is overlaid with a black circular spot.

The genitalia of *S. paraplesiana* resemble those of *S. constrictana*. Subtle differences in valva shape and sterigma structure are discussed under the latter taxon.

*Sonia paraplesiana* is broadly distributed in eastern North America. We examined 315 specimens (236 ♂, 79 ♀) which, together with various literature records, document a range extending from Minnesota to New York, south to the Gulf Coast. In Florida, adult capture dates vary from 5 February to 21 December; in the northern part of the range, they are concentrated from May through August.

## 55. *Sonia roseomaculana* Wright and Gilligan, new species

(Plate U, 55a–c; Plate 23, 55a–c)

**Diagnosis.** *Sonia roseomaculana* differs from other members of the group in size (mean FWL = 7.9 mm vs. 7.2, 6.3), forewing geometry (AR = 2.47 vs. 2.81, 2.67), shape of the forewing apex (weakly produced vs. falcate in *S. constrictana* and not produced in *S. paraplesiana*), shape of the pretornal mark (rectangular vs. triangular), and in the absence of a clearly defined postmedian fascia. It agrees with *S. constrictana* in the presence of a prominent white spot in the apical quadrant of an otherwise gray ocellus and in the presence of a circular black spot at the apex, but it differs from that taxon in the color of the fasciate markings (rosy brown vs. blackish brown). The ocellus lacks the pink tints usually present in *S. paraplesiana*. *Sonia roseomaculana* is distinguished from members of the *ferruginana* group by the shape of the pretornal mark (rectangular vs. triangular) and by the expression of the discal mark (prominent vs. absent or weakly discernable). The male genitalia resemble those of *S. constrictana* and *S. paraplesiana*, but the SA is somewhat larger (146° vs. 120°, 143°), the neck is slightly broader (NR = 0.62 vs. 0.48, 0.51), and the cucullus is a little more elongate (length-to-width ratio = 1.65 vs. 1.43, 1.49). We encountered only one female (Plate U:55c), which resides in the MEM. It had been dissected by W. E. Miller (slide 1011982), but we were unable to locate the slide.

**Description.** *Head.* Third segment of labial palpus black with white tip; second segment white with two blackish-brown spots on lateral surface and gray-brown scales on ventral margin; lateral surface of first segment whitish with black spot; lower frons whitish; upper frons and vertex whitish to pale rosy brown; antenna brown with conspicuous black mark on pedicel; scape whitish with gray-brown patch on ventral surface. *Thorax.* Dorsal surface brown with whitish transverse line on mesothorax; scales of tegula gray brown with whitish tips, producing weakly speckled effect; fore- and mid-leg with anterior surfaces dark brown, posterior surfaces tan, one or two tan marks on tibia; hind-leg uniformly tan; tarsus with alternating light and dark annulations. Forewing: ♂ FWL 7.2–8.0 mm (mean = 7.7, n = 4), AR = 2.45, CFR = 0.27; ♀ FWL 8.6 mm (n = 1), AR = 2.53; dorsal surface grayish brown with rosy-brown markings; apex weakly falcate, with blackish-brown apical spot; interfascial spot concolorous with ground; subbasal mark quadrate, extending from inner margin to cell, with thin white edging on distal margin; discal mark prominent, with black radial dash on anterior margin and thin white edging on proximal margin; pretornal mark quadrate, narrowly joined with or approximate to discal mark, thinly edged with white on proximal margin; postmedian fascia not discernable; subapical area with two black radial dashes anterior to ocellus; ocellus gray except for conspicuous white spot in apical quadrant and two short black dashes on white spot; costal strigulae white, inconspicuous, delimited by rosy-brown streaks and short black dashes; fringe gray. Hindwing: Dark gray brown. *Abdomen.* Male genitalia (n = 4): Typical of the genus; uncus with weakly developed pair of rounded projections located medially on common boundary of unscaled ventral surface and densely scaled dorsal surface; socius with apex truncated; vesica with 55–75 cornuti (mean = 64, n = 4); valva with NR = 0.62, SA = 146°; cucullus with length-to-width ratio = 1.65. Female genitalia unknown.

**Holotype** (Plate U:55a; Plate 23:55b). ♂, Franklin County, Apalachicola National Forest, H. D. Baggett, 1 September 1991, slide DJW 4082, USNM.

**Paratypes** (2 ♂, 1 ♀). Same collection data as holotype (2 ♂, slides DJW 4273, 4279; 1 ♀, slide WEM 1011982), MEM.

**Etymology.** The specific epithet refers to the color of the fasciate markings.

**Distribution.** This species is currently known only from the type locality southeast of Tallahassee in the Florida panhandle.

## The *ferruginana* group (species 56–58)

We examined more than 300 *Sonia* specimens from eastern North America that are sufficiently similar in forewing pattern and genitalia to form a cohesive group but which are not accommodated by any of the available names. Many were misidentified in collections as *S. constrictana*, *S. paraplesiana*, or *S. canadana*; others were set aside as undetermined. From this material we recognize two subgroups in which stable forewing phenotype combines with subtle genitalia features to support diagnoses of the two new species described below as *S. ferruginana* and *S. griseolana* (represented by about 210 and 25 specimens, respectively). The remaining individuals (numbering about 65) might conceivably be viewed as variants of one or the other of these new taxa, but some exhibit combinations of character states that are distinctive enough to suggest the possibility of two or more additional new species. We refer to them in aggregate as the *ferruginana* complex, to which we assign the number 58. These 65 individuals can be sorted roughly into two forewing color forms: gray with dark brown markings (Plate W:58a–i) and gray brown with reddish markings (Plate W:58j–o). In both forms there is variation in the expression of the discal mark, the presence/absence of a white ocelloid spot, the shape of the forewing apex, and the amount of white edging on the fasciate markings. We found no stable genitalia distinctions that would combine with forewing features to produce clear diagnoses of additional taxa, so we defer further refinement of the group to a time when additional information is available. In the interim, we illustrate examples of the various phenotypes we encountered. Group characteristics include:

*Head.* Third segment of labial palpus black or brown, often with whitish tip; second segment whitish, with long brown scales on ventral margin and dark longitudinal bar (sometimes reduced to two spots) on lateral surface; antenna with black scaling on pedicel and patch of grayish/brownish scales on ventral surface of scape.

*Forewing.* Mean FWL: 6.8–7.4 mm; AR: 2.56–2.64; CFR: 0.27–0.30; dorsal surface medium brown to dark gray; fasciate markings reddish brown to blackish brown; subbasal mark extending from inner margin to radius, with thin white edging along distal margin; mid-costal bar absent to barely detectable; discal mark absent to weakly expressed, in the latter case disjunct from pretornal mark; pretornal mark triangular, with white edging on proximal margin; postmedian fascia usually inconspicuous; apex with prominent circular black or brown spot; ocellus grayish, often obscure, sometimes with white spot in apical quadrant; termen with narrow salt-and-pepper colored band from tornus to apex; fringe gray.

*Genitalia.* Typical of the genus; vesica with 30–68 cornuti; neck well defined; NR: 0.50–0.56; saccular corner angular; SA: 105°–134°; sterigma length-to-width ratio: 1.25–1.46.

## 56. *Sonia ferruginana* Wright and Gilligan, new species

(Plate V, 56a–i; Plate 24, 56a–f)

**Diagnosis.** This species is distinguished from its congeners by the following combination of forewing features: ground color brown, markings rusty brown, pretornal mark triangular, discal mark absent to barely discernable. It often is confused with *S. constrictana* and *S. paraplesiana* but differs from the first in the lack of a falcate forewing apex and from the second in the presence of an apical spot. The genitalia of *S. ferruginana* resemble those of its congeners in the *constrictana, ferruginana*, and *canadana* groups to the extent that species-level determinations on this basis are problematic. Nevertheless, *S. ferruginana* seems to resemble most closely *S. nigrana* in valva shape and sterigma-sternum 7 structure. These two species are readily separated by forewing color (rusty brown vs. blackish) and expression of the discal mark (absent vs. prominent).

**Description.** *Head.* Third segment of labial palpus blackish-brown with white tip; lower frons white; upper frons and vertex brown; antenna scales pale brown basally, dark brown distally, producing fine alternating light and dark annulations. *Thorax.* Dorsal surface brown; fore- and mid-leg with anterior surfaces dark brown, posterior surfaces tan; hind-leg uniformly tan; tibia with alternating tan and brown annulations. Forewing: ♂ FWL 4.0–8.2 mm (mean = 6.8, n = 125), AR = 2.63, CFR = 0.29; ♀ FWL 5.1–9.1 mm (mean = 7.5, n = 37), AR = 2.68; costa weakly arched; apex slightly acute; termen nearly straight, sometimes weakly concave from $M_3$ to apex; dorsal surface medium brown with rusty-brown markings; subbasal mark broad, moderately quadrate, with proximal margin vaguely defined and distal margin weakly edged with white; pretornal mark prominent, triangular, with proximal margin weakly edged with white; discal mark absent (Plate V:56a) to faintly discernable (Plate V:56c); interfascial spot slightly paler than ground color, with inconspicuous whitish reticulations; ocellus obscure, vaguely divided by thin brown line segments into three gray components (proximal transverse bar, apical quadrant, and tornal quadrant); costal strigulae white to gray, finely delimited by orange-brown streaks and gray-brown striae; apical spot brown, often with blackish center. Hindwing: Dark gray brown. *Abdomen.* Male genitalia (n = 13): Uncus often with weakly developed pair of rounded projections medially located on boundary between unscaled ventral surface and densely scaled dorsal surface; socius with apex rounded or truncated; vesica with 30–61 cornuti (average = 42, n = 11); valva with costal margin moderately concave, NR = 0.56, SA = 132°; cucullus with dorsal lobe well developed, apex rounded, distal margin nearly straight, ventral lobe not developed. Female genitalia (n = 10): Length of apophyses posteriores approximately three-fourths that of apophyses anteriores; lamella postvaginalis nearly rectangular, with posterior width slightly greater than ostium diameter; otherwise, typical of the genus.

**Holotype** (Plate V:56a; Plate 24:56a). ♂, Kentucky, Bracken County, Meldahl Dam, L. D. Gibson, 21 June 2007, slide DJW 4064, USNM.

**Paratypes** (35 ♂, 15 ♀). ALABAMA. Lawrence County, Prairie Grove Glade, S. M. Lee, E. Martinez, and N. Pinkaew, 24 May 2004 (1 ♀). FLORIDA. Alachua County, Gainesville, E. C. Knudson, 21 June 1980 (1 ♂); Dade County, Long Pine Key, J. B. Heppner, 26 April 1975, (1 ♀, slide DJW 4129). IOWA. Johnson County, Williams Prairie, D. J. Wright, 15 June 1992 (1 ♂; 1 ♀, slide DJW 4084). KENTUCKY. Boone County, Boone Cliffs Nature Preserve, L. D. Gibson, 5 July 1984 (1 ♀, slide WEM 263903); Boone County, Big Bone Lick State Park, L. D. Gibson, 18 June 1990 (1 ♂), 26 June 1990 (2 ♂, slide WEM 207904); Bracken County, Meldahl Dam, L. D. Gibson, 21 June 2007 (2 ♂); Bullitt County, 7 mi. E of Shepherdsville, D. J. Wright, 8 July 1988 (1 ♂, slide TD 97905); Campbell County, A. J. Jolly County Park, L. D. Gibson, 15 July 1982 (1 ♀); Laurel County, Bolton Branch, D. J. Wright, 17 June 1994 (1 ♂); McCracken County, Paducah Zip Track, D. J. Wright, 12 August 2008 (1 ♂, slide DJW 3801). LOUISIANA. Bossier Parish, Barksdale AFB, R. Brown and D. Pollock, 8 August 1996 (1 ♂); Calcasieu Parish, S. H. Jones State Park, R. Brown and D. Pollock, 19–23 May 1993 (2 ♀, slide DJW 4065). MARYLAND. Harford County, Bush River Marsh, J. Glaser, 30 May 2002 (1 ♀). MISSISSIPPI. Oktibbeha County, 4 mi. S of Starkville, D. J. Wright, 19 June 2008 (1 ♀); 6 mi. SW of Starkville, R. L. and B. B. Brown, 11 June 1984 (1 ♂), 30 July 1984 (1 ♂). MISSOURI. Lewis County, 1 mi. S of Canton, A. M. Dykstra, 28 June 1975 (1 ♀). NEW YORK. Tompkins County, Ithaca, Snyder Heights, J. G. Franclemont, 15 July 1979 (1 ♂), 6 August 1976 (1 ♂, slide DJW 4069). OHIO. Adams County, 1 mi. SE of Lynx, D. J. Wright, 27 June 1998 (7 ♂, slide DJW 3800, wing slide DJW 3; 2 ♀, slide DJW 3802), 3 August 1998 (1 ♂), 3 August 2000 (1 ♂); Lynx Prairie Preserve, D. J. Wright, 6 July 1989 (1 ♂, slide TD 97901); Delaware County, Delaware Reservoir Wildlife Area, L. D. Gibson, 29 June 1990 (1 ♂, 1 ♀); Delaware County, State Route 229, 3 mi E of US 23, D. J. Wright, 29 June 1990 (3 ♂); Hamilton County, Miami Whitewater County Park, L. D. Gibson, 6 August 1986 (1 ♀); Erie County, Resthaven Wildlife Area, D. J. Wright, 20 July 1990 (1 ♂); Wyandot County, County Road T-108, 2.5 mi. S of

Route 294, D. J. Wright, 30 June 1990 (2 ♂, slides WEM 139904, 189904). TEXAS. Brazoria County, Churchill Bridge, E. C. Knudson, 9 May 1982 (1 ♂); Hardin County, Big Thicket National Preserve, E. C. Knudson, 25 June 1994 (1 ♀), Harris County, Spring Valley, E. C. Knudson, 20 April 1991 (1 ♂); Sabine County, Six Mile, E. C. Knudson, 18 July 1987 (1 ♂). Paratype depositories: LDG, MEM, MGCL, USNM.

**Etymology.** The specific epithet derives from the Latin adjective *ferruginus*, meaning rust colored.

**Distribution.** *Sonia ferruginana* is broadly distributed through much of eastern United States. The material we examined documents a range that includes the region extending south of a line from Iowa to Massachusetts to the Gulf Coast (Texas to Florida).

## 57. *Sonia griseolana* Wright and Gilligan, new species

(Plate V, 57a–f; Plate 24, 57a–h)

**Diagnosis.** *Sonia griseolana* resembles *S. ferruginana* in forewing pattern, but the ground color is gray rather than brown, the costa is more strongly arched, the apex is weakly falcate, and the apical quadrant of the ocellus contains a conspicuous white spot. The male genitalia of *S. griseolana* differ from those of *S. ferruginana* in the size of the saccular angle (105° vs. 132°).

**Description.** *Head.* Labial palpus typical of the group, with third segment entirely black; lower frons white; upper frons and vertex brown; scales along anterior margin of eye white; scales of antenna pale brown basally, dark brown distally; pedicel black; ventral surface of scape with dark gray patch. *Thorax.* Dorsal surface brownish gray; fore- and mid-leg with anterior surfaces dark brown, posterior surfaces tan; hind-leg uniformly tan; tibia with alternating tan and brown annulations. Forewing: ♂ FWL 5.5–7.3 mm (mean = 6.6, n = 20), AR = 2.65, CFR = 0.27; ♀ FWL = 7.2–8.2 (mean = 7.8, n = 3), AR = 2.58; costa moderately arched; apex weakly falcate; dorsal surface brownish gray with dark brown markings, the latter frequently with faint rusty tint; subbasal mark broad, moderately quadrate, with proximal margin vaguely defined; mid-costal bar barely detectable; discal mark absent or vaguely discernable; pretornal mark prominent; postmedian fascia dark brown, well defined but inconspicuous; apical spot blackish brown; interfascial spot slightly paler than ground, with lateral margins thinly accentuated with white; ocellus mostly gray, with bright white spot in apical quadrant. Hindwing: Dark brownish gray, concolorous with interfascial areas of forewing. *Abdomen.* Male genitalia (n = 14): Typical of the genus; uncus with apex rounded; socius with apex rounded or truncated; vesica with 30–60 cornuti (average = 42, n = 12); valva with costal margin strongly concave, NR = 0.51, SA = 105°; cucullus with apex rounded, distal margin weakly convex, ventral lobe not developed. Female genitalia (n = 3): As in *S. ferruginana* except sterigma slightly more elongate (length-to-width ratio = 1.46 vs. 1.28); one specimen with lateral margins of lamella postvaginalis weakly concave (Plate 24:57h).

**Holotype** (Plate V:57a; Plate 24:57a). ♂, Mississippi, Oktibbeha County, 6 mi. SW of Starkville, R. L. and B. B. Brown, 23 July 1984, slide DJW 4119, USNM.

**Paratypes** (21 ♂, 3 ♀). FLORIDA. Highlands County, Archbold Biological Station, J.-F. Landry and P. Hebert, 19 June 2006 (1 ♀; slide TOR 1603). KENTUCKY. McCracken County, Paducah Zip Track, L. D. Gibson, 16 June 2007 (1 ♂, slide DJW 4125), 12 August 2008 (5 ♂, slide DJW 4061). LOUISIANA. Bossier Parish, Barksdale Air Force Base, D. M. Pollock, 21 August 1996 (1 ♂, slide DJW 4118); Cameron Parish, Cameron Prairie National Wildlife Refuge, R. L. Brown and D. M. Pollock, 15 September 1993, (3 ♂, slide USNM 114336). MISSISSIPPI. Hancock County, 2 mi. N of Pearlington, T. L. Schiefer, 23 July 1989 (1 ♂, slide WEM 308854); Hinds County, Jackson, B. Mather, 12 August 1961 (1 ♂, slide JAP 1276); Oktibbeha County, 6 mi. SW of Starkville, R. L. and B. B. Brown, 18 July 1984 (1 ♂), 2 August 1984 (1 ♂, slide DJW 4277), 3 August 1986 (2 ♂, slide DJW 4126), 21 August

1984 (1 ♀, slide DJW 4123); Webster County, 7 mi W of Walthall, R. L. Brown, 15 June 1988 (1 ♂, slide DJW 4124); Winston County, Noxubee National Wildlife Refuge, T. L. Schiefer, 14 June 1992 (1 ♂, slide WEM 308853), 20 July 1991 (1 ♂, slide WEM 1310995); TEXAS. Harris County, Houston, A. and M. E. Blanchard, 6 June 1968 (1 ♂, slide USNM 68984); Newton County, Deweyville, E. C. Knudson, 18 August 1979 (1 ♂, slide DJW 4118); Tyler County, Big Thicket National Preserve, R. Leuschner, 20 June 1996 (1 ♀, slide DJW 4280). Paratype depositories: CNC, EME, LDG, MEM, MGCL, USNM.

**Etymology.** The specific epithet derives from the Latin adjective *griseolus*, meaning grayish.

**Distribution.** All but one of the paratypes were collected in the region extending from eastern Texas to central Mississippi and north to western Kentucky. The exception is a female from central Florida.

## 58. *ferruginana* complex

(Plate W, 58a–o; Plate 25, 58a–p)

The specimens aggregated under number 58 are closely related to *S. ferruginana* and *S. griseolana*. Some may, in fact, represent subtle intraspecific variation that is not accounted for in the definitions of those two taxa. Others, especially the reddish phenotypes, may represent one or more unnamed species. We illustrate the principal forms to promote the collection of additional material that may aid in resolving the taxonomic status of these specimens. They all have a dark circular spot at the apex of the forewing, but in many, the discal mark is more prominent than in other members of the group.

### Grayish phenotype

(Plate W, 58a–i; Plate 25, 58a–k)

The illustrated individuals are representative of 36 specimens (25 ♂, 11 ♀) that share with *S. griseolana* a predominantly gray forewing but differ from that taxon in the absence of a white ocelloid spot and a falcate apex.

Images 58a–c depict three of six similar looking specimens collected in Baldwin County, Alabama on 5 August 2000 (2 ♂, 1 ♀) and in Collier County, Florida on 30 June 1984 (1 ♂, 2 ♀). They are larger than typical *S. ferruginana* and *S. griseolana* (mean FWL = 8.1 vs. 7.0 and 6.8 mm, respectively). Images 58d–e represent an especially gray phenotype collected in Cumberland County, New Jersey (3 ♂ on 29 June 2019; 1 ♀ on 18 July 2020). The specimen in image 58f was collected along with the gray males from Cumberland County and might be conspecific with them despite its brownish appearance and lighter interfascial spot. Images 58g–h are typical of what might be a particularly grayish variant of *S. ferruginana* that has been collected in the region from Ohio and Kentucky northeast to Massachusetts, Connecticut, and New Jersey. Image 58i shows one of two rather large females (mean FWL = 8.3 mm) in the INHS. They were collected by Murray Glenn in Putnam County, Illinois and feature a particularly pale interfascial spot and a prominent discal mark, the latter connecting to the postmedian fascia.

### Reddish phenotype

(Plate W:58j–o, Plate 25:58l–p)

These images and genitalia drawings are representative of 28 specimens (22 ♂, 6 ♀) with three features that are not present in typical *S. ferruginana*: especially reddish markings, a moderately expressed discal mark, and a conspicuous white spot in the apical quadrant of the ocellus. They were collected along the Atlantic and Gulf Coasts, and their forewing and genitalia statistics are similar to those of *S. ferruginana*. Images 58j–k are typical of 6 specimens (5 ♂, 1 ♀) in the USNM from the eastern shore of Maryland;

image 58l depicts a specimen in the JBS Collection from coastal North Carolina; and images 58m–o depict the variation in 12 specimens (8 ♂, 4 ♀) from Florida in the MEM, the MGCL, and the USNM. We also examined 5 ♂ in the MEM from Baldwin County, Alabama and 1 ♂ in the MEM from Cameron Parish, Louisiana. The Maryland specimens have a particularly striking forewing appearance (due to the intensity of their red markings) and may prove to be a distinct species. A few of the individuals from Florida are suggestive of *S. roseomaculana* but differ from that taxon in the shape of the pretornal mark (triangular vs. rectangular).

## The *canadana* group (species 59–63)

This group includes four predominantly black or dark gray species: *S. canadana*, *S. nigrana*, *S. albimacula*, and *S. parva*. In addition, we discuss under number 63 two phenotypes that have similarities with other members of the group but may not be conspecific with any of those taxa. We refer to these later individuals in aggregate as the *canadana* complex. Group characteristics include:

*Head.* Third segment of labial palpus black or brown with pale tip; second segment with medial surface white, lateral surface whitish or brownish, long scales along ventral margin dark gray; lower frons white; upper frons brown; vertex brown; antenna brown, usually with black scales on pedicel.

*Forewing.* Mean FWL: 5.7–7.2 mm; AR: 2.56–2.75; CFR: 0.29–0.31; dorsal surface blackish brown to gray; fasciate markings dark brown; subbasal mark somewhat quadrate, extending from inner margin to cell, with proximal margin vaguely defined; mid-costal bar obscure but often detectable; discal mark present, often vaguely defined; pretornal mark prominent, triangular, and frequently fused with discal mark; postmedian fascia moderately well defined; interfascial spot white to gray, often with dark transverse reticulations and white edging on lateral margins; ocellus gray, sometimes with white speckling; termen with narrow salt-and-pepper colored band from tornus to apex; fringe gray.

*Genitalia.* Typical of the genus; vesica with 22–56 cornuti; valva with neck well defined, NR: 0.52–0.58, saccular corner angulate, SA: 124°–136°.

### 59. ***Sonia canadana*** McDunnough, 1925

(Plate X, 59a–f; Plate 25, 59a–f)

*Sonia canadana* McDunnough 1925b:115; McDunnough 1939:49; MacKay 1959:26; Powell 1983:35; Miller 1987:62; Brown 2005:555; Gilligan et al. 2008:131.

**Holotype.** ♂, Manitoba, Aweme, N. Criddle, 16 July, TOR 4045, CNC.

McDunnough (1925) based this species on the holotype, an allotype collected on 8 July by N. Criddle at Aweme, Manitoba, and four male paratypes collected between 30 July and 4 September by J. Evans at Trenton, Ontario, Canada.

*Sonia canadana* is the largest member of the group (mean FWL = 7.2 vs. 5.7–6.7 mm). The third segment of the labial palpus is black with a few whitish scales at the tip; the lateral surface of the second segment is whitish to grayish brown, with a scattering of black/brown scales that often coalesce into a dark bar or a pair of dark spots near the dorsal margin. The upper frons is brown, the vertex is pale brown to gray brown, and the antenna has black scaling on the pedicel and a gray patch on the ventral surface of the scape. Forewing color varies from brown to blackish brown, with white speckling or reticulations in the interfascial spot and ocellus. The subbasal mark is prominent, its distal margin edged with white and oriented somewhat obliquely with respect to the inner margin; the mid-costal bar is barely discernable;

the discal mark is moderately well defined and usually connected to the postmedian fascia (as in *S. paraplesiana*); and the pretornal mark is separated from the discal mark by a white-edged grayish band that extends distally to the ocellus. The ocellus has a gray band along the proximal margin, a variably expressed white spot in the apical quadrant, a gray spot in the tornal quadrant, and sometimes a few centrally located black spots. The costal strigulae are white and moderately conspicuous.

In males, the uncus frequently has a weakly bilobed apex, the distal extremity of the socius is rounded or truncated, and the vesica has 22–56 cornuti. The shape of the valva varies as in Plate 25:59a–e, with NR = 0.52 and SA = 127°. In females, the length of the apophyses anteriores is about 1.5 times that of the apophyses posteriores, and the posterior margin of the lamella postvaginalis is slightly wider than the ostium.

We examined specimens from Manitoba, Ontario, Illinois, Kentucky, Michigan, New York, and Ohio. Adult capture dates range from 14 June to 10 September, but most records are from August. One specimen in the USNM from Ontario is labeled "host: *Aster novae-angliae*" = *Symphyotrichum novae-angliae* (L.) G.L. Nesom (New England aster). Ĉapek (1971) reported larvae boring in root-stalks of *Solidago* L. (goldenrod) in Ontario. We have been unable to locate specimens from Ĉapek's study and, therefore, cannot confirm his species determinations. It is possible that his specimens are exemplars of *S. nigrana* (described below), which has been confused with *S. canadana*.

## 60. ***Sonia nigrana*** Wright and Gilligan, new species

(Plate X, 60a–f; Plate 26, 60a–f)

**Diagnosis.** *Sonia nigrana* resembles *S. canadana* in forewing color (blackish brown) but is slightly smaller than the latter (mean FWL = 6.4 vs. 7.2 mm) and has considerably fewer white flecks in the forewing. In *S. nigrana* the white scaling is confined to transverse reticulations in the interfascial spot; in *S. canadana* it occurs in the interfascial spot, in the edging of the fasciate markings (especially noticeable between the discal and pretornal marks), and in the ocellus. The pretornal and discal marks in *S. nigrana* are confluent, forming a band from the inner margin nearly to the radius; in *S. canadana* they are disjunct and separated by a narrow gray band with white edging. The distal margin of the subbasal mark is usually perpendicular to the inner margin in *S. nigrana* vs. oblique thereto in *S. canadana*. In males, the cucullus tends to be slightly less elongate in *S. nigrana* than in *S. canadana* (length-to-width ratio = 1.54 vs. 1.68), but we found no stable genitalia differences that would reliably distinguish the two species.

**Description.** *Head.* Third segment of labial palpus black with white tip; second segment with long grayish-brown scales on ventral margin, medial surface whitish, lateral surface whitish to pale gray with two gray-brown spots near dorsal margin; lower frons whitish; upper frons and vertex pale brown to medium brown; antenna brown, with black scaling on pedicel and dark gray patch on ventral surface of scape. *Thorax.* Dorsal surface brownish gray; fore- and mid-leg with anterior surfaces blackish brown, posterior surfaces tan; hind-leg uniformly tan; tibia with alternating tan and blackish-brown annulations. Forewing: ♂ FWL 5.0–7.2 mm (mean = 6.3, n = 66), AR = 2.61, CFR = 0.30; ♀ FWL 6.5–8.0 mm (mean = 7.1, n = 8), AR = 2.57; costa weakly convex; apex slightly acute; distal margin nearly straight, sometimes weakly concave near apex; dorsal surface dark brownish gray to blackish gray with blackish-brown markings; subbasal mark broad, quadrate, extending nearly to radius, with distal margin highlighted by white line perpendicular to inner margin; interfascial spot white with gray transverse reticulations; discal and pretornal marks confluent, forming median band that sometimes connects with narrow mid-costal bar; postmedian fascia well defined but inconspicuous; cubitus and $CuA_2$ weakly emphasized by thin blackish line from base to tornus; ocellus divided by thin dark lines into three dark gray components (proximal transverse band, apical quadrant, and tornal quadrant); costal strigulae white and sharply defined; fringe dark gray. Hindwing: Dark gray brown, nearly concolorous with forewing ground color. *Abdomen.* Male genitalia: Typical of the genus, usually with apex of uncus rounded; apex of socius rounded or truncated; vesica with 22–38 cornuti; valva with NR = 0.58, SA = 131°. Female genitalia: Typical of the genus.

**Holotype** (Plate X:60a, Plate 26:60c). ♂, Ohio, Adams County, 1 mi SE of Lynx, D. J. Wright, 5 July 2002, slide DJW 4077, USNM.

**Paratypes** (32 ♂, 5 ♀). IOWA. Monona County, Loess Hills State Forest, L. D. Gibson, 1 August 1995 (3 ♂). Pocahontas County, Kalsow Prairie, D. J. Wright, 22 June 2000 (1 ♀, slide DJW 4284), 18 July 2003 (1 ♂, DJW 4283), 19 July 1998 (3 ♂, slide DJW 3804; 1 ♀, slide DJW 3806). KANSAS. Riley County, Konza Prairie, D. M. Pollock, 12 June 1999, (1 ♂); L. D. Gibson, 23 July 1995 (3 ♂). MARYLAND. Washington County, Sideling Hill WMA, J. Glaser, 24 August 2000 (1 ♂, slide DJW 3910). OHIO. Adams County, 1 mi SE of Lynx, D. J. Wright, 27 June 1998 (5 ♂; 1 ♀), 2 July 2003 (2 ♂, slide DJW 3803), 5 July 2002 (4 ♂, slides DJW 4077, 4285), 25 July 1997 (4 ♂; 1 ♀, slide DJW 4074), 25 July 1998 (1 ♂, slide DJW 4076; 1 ♀, slide DJW 3805), 3 August 1998 (1 ♂, slide DJW 4286), 3 August 2000 (2 ♂), 12 August 1998 (1 ♂). Paratype depositories: LDG, MEM, USNM.

**Etymology.** The specific epithet derives from the Latin adjective *niger*, meaning black or dark.

**Distribution.** We examined 72 specimens (64 ♂; 8 ♀) that we refer to this taxon. One was collected at Pontiac, Quebec; two on Manitoulin Island, Ontario; one in Washington County, Maryland; four in Pennsylvania, two in Michigan, and the rest in remnant prairie habit in Iowa, Kansas, and Ohio.

## 61. ***Sonia albimacula*** Wright and Gilligan, new species
(Plate X, 61a–c; Plate 26, 61a–d)

**Diagnosis.** *Sonia albimacula* is distinguished from other members of this group by the following combination of characteristics: head with white scaling on medial portion of vertex; interfascial spot large, white, and prominent; basal fascia flanked laterally by moderately contrasting whitish or grayish transverse bands. This species most closely resembles the undescribed morphospecies illustrated in Plate Y:63a, but differs from that taxon in size (mean FWL = 6.4 vs. 6.9 mm), forewing geometry (AR = 2.58 vs. 2.70), size and shape of the interfascial spot (larger, with proximal margin convex vs. concave), expression of the costal strigulae (brighter and more conspicuous), and prominence of the postmedian fascia (accentuated by pale gray lateral edging). We found no male genitalia features that clearly distinguish *S. albimacula* from other members of the group (NR = 0.58 vs. 0.52–0.58; SA = 131° vs. 124°–136°). The female genitalia are typical of the genus.

**Description.** *Head.* Third segment of labial palpus black with pale tip; second segment with medial surface white, dorsal margin white, long scales along ventral margin brownish gray, lateral surface whitish and variably suffused with black/brown; first segment whitish, sometimes with black spot on lateral surface; frons white; vertex white medially, pale brown to brownish gray laterally; antenna grayish brown, with some black scaling on pedicel; scape pale, with gray patch on ventral surface. *Thorax.* Dorsal surface dark brownish gray, often with white-tipped scales; fore- and mid-leg with anterior surfaces dark gray, each with three whitish marks, posterior surfaces pale tan; hind-leg mostly tan; tarsus with white and dark gray annulations. Forewing: ♂ FWL 5.4–7.1 (mean = 6.4, n = 25), AR = 2.58, CFR = 0.30; ♀ FWL 6.4–6.6 (mean = 6.5, n = 4), AR = 2.61; costa moderately arched; apex slightly acute; distal margin weakly concave from $M_3$ to apex; dorsal surface mostly brownish gray, with fasciate markings dark brown; basal fascia narrow, weakly expressed, bordered laterally with whitish bands from inner margin to cell; subbasal mark prominent, extending from inner margin into cell, with distal margin weakly concave; interfascial spot large, bright white, with a few brownish-gray transverse reticulations; discal and pretornal marks fused, forming band from inner margin to radius; mid-costal bar well defined, often connecting with discal mark;

postmedian fascia conspicuous, edged laterally with gray striae; cubitus and $CuA_2$ accentuated by gray line from base to tornus (weakly expressed across interfascial spot); costal strigulae white and prominent; ocellus mostly gray with some white suffusion, bisected transversely by line of brownish-gray spots; termen with thin salt-and-pepper colored band from tornus to apex; fringe gray. Hindwing: Gray brown. *Abdomen.* Male genitalia: Typical of genus; uncus rounded; socius with apex rounded or truncated; vesica with 26–42 cornuti (mean = 32, n = 11); valva with costal margin concave, ventral emargination shallow, NR = 0.58, SA = 131°; cucullus with dorsal lobe strongly produced and tapering slightly to rounded apex, distal margin weakly convex, ventral lobe not developed, anal angle obtuse, setae on medial surface coarse, 5–8 spiniform setae on distal margin. Female genitalia: Typical of the genus.

**Holotype** (Plate X:61a; Plate 26:61a). ♂, Ohio, Adams County, 1 mi SE of Lynx, D. J. Wright, 27 June 1998, slide DJW 4394, USNM.

**Paratypes** (20 ♂, 5 ♀). ILLINOIS. Saline County, Garden of Gods, J. A. Powell, 21 July 1985 (1 ♂); Pope County, Burden Falls, J. A. Powell, 22 July 1985 (1 ♂, slide DJW 4361). MAINE. Penobscot County, Lincoln, 15 July, (1 ♀, slide DJW 4593). MASSACHUSETTS. Barnstable County, Barnstable, C. P. Kimball, 19 June 1949 (1 ♂), 20 June 1949 (1 ♂), 25 June 1949 (1 ♂), 26 June 1949 (1 ♂, 1 ♀), 30 June 1951 (1 ♂), 1 July 1949 (2 ♀, slide DJW 4591), 2 July 1949 (1 ♂, slide DJW 4592), 3 July 1960 (1 ♂, slide DJW 4412), 4 July 1949 (1 ♀), 6 July 1950 (1 ♂, slide WEM 308032), 11 July 1949 (2 ♂, slide DJW 4594), 13 July 1949 (1 ♂), 17 July 1950 (1 ♂). NEW YORK. Suffolk County, Orient, Long Island, 29 May 1934, R. Latham (1 ♂, slide DJW 4390). OHIO. Same data as holotype (1 ♂, slide DJW 4282). VERMONT. Caledonia County, Peacham, M. Sabourin, 26 June 2007 (1 ♂, slide DJW 4534), 5 July 2007 (1 ♂, slide DJW 5435), 11 July 2006 (1 ♂, slide DJW 4536), 20 July 2006 (1 ♂, slide DJW 4538). Paratype depositories: CMNH, CUIC, EME, MGCL, USNM.

**Etymology.** The specific epithet is a combination of the Latin *albus* and *macula*, meaning white and spot, respectively, and refers to the conspicuous interfascial spot.

**Distribution.** We examined 31 specimens (25 ♂; 6 ♀) which document a geographic range that includes southern Illinois; southeastern Ohio; northeastern Vermont; Long Island, New York; Cape Cod, Massachusetts; and coastal Maine.

## 62. *Sonia parva* Wright and Gilligan, new species

(Plate Y, 62a–e; Plate 26, 62a–d)

**Diagnosis.** *Sonia parva* differs from other members of the group in size (mean FWL = 5.7 vs. 6.4–7.2 mm) and general forewing appearance (dull grayish brown with weakly contrasting interfascial spot). It resembles *S. hodgesi* in size (mean FWL = 5.7 vs. 5.8 mm) but differs from that species in forewing color (grayish brown vs. gray and white), male genitalia (Plate 26:62a–c vs. Plate 28:68a–c), and geographic distribution (Iowa, Kentucky, Missouri, and Ohio vs. Arizona). *Sonia parva* superficially resembles the grayish phenotype in the *ferruginana* complex (Plate W:58a–i) but differs from those specimens in size (mean FLW = 5.7 vs. 7.4 mm) and in the expression of the discal mark (small but clearly discernable vs. absent to barely detectable). In males of *S. parva*, the saccular angle (124°) is smaller than in all members of the *ferruginana* and *canadana* groups except *S. griseolana* (127°–136° and 105°, respectively). Otherwise, *S. parva* agrees substantially in genitalia with the members of those groups.

**Description.** *Head.* Third segment of labial palpus black or brown with whitish tip; second segment with long scales on ventral margin gray, medial surface white, lateral surface a mixture of grayish brown and white; lower frons white; upper frons tan; vertex grayish brown; antenna grayish brown, usually with some black scaling on pedicel and considerable gray-brown suffusion on whitish scape. *Thorax.* Dorsal surface grayish brown; fore- and mid-leg with anterior surfaces dark gray, posterior surfaces tan; hind-leg mostly tan; tarsus with alternating tan and dark gray annulations. Forewing: ♂ FWL 5.0–6.2 (mean = 5.7, n = 20), AR = 2.69, CFR = 0.30; ♀ FWL = 6.3 mm (n = 1), AR = 2.62; costa weakly arched; apex slightly acute; distal margin nearly straight; dorsal surface grayish brown with dark brown markings; basal fascia represented by weakly expressed transverse line; subbasal mark prominent, extending from inner margin into cell, with grayish-white edging on distal margin; interfascial spot grayish brown, with some whitish suffusion; mid-costal bar thin and variably expressed; discal mark obscure but discernable; pretornal mark prominent, triangular, and disjunct from discal mark; postmedian fascia well defined, sometimes weakly connected to discal mark, edged by gray stria emanating from strigula 7; ocellus gray, divided by black/brown spots and dashes into three components (proximal band, apical quadrant, and tornal quadrant); costal strigulae whitish to grayish, well delimited from mid-costa to apex; termen with thin salt-and-pepper colored band from tornus to apex; fringe gray. Hindwing: Gray brown. *Abdomen.* Male genitalia: Typical of the genus; uncus with apex flat to weakly indented medially; socius with apex truncated; vesica with 30–42 cornuti (n = 7); valva with costal margin moderately concave, ventral emargination shallow, NR = 0.53, SA = 124°; cucullus with dorsal lobe strongly produced and tapering evenly to rounded apex, distal margin straight to weakly convex, setae on medial surface coarse. Female genitalia: Typical of the genus.

**Holotype** (Plate Y:62a; Plate 26:62a). ♂, Iowa, Pocahontas County, Kalsow Prairie, D. J. Wright, 18 July 2003, slide DJW 4088, USNM.

**Paratypes** (19 ♂, 1 ♀). IOWA. Same data as holotype (5 ♂, slide DJW 4287). KENTUCKY. Menifee County, Leatherwood Fork, Indian Creek, Road 9A, 6 July 1991, L. D. Gibson (3 ♂, slides DJW 4387, 4388, 4552; 1 ♀, slide DJW 4551). MISSOURI. Mississippi County, Big Oak Tree State Park, 28 May 1985, G. J. Balogh, (5 ♂, slide DJW 4632). OHIO. Erie County, Resthaven Wildlife Area, D. J. Wright, 16 July 1996 (2 ♂, slide DJW 4389), 20 July 1990 (4 ♂, slides DJW 4089, WEM 189901). Paratype depositories: LDG, USNM.

**Etymology.** The specific epithet is the nominative singular feminine form of the Latin adjective *parvus*, meaning small.

### 63. *canadana* complex

(Plate Y, 63a–b; Plate 26, 63a–b)

The specimens illustrated under this number resemble the other members of the group in male genitalia but differ sufficiently from those taxa in forewing appearance to raise the prospect that they may represent unnamed species. In each case, the female is unknown.

**Plate Y:63a, Plate 26:63a.** This specimen is one of four males in the CNC collected on 26 June 1995 at Spring Green Nature Conservancy Preserve in Sauk County, Wisconsin. Its most distinctive feature is the nearly immaculate white interfascial spot.

**Plate Y:63b, Plate 26:63b.** This is one of two males in the CNC with a particularly dark interfascial spot. They were collected 17–18 July at Manitoba Tall Grass Prairie Preserve, Manitoba, Canada.

## Species not assigned to a group

### 64. ***Sonia divaricata*** Miller, 1990
(Plate Y, 64a–e; Plate 27, 64a–e)

*Sonia divaricata* Miller 1990:88; Brown 2005:555; Gilligan et al. 2008:132.

**Holotype.** ♂, Missouri, Barton County, Prairie State Park, J. R. Heitzman, 10 June 1980, slide WEM 243893, AMNH.

The type series for *S. divaricata* includes three paratypes: 2 ♂ with the same collection data as the holotype (slides DH 225811 and DH 303821), one in the University of Minnesota Saint Paul Entomology Museum, the other initially retained in the J. R. Heitzman Collection; and 1 ♀ from Powell County, Kentucky (slide WEM 243894) in the LDG Collection.

This species is a slightly larger than average *Sonia* (mean FWL = 8.0 mm). The third segment of the labial palpus is two-toned, with distal one-half black (apex often somewhat paler) and basal one-half tan; the lateral surface of the second segment is dark gray except for a patch of tan scales at the distal extremity; and the first segment is tan. The lower frons is tan, the upper frons and vertex are brown, and the basal portion of the antenna has some blackish scaling but lacks a well-defined black mark on the pedicel or the scape. The forewing has dark brown fasciate markings and white to pale brownish-gray interfascial areas, the latter with dark brown transverse reticulations. *Sonia divaricata* is distinguished from its congeners by the following maculational features: the subbasal fascia and the interfascial spot are bandlike, extending from the inner margin to the costa, the former nearly uniform in width, the latter abruptly constricted at the radius; the mid-costal bar is broad and quadrate; and the prominent pretornal mark is usually confluent with the discal mark. The postmedian fascia is broad anterior to the ocellus but constricted near the costa, and the gray ocellus is divided into three components by brown line segments forming an inverted Y. The costal strigulae are conspicuous and whitish (sometimes with gray suffusion), the termen has a narrow salt-and-pepper colored band from tornus to apex that is edged proximally by a white line from $M_1$ to the apex, and the fringe is gray.

Miller (1990) chose the name *divaricata* to call attention to what he considered to be a "bifid" uncus in the three male specimens he examined. He stated that this feature distinguishes *S. divaricata* from all other known *Sonia* except *S. comstocki*. In our experience, this character is variable and, to some extent, present in several species of *Sonia*, including *S. canadana*, *S. ferruginana*, *S. filiana*, and *S. comstocki*. Slide mounting the male genitalia often contorts the tegumen and its appendages, affecting the perceived shape of the uncus. The aforementioned taxa often have a pair of small semicircular projections on the boundary between the unscaled ventral and scaled dorsal surfaces of the uncus. These may give the impression of a bilobed uncus but can be difficult to discern, depending on the orientation of the genitalia on the slide. The most distinguishing characteristics of the male genitalia are: the length of the valva neck (relatively short), the size of the cucullus (large compared to the basal portion of the valva), and the shape of the distal margin of the cucullus (strongly curved, almost semicircular). In these respects, *S. divaricata* differs from all its congeners except *S. hodgesi*, a taxon that is easily separated from *S. divaricata* by size, color, maculation, and geographic distribution. We found no characters in the female genitalia that reliably diagnose this species.

We examined 29 adults (25 ♂, 4 ♀) from Arkansas, Florida, Kentucky, Louisiana, Missouri, and Tennessee. Most were collected between 31 March and 11 June, but three from Tampa, Florida are dated 25 September.

## 65. *Sonia comstocki* Clarke, 1952
(Plate Y, 65a–c; Plate 27, 65a–e)

*Sonia comstocki* Clarke 1952:62; Powell 1983:35; Brown 2005:555; Powell and Opler 2006:192; Powell and Opler 2009:136.

**Holotype.** ♂, California, San Diego County, San Felipe Wash, J. A. Comstock, 9 February 1938, slide 72842, USNM.

Clarke (1952) described this species from the holotype and seven paratypes (6 ♂, 1 ♀), all with the same collection data. We examined 13 specimens (9 ♂, 4 ♀), three from the type series and ten collected by E. C. Knudson at Big Bend National Park in Brewster County, Texas.

*Sonia comstocki* is one of the largest members of the genus (mean FWL = 10.1 mm), exceeded in size only by *S. filiana* (mean FWL = 11.1 mm). The head is uniformly yellowish white except for two black spots on the labial palpus, one on the dorsal surface of the third segment, the other on the lateral surface of the second segment. Forewing color varies from pale brownish yellow (Plate Y:65a–b) to pale brown (Plate Y:65c). Fasciate markings are sometimes faintly detectable as slightly darker shades of the ground color. They include a subbasal mark that extends from the inner margin to the cell and a fragmented median fascia represented by a discal mark, and a pretornal mark. The ocellus is obscure, and the whitish costal strigulae are defined by numerous short brown dashes. The termen has a narrow inconspicuous salt-and-pepper colored band from tornus to apex that is edged proximally by a pale whitish bar from $M_1$ to the apex. The pale yellowish coloration distinguishes this species from its congeners.

In males, the uncus sometimes has a pair of weakly developed projections located medially on the boundary between its unscaled ventral and scaled dorsal surfaces (Plate 27:65c); the socius tapers to a rounded apex; and the vesica has 70–88 cornuti (mean = 80, n = 4), their sockets overlapping to the extent that an exact count is often impossible to obtain. The neck of the valva is relatively wide compared to other members of the genus (NR = 0.69 vs. 0.48–0.66), the saccular corner is broadly rounded, with SA = 161°, the dorsal lobe of the cucullus tapers slightly to a rounded apex, and the distal margin of the cucullus is straight. The female genitalia are typical of the genus.

We examined a rather small sample of specimens, as mentioned above. The geographic range of this species seems to be restricted to the desert Southwest, but its extent is poorly documented. It was described from San Diego County, California; Powell and Opler (2009) reported its presence in the western Mojave Desert; and E. C. Knudson's specimens (in the MGCL) were collected in west Texas (Big Bend National Park). The specimens we examined were captured in August and early September. Powell and Opler (2006) reported rearing this species from larvae boring in *Isocoma acradenia* (Greene) Greene (alkali goldenbush) in Los Angeles County, California. This plant occurs in desert areas of California, Nevada, Utah, and Arizona but apparently is not known from Texas.

## 66. *Sonia filiana* (Busck, 1907)
(Plate Z, 66a–c; Plate 27, 66a–e)

*Hendecaneura* (?) *filiana* Busck 1907:135.
*Eucosma filiana*: Barnes and McDunnough 1917:170.
*Sonia filiana*: Heinrich 1923:161; McDunnough 1939:49; Powell 1983:35; Brown and Lewis 2000:1035; Brown 2005:555; Powell and Opler 2006:193; Powell and Opler 2009:136.

**Holotype.** ♂, California, [Riverside County], West Riverside, 26 October 1905, slide 72866, USNM.

Busck (1907) tentatively described this species in the genus *Hendecaneura* Walsingham but expressed misgivings as to the propriety of this placement. He did not indicate the number of specimens supporting the description but gave the "alar expanse" as 26 mm (as opposed to a range of values), suggesting that a single specimen may have been involved. Busck (1907) reported a male type in the USNM (No. 10318) from West Riverside, California whose wing slide had been deposited in Lord Walsingham's collection. Barnes and McDunnough (1917) treated *H. filiana* as a synonym of *Eucosma palousana* Kearfott, and Heinrich (1923) transferred it to *Sonia*, illustrating the genitalia of the specimen Busck considered to be the type. We follow Brown and Lewis (2000) in interpreting that specimen as a holotype.

*Sonia filiana* is the largest member of the genus (mean FWL = 11.1 vs. 5.7–10.1 mm). The third segment of the labial palpus is blackish with a pale tip; the medial and lateral surfaces of the second segment are tan and brown, respectively; the lower frons is tan; the upper frons and vertex are brown; and the antenna is brown, with a black spot on the pedicel. The dorsal surface of the forewing is brown with dark brown markings, the latter including a well-defined subbasal mark extending from the inner margin to the radius, a fragmented median fascia comprised of three disjunct components (a faint mid-costal bar, a discal mark, and a pretornal mark that straddles $CuA_2$), and a postmedian fascia that is broad anterior to the ocellus but severely constricted near the costa. The discal mark is disjunct from the pretornal mark and is sometimes connected with the postmedian fascia. The ocellus is vaguely defined and mostly tan, with a narrow brown central field that is sometimes marked with two or three dark spots. The termen has a narrow salt-and-pepper colored band from tornus to apex and a whitish/grayish bar from $M_1$ to the apex.

The male genitalia resemble those of *S. comstocki*, but the number of cornuti in the vesica is smaller by half (28–32, n = 3 vs. 70–88, n = 4). The female genitalia are typical of the genus.

We examined ten specimens (8 ♂, 2 ♀) including the holotype, all from Los Angeles, Riverside, and San Diego counties in southern California. Powell and Opler (2009) reported the range of this species as coastal southern California, extending as far north as Morro Bay in San Luis Obispo County. *Isocoma acradenia* (Greene) Greene (alkali goldenbush) has been documented as a larval host in San Diego County (Powell and Opler, 2006).

## 67. ***Sonia vovana*** (Kearfott, 1907)
(Plate Z, 67a–f; Plate 28, 67a–f)

*Eucosma vovana* Kearfott 1907:33; Barnes and McDunnough 1917:170.
*Sonia vovana*: Heinrich 1923:161; McDunnough 1939:49; Powell 1983:35; Brown 2005:555; Gilligan et al. 2008:132; Powel and Opler 2009:136.
*Eucosma typicodes* Meyrick 1912:35, unnecessary replacement name for *vovana*.
*Hendecaneura* (?) *fraternana* Busck 1907:134; Brown and Lewis 2000:1036.
*Eucosma fraternana*: Barnes and McDunnough 1917:170.

**Types.** *Eucosma vovana*. Lectotype (designated by Heinrich 1923). ♂, Utah, Stockton, Tom Spalding, 8 August 1904, AMNH. *Hendecaneura fraternana*. Holotype. ♂, California, [Riverside County], West Riverside, 31 October, Type No. 10317, slide 72865, USNM.

Kearfott (1907) described *E. vovana* from 13 syntypes (sex not recorded) with the following collection data: Arizona, Yuma County, 26 March, J. B. Smith; Colorado, Bear Creek, H. K. Morrison; Colorado, Glenwood Springs, 24 August, Dr. Barnes; Texas, Amarillo, 30 August, Cockerell; and Utah, Stockton, 2 August–3 September, T. Spalding. Klots (1942) reported six syntypes in the AMNH, including a male lectotype from Stockton, Utah whose designation he credited to Heinrich (1923). That specimen is the only male syntype in the AMNH from Stockton, Utah, so Heinrich's remarks regarding the type seem to be sufficient to designate it as the name-bearing specimen. It bears a blue-green "LECTOTYPE" label that probably was attached by Klots. Busck (1907) described *fraternana* in the genus *Hendecaneura* but expressed uncertainty as to whether that was the proper generic placement. He did not report the number of syntypes, but he gave the "alar expanse" as 19–22 mm and mentioned October and November (presumably referring to adult captures), indicating that there were at least two specimens in the type series. He stated that the type was a male in the USNM from Riverside, California and that a wing slide associated with that specimen had been deposited in Lord Walsingham's collection. We located eight male specimens in the USNM that were collected at Riverside, California between 26 October and 2 November, 1905. The one indicated above as the holotype resides in the type collection at the USNM, and its right fore and hindwings are missing. It was reported as a holotype by both Brown and Lewis (2000) and Brown (2005). Heinrich (1923) recognized *fraternana* as a synonym of *vovana* and placed both taxa in *Sonia*.

*Sonia vovana* is variable in size (FWL: 6.9–11.3 mm, mean = 9.2). The third segment of the labial palpus is black with a white tip; the second segment has a whitish medial surface and a tan to brownish-gray lateral surface, the last with a variable scattering of blackish scales that often coalesce into a dark mark near the dorsal margin. The lower frons is tan; the upper frons and vertex are brown or gray, often with whitish patches on the lateral margins; and the antenna has a black mark on the pedicel and some dark scaling on the scape. The forewing has dark brown fasciate markings and whitish interfascial areas, the latter with brownish-gray suffusion and often with transverse brownish-gray reticulations. The subbasal fascia is usually complete though often weakly expressed near the costa, and the median fascia is fragmented into a broad mid-costal bar, a prominent discal mark, and a triangular pretornal mark. The postmedian fascia arches over the apical quadrant of the ocellus and is strongly constricted at the radius. The ocellus has pinkish-gray transverse bands on the lateral margins and a narrow whitish central field, the latter variably marked with dark dots and/or dashes. The costal strigulae between the median fascia and the apex are white and conspicuous, the termen has a narrow salt-and-pepper colored band from the tornus to the apex and a white dash from $M_1$ to the apex, and the fringe is white to pale grayish tan except for a dark section from $M_1$ to $M_3$ and a dark patch at the apex.

The male genitalia are typical of the genus. The most distinctive feature is the shape of the socius, which is broad basally, narrows abruptly at the midpoint, and attenuates from there to a narrowly rounded apex. The vesica has 35–76 cornuti (mean = 54, n = 7); the neck is moderately wide, with NR = 0.66; the saccular corner is angulate, with SA = 151°; and the cucullus has a somewhat elongate dorsal lobe, with lateral margins nearly parallel. The female genitalia are typical of the genus, except that the lamella postvaginalis narrows rather abruptly near the ostium.

This species is common in arid regions of the Southwest. We examined 46 specimens (28 ♂, 18 ♀) from Arizona, California, Colorado, southwest Kansas, New Mexico, Utah, and Texas. Nearly all adult capture dates are from August to mid-October, but we encountered two records from south Texas dated 29 May and 4 June. The larva is a root-borer in species of *Gutierrezia* Lag. (snakeweed) and *Isocoma* Nutt. (goldenbush). Powell and Opler (2006) reported rearing *S. vovana* in California from the following hosts: *G. californica* (DC.) Torr. & A. Gray (San Joaquin snakeweed) (in San Diego and Riverside counties), *G. microcephala* (DC.) A. Gray (threadleaf snakeweed) (in Los Angeles, San Bernardino, San Diego, and Contra Costa counties), *G. sarothrae* (Pursh) Britton & Rusby (broom snakeweed) (in San Diego County), *Gutierrezia* sp. (in Los Angeles, San Diego, and San Luis Obispo counties), and *I. menziesii* (Hook. & Arn.) G.L. Nesom (Menzies' goldenbush) (in San Diego County). They also reared it in Arizona from *Gutierrezia* sp. in Yavapai County and from *I. tenuisecta* Greene (burroweed) in Pima County.

## 68. ***Sonia hodgesi*** Wright and Gilligan, new species

(Plate Z, 68a–e; Plate 28, 68a–d)

**Diagnosis.** *Sonia hodgesi* resembles *S. vovana* in forewing appearance but differs from that species in size (mean FWL = 5.8 vs. 9.2 mm), head color (white vs. brown or gray), subtle aspects of forewing appearance, and genitalia characteristics. Differences in forewing markings include the expression of the mid-costal bar (weak vs. strong) and the location of the pretornal mark (resting on $CuA_2$ instead of the inner margin). The male genitalia resemble those of *E. divaricata* in that the valval neck is relatively short and the cucullus is rather large. There are substantially fewer cornuti in the vesica of *S. hodgesi* than in any of its congeners (7–12 vs. 22–88). Females of *S. hodgesi* are unique in the genus in that the lamella postvaginalis lacks microtrichia, the ductus bursae is encircled near the ductus seminalis by an obliquely oriented sclerotized band whose inner surface lacks microtrichia, and the corpus bursae has only one signum.

**Description.** *Head.* Labial palpus white except for some blackish scaling on ventrodistal corner and lateral surface of second segment; frons and vertex white; antenna white. *Thorax.* Dorsal surface white, with blackish band across mesothorax and blackish spot at base of tegula; fore- and mid-leg with anterior surfaces blackish brown, posterior surfaces whitish; hind-leg uniformly whitish; tibia with alternating white and blackish annulations. Forewing: ♂ FWL 4.7–6.7 mm (mean = 5.7, n = 33), AR = 2.92, CFR = 0.27; ♀ FWL 4.8–7.0 mm (mean = 6.0, n = 21), AR = 2.90; costa weakly arched; apex acute; termen weakly convex; dorsal surface pale grayish brown with extensive white suffusion and dark brown markings; basal fascia weakly expressed as narrow (often fragmented) line from inner margin to costa; subbasal fascia extending from inner margin to cell, separated from costa by whitish streak along radius; median fascia with mid-costal bar barely discernable, discal mark slightly larger than pretornal mark, posterior margin of pretornal mark on $CuA_2$; postmedian fascia well defined from distal end of cell to mid-termen, strongly constricted at radius; ocellus mostly gray, often suffused with pink on lateral margins, frequently with black ">" shaped mark at costal margin surmounting one or two black dots in narrow central field; costal strigulae from mid-wing to apex whitish, separated by gray-brown wedges; termen with inconspicuous salt-and-pepper colored band from tornus to apex and with white dash from $M_1$ to apex; fringe white from tornus to $M_3$, gray from $M_3$ to apex. Hindwing: Gray brown; fringe white with narrow gray band along distal and posterior margins. *Abdomen.* Male genitalia (n = 5): Uncus triangular, barely differentiated from dorsolateral shoulders of tegumen; socius broad basally, abruptly narrowing at midpoint, attenuating thereafter to narrowly rounded apex; vesica with 7–12 cornuti; valva with costal margin concave, ventral emargination shallow, NR = 0.66, saccular corner weakly defined, SA = 166°; cucullus with dorsal lobe strongly developed, apex semicircular, distal margin convex, ventral lobe not developed, setae on medial surface moderately course. Female genitalia (n = 2): Papillae anales Type 3; length of apophyses posteriores approximately three-fourths that of apophyses anteriores; sterigma U-shaped; lamella postvaginalis lacking microtrichia, rectangular but narrowing abruptly at ostium, with lateral margins curling inward; lamella antevaginalis ringlike, weakly sclerotized, with quarter-round cross-sections; sternum 7 with posterior margin concavely emarginated to one-third length of sterigma and diverging laterally from sterigma; sclerotization of ductus bursae ringlike, obliquely oriented with respect to central axis, and lacking microtrichia on inner surface; corpus bursae with one small signum located near juncture with ductus bursae.

**Holotype** (Plate Z:68a, Plate 28:68b). ♂, Arizona, Santa Rita Mountains, Madera Canyon, 5 August 1959, R. W. Hodges, slide DJW 4036, USNM.

**Paratypes** (23 ♂, 14 ♀). ARIZONA. Madera Canyon, Santa Rita Mountains, R. W. Hodges, 3 July 1959 (1 ♂), 27 July 1959 (1 ♀), 29 July 1959 (1 ♀), 31 July 1959 (1 ♂), 4 August 1959 (4 ♂), 6 August 1959 (1 ♀), 26 August 1959 (1 ♂), 27 August 1959 (1 ♂); Cochise County, Southwest Research Station, Chiricahua Mountains, J. W. Brown, 7 August 2010 (1 ♂), 8 August 2010 (2 ♀), 9 August 2010 (1 ♂), 10

August 2010 (1 ♂, slide USNM 144044), 14 August 2011 (1 ♂), 15 August 2011 (2 ♂, slide DJW 4037); R. L. Brown and S. Lee, 14 August 2011 (4 ♀, slide DJW 4025); Coronado National Forest, R. L. Brown and S. Lee, 14 August 2011 (3 ♂, slides DJW 4024, 4087, wing slide W-5; 1 ♀); Santa Cruz County, Pina Blanca Canyon, R. W. Hodges, 8 August 1959 (1 ♂), 11 August 1959 (5 ♂, 2 ♀), 24 August 1959 (1 ♀, slide DJW 4035), 26 August 1959 (1 ♀). Paratype depositories: CUIC, MEM, USNM.

**Etymology.** The specific epithet honors the late R. W. Hodges, who collected many of the specimens in the type series.

**Distribution**. We examined 60 specimens (37 ♂, 23 ♀), all but one of which were collected in southeastern Arizona. The exception is a male in the USNM that was taken at the intersection of State Route 90 and Gold Gulch Road in Grant County, New Mexico.

## 69. ***Sonia gibsoni*** Wright and Gilligan, new species

(Plate Z, 69; Plate 28, 69)

**Diagnosis.** This species is known only from the holotype. It superficially resembles *S. filiana*, but the two taxa are substantially different in size (FWL = 8.1 vs. 10.1–12.2 mm). Both species are brown with dark brown markings, but the interfascial areas and the ocellus in *S. gibsoni* are more strongly suffused with white, producing a much paler overall appearance. The male genitalia differ from those of *S. filiana* in that the neck of the valva is longer and narrower, NR = 0.54 vs. 0.66, the socius is more evenly tapered, and the dorsal lobe of the cucullus tapers to a more narrowly rounded apex. The female of *S. gibsoni* is unknown. *Sonia gibsoni* and *S. filiana* might differ in geographic distribution. The former was collected at an elevation of 6,144 feet in Grant County, New Mexico, whereas the latter appears to be restricted to coastal southern California.

**Description.** *Head.* Third segment of labial palpus brown with whitish tip; second segment with medial surface and dorsal margin whitish, lateral surface tan; lower frons whitish; upper frons whitish; vertex tan with reddish-brown tints; antenna pale brown, with conspicuous black mark on pedicel; scape whitish with patch of gray-brown scales on ventral surface. *Thorax.* Dorsal surface tan; fore-leg with anterior surfaces brown, posterior surfaces tan; mid- and hind-legs mostly tan; tarsus with alternating tan and brown annulations. Forewing: ♂ FWL 8.1 mm, AR = 2.84, CFR = 0.25 (n = 1); ♀ unknown; costa weakly arched; apex acute; termen straight; dorsal surface pale brown with medium-brown markings; basal fascia present as weakly expressed line; subbasal mark quadrate, extending from inner margin to radius, with lateral margins thinly accentuated with dark brown; interfascial spot large, tan, with considerable white suffusion, extending from inner margin to cell; mid-costal bar inconspicuous; discal mark well defined, with blackish-brown edging on proximal and tornal margins; pretornal mark somewhat triangular, disjunct from discal mark, with dark edging on lateral margins; postmedian fascia broad anterior to ocellus, narrow near costa; ocellus mostly tan, with faint rosy tints; costal strigulae whitish and moderately well defined from distal end of fold to apex; termen with narrow salt-and-pepper colored band from tornus to apex; fringe tan from tornus to $M_2$, dark gray from $M_2$ to apex. Hindwing: Gray brown. *Abdomen.* Male genitalia: Typical of the genus; uncus with apex rounded; socius tapering evenly toward rounded apex; vesica with 32 cornuti; valva with distal margin of basal excavation sinuous, costal margin weakly concave, ventral emargination long and shallow, neck long and narrow, NR = 0.54, SA = 149°. Female genitalia unknown.

**Holotype** (Plate Z:69; Plate 27:69). ♂, New Mexico, Grant County, W side State Route 90, 14 mi SW of Silver City, elevation 6,144 ft., 32.5508° N, 108.3784° W, L. D. Gibson, 2 August 2005, slide DJW 4059, USNM.

**Etymology.** This species is named after Loran D. Gibson, the collector of the holotype.

# *Suleima* Heinrich

*Suleima* Heinrich, 1923, Bull. U.S. natn. Mus. 123: 155. Type species: *Semasia helianthana* Riley, 1881.

Heinrich (1923) proposed the genus *Suleima* for four species that previously had been placed in *Thiodia* (*T. helianthana*), *Eucosma* (*E. baracana* and *E. daracana*), and *Enarmonia* (*E. lagopana*). He designated *S. helianthana* (Riley) as the type and described two additional species (*S. skinnerana* and *S. cinerodorsana*). Sixty years later, Blanchard and Knudson (1983) added *S. mendaciana* to the list, bringing the species count to seven.

Placement in this genus can be inferred from distinctive features of the genitalia, but the uniformity in those features greatly reduces their effectiveness in making species level determinations. Consequently, species circumscriptions in *Suleima* rely heavily on differences in forewing appearance and, to some extent, geographic distribution.

**Diagnosis.** *Suleima* resembles *Sonia* in forewing venation (with $R_4$ and $R_5$ are united), but differs from that genus in that veins $M_3$ and $CuA_1$ of the hindwing are united (Fig. A3b) instead of stalked (Fig. A2b). Males of *Suleima* differ from those of *Sonia* in the absence of a forewing costal fold, the shape of the valva (Fig. B14 vs. B13), and the absence of a clasper on the medial surface of the valva. Females of *Suleima* have a distinctive pattern of sclerotization on the ductus bursae (an elongate semi-helical band, Fig. C14) that sets them apart from all other females treated in this volume.

**Description.** *Forewing* (Fig. A7). Mean FWL: 6.2–9.6 mm, average = 7.8; AR: 2.86–3.28, average = 3.06; male without costal fold; venation as in Fig. A3a ($R_4$ and $R_5$ united); costal margin nearly straight; apical angle slightly acute; termen straight; maculation usually semifasciate (Fig. A7), sometimes with longitudinal streaking; ocellus conspicuous, with whitish central field flanked laterally by pale pink or gray transverse bands; costal strigulae usually sharply delimited by small dark quadrate marks and fine striate dashes.

*Hindwing.* Venation as in Fig. A3b ($M_3$ and $CuA_1$ united); coloration uniform in shades of gray or brown.

*Male genitalia.* Tegumen + appendages with mixture of Type 1 and Type 2 characteristics; gnathos arising from tegumen (as in Type 1, Fig. B2); uncus not clearly differentiated from dorsolateral shoulders of tegumen (as in Type 2, Fig. B3); socius short and often stubby; phallus attenuating distally, with base loosely surrounded by anellus; vesica with 12–40 deciduous cornuti; valva (Fig. B14) usually with distal margin of basal excavation sinuous, infrequently (*S. skinnerana* and *S. flavidana*) with conspicuous basal process, costal margin weakly concave (except in *S. skinnerana*), neck short and broad, with ventral emargination shallow, NR: 0.79–0.90, saccular corner weakly angulate, SA: 140°–160°; cucullus large in comparison to basal portion of valva, with apex and anal angle broadly rounded (the latter often barely produced) and with 8–12 prominent spiniform setae along distal margin from anal angle to apex; setae on medial surface of valva densely spaced and moderately stiff in a band along distal margin of cucullus from apex to anal angle (sometimes extending from anal angle onto neck), otherwise hairlike and more sparsely distributed.

*Female genitalia.* Papillae anales Type 2 (Fig. C2); apophyses posteriores and anteriores nearly equal in length; sterigma cylindrical or U-shaped and coarsely microtrichiate; lamella antevaginalis ringlike, with central portion conspicuously unsclerotized; posterior margin of sternum 7 concavely emarginated (except in *S. skinnerana*), medially approximate to lamella antevaginalis, laterally divergent from sterigma; ductus bursae distorted by elongate semi-helical sclerotized band extending from ductus seminalis nearly to constriction anterior to ostium; corpus bursae with two signa located at or slightly posterior to mid-bursa, one somewhat larger than the other.

**Immature stages.** MacKay (1959) provides a detailed description of *Suleima* larvae based on four species. She separated larvae of *Suleima* from those of *Epiblema* and *Sonia* by the structure of the spinneret (blunt or indented versus rounded or somewhat broadened).

*Egg*. As described for *Epiblema*.

*Larva*. Head with stemmatal area strongly rounded and the vertical angle at a right angle or acute; stemmata well defined in outline and convex; spinneret with blunt tip, often broadened but not rounded or tapered, or with a medial indentation; integument with slender, easily observed spinules; spiracles circular; setal pinacula minute, and setae of groups, e.g., SV group, often on their own individual pinacula; L-group on T1 trisetose, never enlarged or extending under the spiracle, L1 never in a straight line with L2 and L3; SD1 and SD2 on T1 close together; L1 and L2 on A1–8 with a tendency to be posterior to a vertical line through spiracle; SV-group setal counts on A1, 2, 7, 8, 9 are 2:2:2:2(1):1; SD1 on A8 anterior or anterodorsal to spiracle, with no tendency to be anteroventral in position; V1s of A9 usually about twice as far apart as those on A8; anal shield rounded posteriorly; crotchets few in number and always uniordinal although often unevenly so; dorsal setae of claw usually of unequal length, with inner one shorter; anal comb absent.

*Pupa*. As described for *Epiblema*.

**Biology.** All *Suleima* treated here with known larval hosts feed on sunflowers in the genera *Helianthus* L. (sunflower), *Helianthella* Torr. & A. Gray, or *Tithonia* Desf. ex Juss. Those with recorded life histories are borers in stalks and stems. The best-documented species is *Suleima helianthana*, which is referred to in the economic literature as the "sunflower bud moth" for the damage it causes to sunflowers. We summarize its life history here on *Helianthus annuus* L. (common sunflower) from Satterthwait (1948) and Knodel et al. (2010). Females lay as many as 130 eggs on buds, bracts, leaves, or flower heads. Early instar larvae mine leaves by boring into the leaf axil or feeding under webbing on the leaf surface, and later instars bore into the buds, bracts, stems, or stalks. The larva spins a cocoon close to the entrance hole of its burrow and pupates inside. The complete life cycle (from egg to adult emergence) takes approximately 50 days, and two or more generations are completed per year. The last generation overwinters as a late instar larva and pupates in the spring. Although as many as 24,000 larvae per acre have been reported in commercial sunflower fields in Texas, economic loss is minimal. Larval feeding may cause abnormal plant growth and deformed flower heads, but larvae do not typically feed on developing seeds.

## The *helianthana* group (species 70–72)

This group is composed of *S. helianthana*, *S. lagopana*, and *S. powelli*. The first two species are similar in forewing appearance and often are confused with one another in collections; all three are similar in genitalia. *Suleima helianthana* is broadly distributed in central and southwestern United States, *S. lagopana* appears to be restricted to California, and *S. powelli* is known only from the Sonoran Desert in southeastern California. Group characteristics include:

*Forewing* (excluding *S. powelli*). Mean FWL: 7.7, 9.6 mm; AR: 2.97, 2.91; predominant color brownish gray, sometimes extensively suffused with white; semifasciate markings dark brown, including prominent subbasal and pretornal marks, a variably expressed mid-costal bar, and a narrow postmedian fascia arching over the apical margin of the ocellus to approximately mid-termen; ocellus conspicuous, with gray or pale pink bands along lateral margins and one or two black dashes or spots on a white central field; costal strigulae white. *Suleima powelli* resembles *S. helianthana* and *S. lagopana* in forewing pattern but is uniformly pale in coloration, with greatly reduced (or absent) fasciate markings that weakly contrast with the forewing ground.

*Male genitalia.* Typical of the genus; NR = 0.81–0.82; SA: 141°–146°; intraspecific variation in valva shape sufficient to render species determinations on this basis problematic.

*Female genitalia.* Sterigma cylindrical; lamella postvaginalis moderately developed but not clearly differentiated, with variably sclerotized microtrichiate lobes projecting posteriorly from posterolateral corners; otherwise, typical of the genus.

## 70. ***Suleima helianthana*** (Riley, 1881)
(Plate AA, 70a–o; Plate 29, 70a–s)

*Semasia helianthana* Riley 1881:319; Fernald 1882:44.
*Thiodia helianthana*: Fernald 1903:462; Heinrich 1921:824.
*Eucosma helianthana*: Barnes and McDunnough 1917:172; Heinrich 1921:824.
*Suleima helianthana*: Heinrich 1923:156; McDunnough 1939:49; Satterthwait 1948:726; MacKay 1959:40; Powell 1983:35; Miller 1987:61; Brown 2005:576; Gilligan et al. 2008:130; Powell and Opler 2009:136.

**Lectotype** (designated by Heinrich 1923, Plate AA:70a). ♂, Texas, 21 August 1873, C. V. Riley, Type No. 387, slide 72870, USNM.

Riley (1881) described *S. helianthana* from two males that were reared from galls on what he called the "commoner Western Sunflower in Texas." He mentioned dates of 12 and 17 August 1873 (presumably referring to gall collection) and "expanses" of 15 and 20 mm (i.e., FWL = 6.9 and 9.2 mm). Heinrich (1923) illustrated the genitalia of a syntype in the USNM that he called the type, thereby designating a name-bearing specimen. The date 21 August 1873 is inscribed on two of that specimen's pin labels but does not agree with either of the "capture dates" reported by Riley (1881). The fate of the other syntype seems to be unknown.

*Suleima helianthana* exhibits a fair amount of variation in size (FWL: 5.3–10.8 mm, mean = 7.6, n = 40) and forewing appearance (Plate AA). The third segment of the labial palpus tends to be grayish brown (at least on the dorsal surface) with a pale (usually whitish) tip. The second segment has a white medial surface and a brownish-gray lateral surface, the latter often suffused with white. The lower frons is white, the upper frons and vertex vary from brownish gray to white, and the antenna is brownish gray except for whitish scaling on the ventral surface of the scape. Predominant forewing color ranges from dark brownish gray (Plate AA:70a–c, j–l) to fawn (Plate AA:70g–h), with prominent dark-brown semifasciate markings. The basal area usually has fine, brown, weakly discernable, transverse reticulations, and the interfascial spot varies from nearly white (Plate AA:70i) to concolorous with adjacent interfascial areas (Plate AA:70a–c, e–f, j–m, o). The subbasal mark is somewhat quadrate, extends from the inner margin to the cell, and sometimes is interrupted on $A_{1+2}$ (Plate AA:70m–o). The median fascia is represented by a mid-costal bar that varies from moderately prominent (Plate AA:70b, i) to undetectable (Plate AA:70e) and a dark pretornal mark from the inner margin to $CuA_2$, the latter often with a pale (sometimes barely discernable) extension along the proximal margin of the ocellus. The postmedian fascia is narrow, inconspicuous, and usually interrupted near the costa. The ocellus has a prominent white central field that is flanked laterally by gray or pinkish bands, and the common boundary of the ocellus and the pretornal mark often is accentuated by a dark brown line. The fringe is brownish gray and frequently pale near the tornus.

In the male genitalia, the shape of the valva, the setation of the medial surface of the valval neck, and the shape and setation of the socius are variable. Plate 29:70a is representative of material from the Southwest, including the lectotype. In these specimens, the ventral emargination of the valval neck is moderate, NR: 0.70–0.84, the setae on the medial surface of the neck are hairlike and either sparsely distributed or entirely absent, and the socius is short and stubby, with a small patch of densely spaced setae at the apex (Plate 29:70a, c). Plate 29:70b is representative of some specimens from Illinois, Kentucky, and

Ohio in which the ventral emargination of the valval neck is shallow to barely detectable, NR: 0.84–0.96, the band of moderately stiff setae along the ventrodistal margin of the cucullus extends across the medial surface of the neck, and the socius is moderately fingerlike, with setae more broadly distributed (Plate 29:70b, f, m). In the eastern part of the range, both forms (70a and 70b) occur in some local populations, and in the central plains we encountered intermediate variations on these character states (e.g., Plate 29:70e). In many respects, the first form (70a) resembles that of *S. lagopana* and *S. powelli*, and the second (70b) resembles that of *S. cinerodorsana* and members of the *S. baracana* group.

In females, the anterior one-half of the sterigma is cylindrical and the posterior portion is moderately flared into an undifferentiated lamella postvaginalis. Variation in this structure is illustrated in Plate 29:70p–s. In each case, there are two tonguelike microtrichiate lobes projecting posteriorly from the lamella postvaginalis. The membrane associated with those lobes lacks sclerotization in eastern specimens but is weakly sclerotized in specimens from the Southwest.

We examined 62 specimens (45 ♂, 17 ♀) that document a geographic range extending from southern California, Arizona, New Mexico, and south Texas north and east to southern Saskatchewan, southern Manitoba, Ohio, Kentucky, and Arkansas. Literature records place this taxon in Minnesota (Miller 1987) and Maryland (Satterthwait 1948). Specimen size seems to be weakly related to geographic location. In our sample, the variation in FWL was 5.3–7.9 mm (n = 26) in the Midwest, 7.4–9.7 mm (n = 8) in the Great Plains, and 8.3–10.8 mm (n = 10) in California and the Rocky Mountain region. Particularly dark specimens seem to be common in the Southwest (Plate AA:70a–f) but also occur in Illinois, Kentucky, and Ohio (Plate AA:70j–m). The pale phenotype illustrated in Plate AA:70g–h was collected in sandy habitat in Colorado and Kansas.

Known as the sunflower bud moth, *S. helianthana* causes minor economic damage in commercial sunflower fields. Its complete life history is outlined under the biology section in the generic description. In addition to *H. annuus*, this species has been recorded on *Helianthus tuberosus* L. (Jerusalem artichoke), *Hymenoxys odorata* (DC.) (bitter rubberweed), and *Coreopsis grandiflora* Hogg ex Sweet (large-flowered tickseed) (Satterthwait 1948). Capture dates for wild-caught adults ranged from 26 March to 16 September, but most records are from May through August.

*Suleima helianthana* often has been confused with *Eucosma tomonana* (Kearfott) (Wright and Gilligan 2015: species 116) and with some members of the *Pelochrista pulveratana* group (Wright and Gilligan 2017: species 101–113) due to similarity in forewing appearance. It is separated from these species by the color of the ocellus (central field white vs. brown or gray) and by genitalia.

## 71. ***Suleima lagopana*** (Walsingham, 1879)

(Plate BB, 71a–f; Plate 30, 71a–c)

*Steganoptycha lagopana* Walsingham 1879:71; Walsingham 1884:145; Fernald 1882:47.
*Epinotia lagopana*: Fernald 1903:465.
*Enarmonia lagopana*: Barnes and McDunnough 1917:173.
*Suleima lagopana*: Heinrich 1923:158; McDunnough 1939:49; Powell 1983:35; Brown 2005:576; Powell and Opler 2009:136 [photo = *S. baracana*].

**Lectotype** (here designated). ♂, California, Colusa County, 3 mi. from Indian Valley, Walsingham, 28 June 1871, slide 11605, BMNH.

Walsingham (1879) mentioned five syntypes (3 ♂, 2 ♀), presumably all with the same collection data. In the 1960s, Obraztsov chose a name-bearing specimen, prepared its genitalia (BMNH slide 11605), and photographed both items, but he did not publish his selection. We examined his photographs (now on long term loan to the MEM) and include here a formal designation of that specimen as the lectotype. The designation includes some information on the type locality (3 mi. from Indian Valley) that was not reported by Walsingham; we found it in notes written by Obraztsov on the glassine envelopes containing his 35 mm slides.

Historically, there has been considerable confusion regarding the application of the name *lagopana*. Walsingham (1884) reported six specimens under this name (1 ♀ from Arizona, 1 ♀ from Montana, 2 ♂ and 2 ♀ from California), but he noted that, unlike typical *lagopana*, they have a dark basal patch on the forewing and, therefore, might well represent some other species. We did not examine those individuals, but the presence of a dark basal patch indicates that they are likely *S. baracana* and not *S. lagopana*. Heinrich (1923) did not examine the syntypes of *S. lagopana* (in the BMNH), but he did report examining three specimens from California in the Fernald Collection that he considered to have been authentically determined by Walsingham. We found a female of *S. lagopana* in the USNM (Plate BB:71c) labeled "Fernald Collection." It bears no collection data and has no abdomen, but we are confident it is accurately identified. Heinrich's illustration of the male genitalia for *S. lagopana* (1923: fig. 296) is based on a specimen collected by O. C. Poling in southern Arizona (slide USNM 70883). It does not agree with Obraztsov's photograph of the genitalia of the *S. lagopana* lectotype, and we believe that specimen is instead a particularly pale form of *S. baracana*. In our study sample, we found numerous specimens of *S. baracana* that had been determined as *S. lagopana* and some specimens of *S. lagopana* that had been misidentified as *S. helianthana*. In particular, the specimen illustrated in Powell and Opler (2009) as *S. lagopana* is *S. baracana*.

*Suleima lagopana* is larger than *S. helianthana* (mean FWL = 9.6 vs. 7.7 mm) and has a slightly broader forewing (AR = 2.91 vs. 2.97). The labial palpus is white, with a longitudinal brown streak centrally located on the lateral surface of the second segment. The frons and vertex are white, the dorsal surface of the head is pale brown, and the antenna has thin alternating brown and white annulations. The forewing has pale interfascial areas, varying from pale brown to nearly pure white, often with thin brown transverse reticulations between the base and the subbasal fascia. The subbasal mark is broad and prominent. It extends from the inner margin to the cell and sometimes has a weakly expressed extension as a pale brownish shade from the cell to the costa. The median fascia is represented by a conspicuous mid-costal bar and a somewhat hourglass-shaped pretornal mark, the latter being especially dark from the inner margin to $CuA_2$ and often fading somewhat toward the cell. The mid-costal bar and pretornal mark are usually separated by a thin pale streak on the radius. The ocellus is conspicuous, with pinkish bands along the lateral margins. It has a black dash and black dot, respectively, on the upper and lower extremities of the white central field. This species differs from *S. helianthana* in the size of the subbasal mark (broader and more rectangular), the expression of the median fascia (nearly complete from the inner margin to the costa), and the color of the interfascial spot (mostly white vs. brownish gray).

The genitalia resemble those of *S. helianthana* from the southwestern United States. We are unable to adequately assess variation in the male genitalia due to a lack of material, but in females, the shape of the sterigma seems to be stable. We suspect that this taxon cannot reliably be distinguished from *S. helianthana* based solely on genitalia characteristics.

We examined 12 specimens (3 ♂, 9 ♀) from Colusa, Los Angeles, Stanislaus, Tulare, and Yolo counties, California. As far as we know, this species is not present in any other state. Adult capture dates range from 28 June to 31 August. Six of the specimens were labeled "sunflower," "ex sunflower stem," or "reared from sunflower," but no particular species of *Helianthus* L. (sunflower) was indicated as a larval host.

## 72. ***Suleima powelli*** Wright and Gilligan, new species
(Plate BB, 72a–i; Plate 30, 72a–j)

**Diagnosis.** *Suleima powelli* is distinguished from its congeners by its pale uniformly grayish-brown forewing and its lack of prominent fasciate markings. Compared to the other members of the group, it is intermediate in size (mean FLW = 9.0 mm vs. 7.7 and 9.6) and has a somewhat more elongate forewing geometry (AR = 3.18 vs. 2.97 and 2.91). The genitalia resemble those of *S. helianthana*, and we are unable to distinguish between the two on that basis.

**Description.** *Head.* Whitish; labial palpus white, with pale tan tint on lateral surface of second segment; scales on vertex pale tan basally with white vertices, producing pale faun effect; antenna pale tan. *Thorax.* Dorsal surface concolorous with vertex; fore- and mid-legs with anterior surfaces pale tan, posterior surfaces white; hind-leg white; tarsi with subdued pale brown and white annulations. Forewing: ♂ FWL 7.9–10.2 mm (mean = 9.0, n = 10), AR = 3.20; ♀ FWL 7.3–10.5 mm (mean = 8.9, n = 8), AR = 3.16; costal margin nearly straight, apical angle acute, termen straight; dorsal surface uniformly pale grayish brown to pale faun, with brown line on cubitus from base to mid-wing; fasciate markings pale brown, weakly expressed, often undetectable, including narrow subbasal mark from inner margin to cubitus and dark spot on inner margin in pretornal position; ocellus obscure but discernable, with whitish central field flanked laterally by pale pink transverse bands and crossed by 1–3 black dashes; costal strigulae white, defined from base to apex by numerous thin brown dashes; fringe scales whitish with pale brown cross-bars. Hindwing: Pale tan, usually paler than forewing. *Abdomen.* Male genitalia (n = 7): Uncus barely differentiated from dorsolateral shoulders of tegumen, with basal width about two times height and apex rounded or weakly indented; socius short and stubby; base of phallus loosely surrounded by anellus; vesica with 14–24 cornuti; valva with distal margin of basal excavation sinuous, neck short and broad, ventral emargination moderate, NR = 0.81, saccular corner angulate, SA = 141°; cucullus with apex rounded to somewhat flattened, distal margin evenly convex, 8–10 spiniform setae on distal margin; medial surface of cucullus with band of moderately stiff, densely spaced setae along distal margin from apex to anal angle but not extending onto neck. Female genitalia (n = 6): Sterigma cylindrical, usually with two microtrichiate lobes projecting posteriorly from lamella postvaginalis; otherwise, typical of the genus.

**Holotype** (Plate BB:72a; Plate 30:72a). ♂, California, Imperial County, 17 mi. NW of Glamis, J. A. Powell, 27 June 1978, reared from *Helianthus niveus* (Benth.) Brandegee, emerged 1 August 1978, slide DJW 4466, EME.

**Paratypes** (10 ♂, 9 ♀). CALIFORNIA. Imperial County, 17 mi. NW of Glamis, J. A. Powell, 1 April 1978 (6 ♂, slides DJW 4205, 4320; 3 ♀, slides JAP 4235, 4251), 27 June 1978 (3 ♂, slides JAP 4243, DJW 4466, 4467; 6 ♀, slides JAP 4244, DJW 4468, 4469, 4479), 14 mi. NW of Glamis, Dietz and Powell, 31 March 1978 (1 ♂, slide DJW 4477). Paratype depositories: EME, USNM.

**Etymology.** The specific epithet honors J. A. Powell, whose extensive collections in California and neighboring states have been extraordinarily helpful in our understanding of western tortricids.

**Biology**. The EME has 17 specimens (9 ♂, 8 ♀) that were reared by Powell from *Helianthus niveus* (Benth.) Brandegee (showy sunflower). Larvae were collected in two batches labeled 1 April and 27 June, and adults emerged from 10 April to 10 May and from 27 June to 13 August, respectively. Powell's field notes (deposited in the EME) describe larval development in a chamber excavated in the plant stem and adult emergence through a previously prepared "window" in the wall of the chamber.

**Distribution.** The type locality is in the portion of the Sonoran Desert that extends into the southeast corner of California.

**Comments**. The moths illustrated in Plate BB:72h–i and Plate 30:72f–g, j are representative of four specimens (2 ♂, slides Jap 4840, DJW 4475; 2 ♀, slides JAP 4837, DJW 4478) in the EME that were collected at the type locality on 27 August 1978 along with the holotype and ten paratypes of *S. powelli*.

Compared to typical *S. powelli*, they are smaller (mean FWL = 7.6 vs. 9.0 mm), whiter, have a subtly different valva shape, and have a sterigma structure (Plate 30:70j) that resembles that of southwestern specimens of *S. helianthana*. They bear labels with the inscription "collected on *H. niveus*," suggesting that they were taken as adults rather than larvae or pupae. We tentatively refer them to *S. powelli* but do not include them amongst the paratypes and do not include their numerical data with that of *S. powelli* in Appendix A.

## The *baracana* group (species 73–75)

This group is composed of *S. baracana*, *S. leeana*, and *S. blanchardi*, three species that are remarkably similar in genitalia but differ in subtle aspects of forewing appearance. *Suleima baracana* ranges from the Great Plains to the Pacific coast; *S. leeana* occurs in Utah, Arizona, New Mexico, and west Texas; and *S. blanchardi* is known only from Padre Island National Seashore in Texas. Group characteristics include:

*Forewing*. Mean FWL: 6.7–8.5 mm; AR: 3.02–3.23; interfascial spot white, usually with some gray or black speckling and/or irrorations; fasciate markings brown to black; basal patch prominent, with distal margin scalloped; mid-costal bar obscure but usually detectable; pretornal mark prominent, quadrate, and speckled with black; ocellus conspicuous, with pale pink or gray bands on lateral margins and up to four black spots/dashes on white central field; costal strigulae white and sharply defined.

*Male genitalia*. Uncus weakly developed, broad (basal width about 4 times height), and barely differentiated from dorsolateral shoulders of tegumen; socius short, moderately broad, usually uniform in width, sometimes tapering slightly toward apex; vesica with 16–35 cornuti; valva spatulate, with distal margin of basal excavation sinuous, costal margin slightly concave, neck broad and weakly defined by short shallow ventral emargination, NR: 0.83–0.84, saccular corner weakly angulate, SA: 150°–155°; medial surface of valva with band of densely spaced, moderately stiff setae along ventrodistal margin of cucullus and extending across medial surface of neck from anal angle nearly to distal margin of basal excavation; cucullus uniform in width, with apex semicircular and with 6–10 spiniform setae distributed evenly along weakly convex distal margin.

*Female genitalia*. Typical of the genus; sterigma cylindrical or U-shaped and densely microtrichiate; corpus bursae with signa unequal in size, the larger near juncture with ductus bursae, the smaller near mid-bursa.

### 73. ***Suleima baracana*** (Kearfott, 1907)
(Plate CC, 73a–o; Plate 31, 73a–o)

*Thiodia baracana* Kearfott 1907:43.
*Thiodia caracana* Kearfott 1907:43; synonymy by Heinrich 1923:158.
*Thiodia oxyleuca* Meyrick 1912:34, unnecessary replacement name for *baracana*.
*Thiodia famosa* Meyrick 1912:34, unnecessary replacement name for *caracana*.
*Eucosma baracana*: Barnes and McDunnough 1917:172.
*Eucosma caracana*: Barnes and McDunnough 1917:172.
*Suleima baracana*: Heinrich 1923:158; McDunnough 1939:49; MacKay 1959:42; Powell 1983:35; Brown 2005:576.

**Types**. *Thiodia baracana*. Lectotype (here designated, Plate CC:73a). ♂, Utah, Stockton, Tom Spalding, 5 September 1904, AMNH. *Thiodia caracana*. Lectotype (designated by Klots 1942, Plate CC:73b). ♂, Colorado, Denver, Oslar, 3 August 1904, AMNH.

Kearfott (1907) described *T. baracana* from six specimens (sex not reported) that were collected by Tom Spalding at Stockton, Utah, by A. J. Oslar at Denver, Colorado, and by C. P. Gillette in Colorado. We examined what we believe to be the six syntypes, three in the AMNH (1 ♂, 2 ♀; all from Stockton, Utah) and three in the USNM (2 ♂, 1 ♀; all from Colorado). Regarding the syntypes in the AMNH, the collection dates on the pin labels (1 July 1904, 24 August 1904, and 5 September 1904) do not match those reported by Kearfott (1907) (21 August and 5 September). Klots (1942) stated that the male (dated 5 September) is the lectotype that had been designated by Heinrich (1923). Heinrich's comments provide the type locality (Stockton, Utah) and the repository (AMNH) but not the sex. They are, therefore, not sufficient to specify a unique specimen, so we include here a formal designation for the male that Klots interpreted as the lectotype. The data for the three syntypes in the USNM are: Colorado, Denver, Oslar (1 ♂); Colorado, Gillette, (1 ♂, No. 2133, slide 70888; 1 ♀; No. 2147).

Kearfott (1907) proposed the name *T. caracana* for five specimens (sex not recorded) collected by Tom Spalding at Stockton, Utah (29 June), by A. J. Oslar at Denver, Colorado (3 and 6 August), and by H. W. Nash at Pueblo, Colorado (August). We examined four specimens that we consider to be syntypes, two in the AMNH (1 ♀ from Stockton, Utah; 1 ♂ from Denver, Colorado) and two in the USNM (1 ♂ from Denver, Colorado; 1 ♀ from Pueblo, Colorado). Klots (1942) stated that the type material at the AMNH includes a male lectotype from Denver, Colorado (designation attributed to Heinrich 1923) and two female syntypes. We were unable to locate a second female. The male bears a blue-green label with the inscription "LECTOTYPE" (presumably attached by Klots), so we credit Klots with the designation.

The synonymy of *T. caracana* and *T. baracana* was recognized by Heinrich (1923). His photograph of the male genitalia (fig. 297) is based on USNM slide 70888, which was prepared from the syntype of *T. baracana* collected by Gillette in Colorado (No. 2133).

*Suleima baracana* is a somewhat larger than average member of the genus (mean FWL = 8.5 mm) with average forewing geometry (AR = 3.08). The ventral and dorsal surfaces of the third segment of the labial palpus are white and orange brown, respectively, as are the medial and lateral surfaces of the first and second segments. The lower frons is white, the upper frons is tan medially and brown laterally, the vertex is orange brown, and the antenna and dorsal surface of the thorax are concolorous with the vertex. The forewing has brown fasciate markings and a conspicuous white interfascial spot, the latter often with some black speckling. The subbasal fascia is subsumed into a prominent basal patch but is usually distinguishable as a uniformly brown band extending from the inner margin to the cell or costa. The median fascia is composed of a somewhat vaguely expressed mid-costal bar and a prominent brown pretornal mark, the two often separated by an obscure grayish band that extends longitudinally through the cell from the basal patch to the costal margin of the ocellus. The pretornal mark is usually a mottled combination of brown and gray with appreciable black speckling, the last being especially prominent along $CuA_2$ and along the proximal margin of the ocellus. The ocellus is flanked laterally by pale pinkish bands and has up to two black dashes on a white central field. The costal strigulae are white, conspicuous, and separated by brown wedges. The fringe is white from the tornus to $CuA_1$ and dark gray from there to the apex. The hindwing is brownish gray and has a whitish band through the fringe. Intraspecific variation in forewing appearance consists for the most part of subtle differences in the expression of the fasciate markings (Plate CC), but we did encounter some particularly pale specimens (Plate CC:73n–o) from Arizona, New Mexico, and west Texas that might eventually be recognized as a distinct species. Heinrich (1923) misidentified one such USNM specimen from Arizona as *S. lagopana* and illustrated its genitalia under that name (fig. 296), causing subsequent confusion regarding the identity of *S. lagopana*. The two taxa are easily separated by the presence (in *S. baracana*) vs. the absence (in *S. lagopana*) of a dark basal patch on the forewing.

The male genitalia are typical of the group, with 16–35 cornuti, NR = 0.84, and SA = 155°. Valva shape is subtly variable but lacks idiosyncratic features that can be relied upon to distinguish *S. baracana* from other members of the group. In females, the sterigma varies from cylindrical to moderately U-shaped (Plate 31:73k, m–o) and is coarsely microtrichiate. In other respects, the female genitalia are typical of the genus.

*Helianthus annuus* L. (common sunflower) appears to be the preferred larval host. The USNM and the EME both have substantial series of specimens that were reared on *H. annuus* by J. F. G. Clarke from Whitman County, Washington and by J. A. Powell from Los Angeles and Riverside counties, California. According to Powell's field notes (deposited at the EME), the larva is a twig borer that excavates an elongate chamber, causing the death and browning of the terminal portion of the twig. Pupation occurs in the chamber, and the pupa exits through a previously prepared "window" in the side of the chamber. Immatures were collected by Powell between 22 August and 3 September, and the adults emerged between 27 August and 19 September.

We examined 165 specimens (83 ♂, 82 ♀) from the region extending from Washington, Montana, and North Dakota south to southern California, Arizona, New Mexico, and Texas. We encountered only one Canadian record (from Saskatchewan) but expect that the range of this species extends into the Canadian Provinces from British Columbia to Manitoba. Capture dates for wild-caught adults vary from 5 April to 9 September.

## 74. ***Suleima leeana*** Wright and Gilligan, new species

(Plate DD, 74a–i; Plate 32, 74a–f)

**Diagnosis.** *Suleima leeana* resembles *S. baracana* in forewing pattern and genitalia, but the two species differ in head color, forewing color, and forewing geometry. In *S. leeana*, the scales on the lateral surface of the labial palpus and on the dorsal surface of the head are dark brownish gray (often blackish); in *S. baracana* they are orange brown. The interfascial areas of the forewing are white in both species, but the fasciate markings are black in *S. leeana* and brown in *S. baracana*. As a result, the contrast between the pale and dark features is substantially stronger in *S. leeana* than in *S. baracana*. *Suleima leeana* is slightly smaller than *S. baracana* (mean FWL = 7.6 vs. 8.5 mm) and has a noticeably narrower forewing geometry (AR = 3.23 vs. 3.08). We found no genitalia characteristics that reliably separate the two species. *Suleima leeana* is currently known from southern Idaho, central Utah, Arizona, New Mexico, and west Texas, where it is sympatric with *S. baracana*.

**Description.** *Head.* Third segment of labial palpus gray with whitish tip; first and second segments with lateral surfaces brownish gray, medial surfaces whitish, dorsal margins white; long scales on ventral and distal margins of second segment dark brownish gray; lower frons whitish; upper frons and vertex dark brownish gray; antenna with dorsal surface brown, ventral surface tan, scape pale brown. *Thorax.* Dorsal surface dark brownish gray; fore- and mid-leg with anterior surfaces blackish brown, posterior surfaces pale tan; femur and tibia with one or two tan marks; hind-leg uniformly pale tan; tarsus with alternating tan and blackish-brown annulations. Forewing: ♂ FWL 7.2–8.6 mm (mean = 7.7, n = 12), AR = 3.27; ♀ FWL 6.2–8.5 mm (mean = 7.4, n = 8), AR = 3.16; interfascial spot bright white with a few black or dark gray spots, including 1–4 on inner margin; fasciate markings dark brownish gray to blackish gray; basal patch blackish gray with obscure black transverse reticulations and two variably expressed white spots on inner margin; median fascia represented by well-defined mid-costal bar and prominent quadrate pretornal mark, the two usually separated by diffuse grayish band along radius from mid-wing nearly to apex;

ocellus conspicuous, with pale pink or grayish bands on lateral margins and up to four black dots on white central field; distal margin of ocellus edged with narrow brown band; costal strigulae white and sharply delimited by blackish wedges and striae; termen with narrow salt-and-pepper colored band from $CuA_1$ to vertex; fringe white near tornus, gray to blackish gray from $CuA_1$ to vertex. Hindwing: Pale brownish gray. *Abdomen.* Male genitalia (n = 6): Typical of the group; vesica with 20–25 cornuti; NR = 0.84; SA = 153°; anal angle barely detectable; medial surface of valva sometimes with weakly-developed, setose bulge on distal margin of basal excavation (Plate 32:74a–b). Female genitalia: Typical of the genus; sterigma cylindrical and microtrichiate.

**Holotype** (Plate DD:74a, Plate 32:74a). ♂, Arizona, Cochise County, Paradise Road, 5,580 feet, 31.93145 N, 109.20196 W, R. L. Brown, 17 August 2015, slide DJW 4319, ASUHIC.

**Paratypes** (11 ♂, 8 ♀). ARIZONA. Cochise County, Chiricahua Mountains, J. W. Brown, 9 August 2013 (1 ♀, slide DJW 4587); same data as holotype (6 ♂, slides DJW 4205, 4320); Long Park, 9,018 feet, 31.88674 N, 109.27917 W, S. Lee, 19 August 2015, (1 ♀, slide DJW 4208); John Hands Campground, 5,660 feet, 31.88099 N, 109.22246 W, S. Lee, 20 August 2015, (1 ♂, slide DJW 3998; 1 ♀, slide DJW 3999). IDAHO. Oneida County, Curlew National Grassland, D. J. Wright, 1 August 2001 (1 ♀). NEW MEXICO. Bernalillo County, 5 km NW Sandia Park, Rawlins and Zanol, 27 July 1989 (1 ♀, genitalia in vial); Grant County, 7200 feet, 32.909 N, 107.804 W, J. Vargo, 16 August 2015 (1 ♀); Lincoln County, Lincoln National Forest, Road 73, 7000 feet, J. Vargo, 22 August 2008 (1 ♀); Socorro County, Cibola National Forest, 34.0187 N, 107.1137 W, S. Johnson and J. Vargo, 15 August 2015 (3 ♂, slide DJW 4624). TEXAS. Davis Mountains, Mount Locke, A. and M. E. Blanchard, 30 August 1969 (1 ♂, slide USNM 97214). UTAH, Sanpete County, Ephraim Canyon Road, D. J. Wright, 18 July 2006 (1 ♀, slide DJW 4589). Paratype depositories: ASUHIC, CMNH, SJ, USNM.

**Etymology.** This species is named after Sangmi Lee, Collection Manager at ASUHIC, who provided many of the specimens in the type series.

## 75. ***Suleima blanchardi*** Wright and Gilligan, new species

(Plate DD, 75a–f; Plate 32, 75a–f)

**Diagnosis.** *Suleima blanchardi* is the smallest member of the group (mean FWL = 6.7 mm vs. 7.6 and 8.5). Forewing appearance resembles that of *S. baracana*, but the predominant color is darker (blackish with rusty tints rather than brownish), and the interfascial spot is less conspicuous (especially in females). *Suleima blanchardi* differs from *S. leeana* in head color (orange brown vs. dark brownish gray), in the color of the forewing markings (rusty brown vs. blackish brown), and in forewing geometry (AR = 3.02 vs. 3.23). We found no male genitalia characteristics that reliably separate this species from the other members of the group. Females differ from *S. leeana* and *S. baracana* in sterigma structure (U-shaped with well-developed lamella postvaginalis vs. substantially cylindrical). *Suleima blanchardi* seems to be restricted to Padre Island National Seashore, where it appears to be allopatric with *S. baracana* and *S. leeana*.

**Description.** *Head.* Third segment of labial palpus with tip white, ventral surface white, dorsal surface brown; first and second segments with lateral surfaces brown, medial surfaces whitish; lower frons whitish; upper frons and vertex brown; antenna brown dorsally, whitish ventrally, with whitish scales on ventral surface of scape. *Thorax.* Dorsal surface orange brown; fore- and mid-leg with anterior surfaces brown, posterior surfaces whitish; femur of mid-leg with three whitish spots; hind-leg uniformly whitish; tarsus with alternating brown and whitish annulations. Forewing: ♂ FWL 5.8–7.6 mm (mean = 6.7, n = 12), AR = 3.04; ♀ FWL 6.2–7.7 mm (mean = 6.8, n = 11), AR = 3.00; basal patch rusty brown with black or gray transverse reticulations; interfascial spot white, variably marked with black speckles and/or gray transverse reticulations, the latter most strongly expressed in females (Plate DD:75b, d); mid-costal bar weakly detectable; pretornal mark quadrate, weakly defined, with mixture of black, brown, gray, and white scales; postmedian fascia dark gray; ocellus with white central field, pale pink lateral bands, and thin black edging on proximal and distal margins; costal strigulae white and sharply defined; fringe white from tornus to $M_3$, blackish gray from $M_3$ to vertex. Hindwing: Brownish gray. *Abdomen*. Male genitalia (n = 7): Typical of the group; vesica with 22–29 cornuti; NR = 0.83; SA = 150°. Female genitalia: Sterigma U-shaped, with well-developed lamella postvaginalis, otherwise typical of the group.

**Holotype** (Plate DD:75a, Plate 32:75a). ♂, Texas, Padre Island National Seashore, A. and M. E. Blanchard, 29 September 1975, slide DJW 4491, USNM.

**Paratypes** (11 ♂, 11 ♀). TEXAS. Padre Island National Seashore, A. and M. E. Blanchard, 17 May 1976 (1 ♂, slide USNM 97552; 3 ♀, slides USNM 97547, 97554, DJW 4620), 20 May 1979 (1 ♀, slide USNM 97546), 5 June 1978 (4 ♂, slides DJW 4501, 4623), 24 June 1976 (1 ♂; 1 ♀, slide DJW 4500), 19 July 1976 (1 ♂, slide USNM 97551; 1 ♀, slide DJW 4492), 16 August 1976 (2 ♂); 29 September 1975 (2 ♂, slides USNM 97549, DJW 4622; 4 ♀, slides USNM 97545, 97548, DJW 4621), 2 October 1975 (1 ♀, slide USNM 97550). Paratype depository: USNM.

**Etymology.** This species is named after André Blanchard, who together with his wife May, collected all the specimens in the type series.

## The *skinnerana* group (species 76–77)

This group is composed of two poorly known species, *S. skinnerana* and *S. flavidana*. They differ from their congeners in their yellowish-brown forewing color, and they differ from each other in size, genitalia characteristics, and subtle aspects of forewing pattern. Current data indicates that they are sympatric in Arizona and west Texas. Group characteristics include:

*Forewing*. Mean FWL: 9.0, 7.5 mm; AR: 3.05, 2.86; dorsal surface straw yellow with thin longitudinal blackish-brown streaks; fasciate markings absent to vaguely discernable, in the latter case represented by weakly defined brownish patches in the subbasal and pretornal positions; ocellus moderately conspicuous, with pinkish-gray bands on lateral margins and 2–3 black dashes or dots on whitish central field; costal strigulae numbers 4–8 straw yellow, number 9 white.

### 76. *Suleima skinnerana* Heinrich, 1923
(Plate EE: 76a–f; Plate 33, 76a–d)

*Suleima skinnerana* Heinrich 1923:157; McDunnough 1939:49; Powell 1983:35; Brown 2005:576.

**Holotype.** ♂, Arizona, Cochise County, Huachuca Mountains, Carr Canyon, Henry Skinner, August 1905, slide 72885, USNM.

Heinrich (1923) described this species from nine specimens (5 ♂, 4 ♀) with identical collection data. He illustrated the genitalia of a male in the USNM that he referred to as the type, thus designating a name-bearing specimen, and he reported that the eight paratypes were distributed between the AMNH, the USNM, the ANSP, and the Barnes Collection. We examined six of the paratypes: three in the AMNH (2 ♂; 1 ♀, slide DJW 4567) and three in the USNM (1 ♂, slide DJW 4480; 2 ♀, slides DJW 4466, 4499).

*Suleima skinnerana* is poorly represented in collections and often has been confused with the similar looking *S. flavidana*. It is one of the largest members of the genus (mean FWL = 9.0 mm), exceeded in size only by *S. lagopana* (mean FWL = 9.6 mm). The head is whitish, and the dorsal surface of the thorax is pale straw yellow with three brown longitudinal streaks, one located medially, the others on the lateral margins. The forewing is straw yellow with blackish-brown patches in the basal and pretornal positions and thin blackish longitudinal streaks from the base to the ocellus. The subbasal mark is subsumed in a basal patch that is moderately well defined from the inner margin to mid-cell and which is interrupted by a vaguely defined yellowish band along the radius; the interfascial spot is crossed longitudinally by 4–5 thin black streaks; and the pretornal mark is quadrate and extends from the inner margin to the distal end of the cell. The ocellus is edged laterally by pale pinkish-gray bands and has three black dashes or dots on a whitish central field.

In the male genitalia, the uncus is weakly developed (basal width about four times height) and only marginally differentiated from the dorsolateral shoulders of the tegumen; the socius is short and narrow, with apex pointed; and the vesica has 19–24 cornuti (n = 4). The valva is unique in the genus in being substantially symmetric about its central longitudinal axis. The valva has a prominent triangular basal process, a short broad neck, NR = 0.81, an angulate saccular corner, and SA = 140°. The cucullus is nearly circular, with 8–10 moderately long spiniform setae along the distal margin. In females, the sterigma is U-shaped and coarsely microtrichiate. Sternum 7 is substantially longer than it is wide, and its posterior margin is convex (two features that distinguish this species from its congeners). The middle one-third of the posterior part of sternum 7 protrudes outward in a bulgelike manner, forming a shield that covers approximately one-half of the sterigma, including the entire ostium. Moreover, sternum 7 has a broad band of densely spaced scales intermixed with microtrichia that extends over the bulgelike protrusion and along the lateral margins of the sternite to about two-thirds the distance to the anterior margin.

We examined 17 specimens (13 ♂, 4 ♀) from Cochise County, Arizona, Grant County, New Mexico, and Jeff Davis County, Texas. All were collected in August.

## 77. ***Suleima flavidana*** Wright and Gilligan, new species

(Plate EE, 77a–i; Plate 33, 77a–f)

**Diagnosis.** This species previously has been misidentified as *S. skinnerana*, the only other member of the genus with a straw-yellow forewing. It differs from *S. skinnerana* in size (mean FWL = 7.5 vs. 9.0 mm), in forewing geometry (AR = 2.86 vs. 3.05), in the color of the dorsal surface of the head (brown vs. white), and in the lack of the brownish pretornal patch typical of the latter species. Males differ from those of *S. skinnerana* in the degree of differentiation of the uncus from the tegumen (absent vs. weak), the development of the socius (bulgelike and weakly convex vs. short, fingerlike, and tapering to a narrowly rounded apex), the shape of the valva (central axis of cucullus oriented at an angle of approximately 140° with central axis of neck vs. nearly parallel to latter axis), and the size of the basal process (moderate and broadly rounded vs. large and triangular). Females differ from those of *S. skinnerana* in the shape of sternum 7 (width greater than vs. less than length; posterior margin concave vs. convex) and in the extent to which the posterior portion of sternum 7 overlaps the sterigma (covering anterior one-half of lamella antevaginalis vs. anterior one-half of sterigma, including ostium).

**Description.** *Head.* Labial palpus with lateral surface grayish brown, medial surface grayish white; lower frons grayish white; upper frons and vertex brown; antenna brown; ventral surface of scape grayish white. *Thorax.* Dorsal surface tan with three longitudinal brown streaks, one located medially, the others on lateral margins; tegula brown; fore- and mid-leg with anterior surfaces brown, posterior surfaces pale tan; hind-leg pale tan; tarsus with alternating brown and white annulations. Forewing: ♂ FWL 5.7–8.0 mm (mean = 7.2, n = 10), AR = 2.90; ♀ FWL 7.2–8.6 mm (mean = 7.9, n = 6), AR = 2.78; dorsal surface straw yellow, with brown longitudinal streaks and smudges; proximal one-half of wing suffused with brown from inner margin to cubitus, usually with two pale brownish-yellow streaks, one along $A_{1+2}$, the other on inner margin; cubitus highlighted in dark brown from base to mid-wing; $CuA_2$ lightly highlighted in brown from cubitus to tornus; discal cell pale brownish yellow, usually with two faint brown longitudinal streaks; postmedian fascia represented by narrow brown band arching over apical margin of ocellus; termen with narrow salt-and-pepper colored band from tornus to apex; ocellus with pale pinkish-gray bands laterally flanking white central field, the latter with thin black longitudinal streak near anterior margin and black spot on proximal margin; fringe scales from tornus to $M_1$ pale basally with brownish-gray apices, those near apex black. Hindwing: Brownish gray. *Abdomen.* Male genitalia (n = 6): Uncus semicircular, not differentiated from tegumen; socius barely discernable as weakly developed convex projection from tegumen; genital capsule typical of the genus; vesica with 17–25 cornuti; valva with costal margin weakly concave, ventral emargination shallow, NR = 0.82, saccular corner angulate, SA = 152°; basal process moderately developed; cucullus with lateral margins nearly parallel, apex semicircular, 8–10 spiniform setae evenly spaced along distal margin; medial surface of valva with band of stiff densely-spaced setae along distal margin from apex to anal angle but not extending onto medial surface of neck. Female genitalia: Sterigma U-shaped, uniformly microtrichiate, with lateral margins nearly parallel and length slightly greater than width; lamella postvaginalis with weakly developed outward bulges flanking central trough; sternum 7 with width greater than length and posterior margin concave; papillae anales, ductus bursae, and corpus bursae typical of the genus.

**Holotype** (Plate EE:77a; Plate 33:77a). ♂, Arizona, Cochise County, Chiricahua Mountains, Southwest Research Station, D. R. Davis, 31 July–2 August 1964, slide DJW 4481, USNM.

**Paratypes** (9 ♂, 6 ♀). ARIZONA. Cochise County, 1 mi S of Portal, J. H. Davidson, J. M. Davidson, and M. A. Cazier, 17 July 1965 (1 ♀, slide DJW 4355); Chiricahua Mountains, 4 mi W of Portal, D. R. Davis, 3–6 August 1964 (1 ♂; 2 ♀, slide DJW 4482); Chiricahua Mountains, Southwest Research Station, J. W. Brown, 12 August 2012 (1 ♂, slide DJW 4511); Santa Cruz County, Pena Blanca Canyon, R. W. Hodges, 1 September 1959, (1 ♀); Santa Rita Mountains, R. W. Hodges, 3 September 1959 (1 ♂, slide DJW 4568). TEXAS. Brewster County, Alpine, O. C. Poling, 15–22 April 1926 (1 ♂); Brewster County, BIBE Green Gulch, E. C. Knudson, 16–18 May 1999 (1 ♂, slide DJW 4510); Culberson County, Sierra Diablo WMA, A. and M. E. Blanchard, 14 July 1969 (1 ♂, slide 97210 crystalized), 15 July 1969 (1 ♂, slide 70880), E. C. Knudson, 18 August 1984 (1 ♂); Guadalupe Mountains National Park, Dog Canyon, E. C. Knudson, 1 August 1992 (1 ♀, slide DJW 4363); Guadalupe Mountains, Nickel Creek, A. and M. E. Blanchard, 10 July 1968 (1 ♀, slide 146390); Val Verde County, 15 mi W of Del Rio, E. C. Knudson, 19 August 1995 (1 ♂, slide DJW 4464). Paratype depositories: EME, MGCL, USNM.

**Etymology.** The specific epithet derives from the Latin *flavidus* and refers to the yellowish color of the forewings.

**Note.** The USNM has two specimens (1 ♂, slide 146392; 1 ♀, slide 146393) from Puebla, Mexico that are closely related to and possibly conspecific with *S. flavidana*. They are in poor condition and not suitable for photographing, but the forewing appears to resemble that of *S. flavidana* in pattern and color. The female genitalia are virtually identical to those of *S. flavidana*; the male genitalia differ in that the uncus is weakly differentiated from the dorsolateral shoulders of the tegumen and the socius is fingerlike (Plate 33:77e). More material is needed to assess the taxonomic status of these specimens. They were reared from larvae feeding on *Tithonia diversifolia* (Hemsl.) A. Gray (Mexican sunflower), a plant that is native to Mexico and Central America but has been widely introduced in tropical regions around the world and is often considered an invasive species.

## Species not assigned to a group

### 78. *Suleima daracana* (Kearfott, 1907)
(Plate FF, 78a–o; Plate 34, 78a–s)

*Thiodia daracana* Kearfott 1907:44.
*Thiodia profana* Meyrick 1912:34, unnecessary replacement name for *daracana*.
*Eucosma daracana*: Barnes and McDunnough 1917:172.
*Suleima daracana*: Heinrich 1923:156; McDunnough 1939:49; MacKay 1959:42; Powell 1983:35; Brown 2005:576.

**Lectotype** (here designated, Plate FF:78a, Plate 34:78c). ♂, California, Placer County, A. H. Vachell, 1 May, slide DJW 4583, AMNH.

Kearfott (1907) described *T. daracana* from 10 specimens (sex not reported) collected between 20 April and 1 May by A. H. Vachell in Placer County, California. We examined 10 specimens that we believe to be the syntypes, each bearing a red label with the inscription "TYPE, Collection of W. D. Kearfott." One is dated 20 April, eight are dated 1 May, and the tenth has no collection date. This material resides in the AMNH (3 ♂, slide DJW 4583; 1 ♀) and the USNM (4 ♂, slides 70879, DJW 4483, 4584; 2 ♀, slides 145479, DJW 4569). Klots (1942) reported that the four syntypes in the AMNH include a male lectotype that had been designated by Heinrich (1923), but none of the three males are labeled as such. Heinrich's comments about the type and its depository are not sufficient to identify a unique specimen, so we include here a formal designation of a lectotype. The illustration of the male genitalia in Heinrich (1923: fig. 293) is based on slide 70879.

In addition to the types, we examined 46 specimens (33 ♂, 13 ♀) that we refer to *S. daracana*, most of which (40 specimens) were collected by J. F. G. Clarke in southeastern Washington State (Asotin and Whitman counties) or northeastern Oregon (Baker and Wallowa counties). Compared to its congeners, *S. daracana* is average in size (mean FWL = 8.0 mm) and has especially elongate forewing geometry (AR = 3.28 vs. 2.86–3.23). The labial palpus is brownish gray, often with some whitish suffusion on the medial surface. The lower frons, upper frons, and vertex are predominantly brownish gray, sometimes with whitish tints produced by white tips on otherwise brownish-gray scales. The antenna is dark brownish gray, with whitish scales on the ventral surface of the scape. The syntypes (e.g., Plate FF:78a–c) have dull brownish-gray forewings with brown semifasciate markings that weakly contrast with a grayish interfascial spot. The specimens from southeast Washington (e.g., Plate FF:78d–i) and northeast Oregon (e.g., Plate FF:78j–o) vary in forewing appearance from that of the syntypes to that of a more strongly marked phenotype (e.g., Plate FF:78i, o) in which the interfascial spot is white and the fasciate markings are clearly defined. Various intermediate forms are illustrated in Plate FF. The forewing pattern is uniform across the various forms and includes moderately to well-defined subbasal and pretornal marks, a variably expressed mid-costal bar, and a well-defined postmedian fascia. The ocellus is conspicuous, with grayish or pale pinkish bands on the lateral margins and two black longitudinal dashes on a whitish central field.

In males, the uncus is weakly differentiated from the dorsolateral shoulders of the tegumen, the socius is short, broad, and fingerlike, and the vesica has 12–24 cornuti (mean = 17, n = 11). Valva shape varies from that depicted in Plate 34:78a (associated with the syntype illustrated in Plate FF:78c) to that in Plate 34:78b (associated with the phenotype illustrated in Plate FF:78o). In females, the sterigma is U-shaped, with length-to-width ratio varying from about 1.13 to 1.60 (Plate 34:78k–s). The more elongate form (Plate 34:78k–n) appears to be associated with the brownish-gray forewing phenotype (Plate FF:78b, e, k) and occurs in the California, Oregon, and Washington populations. The shorter form (Plate 34:78o–s) seems to be associated with the forewing phenotype featuring greater contrast between the interfascial spot and the fasciate markings (Plate FF:78f, h, l) and occurs in the Oregon and Washington populations.

The USNM has seven specimens (5 ♂, 2 ♀) from Pullman, Washington that were collected by J. F. G. Clarke in the early 1930s and are labeled "reared from *Helianthella douglasii*," which is currently treated as *Helianthella uniflora* (Nutt.) Torr. & A. Gray var. *douglasii* (Torr. & A. Gray) W.A. Weber (Douglas' helianthella). The range of this variety includes British Columbia, Washington, Oregon, Idaho, and Montana. The full range of *H. uniflora* extends from British Columbia to Arizona and as far east as Montana, Wyoming, and Colorado, but apparently does not include California. Consequently, the specimens in the type series of *S. daracana* may have utilized a different larval host than those from Washington and Oregon. Collection dates for wild-caught specimens vary from 20 April to 1 May in Placer County, California, from 15 May to 8 June in southeast Washington, and from 7 to 21 June in northeast Oregon.

The variation observed here in forewing appearance and sterigma shape, together with the possible involvement of more than one larval host, suggests that the material we examined may represent two or more distinct species, but we did not find combinations of the various characteristics that would unambiguously support proposing additional names.

## 79. *Suleima cinerodorsana* Heinrich, 1923

(Plate GG, 79a–g; Plate 35, 79a–d)

*Suleima cinerodorsana* Heinrich 1923:159; McDunnough 1939:49; MacKay 1959:41; Powell 1983:35; Miller 1987:61; Brown 2005:576; Gilligan et al. 2008:130.

**Holotype.** ♂, Pennsylvania, Oak Station, Fred Marloff, 11 August 1911, slide CH 13 February 1920, #15, AMNH.

Heinrich based *S. cinerodorsana* on 16 specimens (12 ♂, 4 ♀) from the Kearfott Collection. They were collected by Fred Marloff at Oak Station, Pennsylvania; by Henry Engel at Pittsburg, Pennsylvania; and by August Busck at Cabin John Bridge, Maryland. Heinrich referred to a male from Oak Station as the type and illustrated its genitalia, thereby designating a unique name-bearing specimen.

*Suleima cinerodorsana* is slightly larger than *S. mendaciana* (mean FWL = 6.5 vs. 6.2 mm) and smaller than the other members of the genus (mean FWL: 6.7–9.6 mm). It is recognized by its two-toned forewing coloration: white from the inner margin to the cubitus and $CuA_2$, black from there to the costa. The head is mostly white, often with some gray-brown shading on the lateral surface of the labial palpus, and the antenna is brown with a white scape. The forewing has no recognizable subbasal or pretornal markings, but a mid-costal bar and a postmedian fascia, though obscure, are often discernable. The ocellus has gray transverse bands flanking a white central field, the latter crossed by one or two black dashes. The subcostal area from mid-wing to the apex has some orange-brown suffusion, the costal strigulae are white, the termen has a narrow salt-and-pepper colored band from tornus to apex, and the fringe is whitish near the tornus but dark gray near the apex.

In males, the uncus is weakly differentiated from the dorsolateral shoulders of the tegumen, the socius is short and fingerlike, and the vesica has 15–32 cornuti. The valva is spatulate, with the distal margin of the basal excavation sinuous, NR = 0.86, and SA = 154°. In females, the sterigma is cylindrical. We found no genitalia characteristics that would reliably distinguish this species from *S. baracana*, *S. leeana*, or some Midwestern specimens of *S. helianthana*.

*Suleima cinerodorsana* appears to be restricted to eastern North America. We examined specimens from Ontario, Illinois, Kentucky, Maryland, Ohio, Pennsylvania, and Wisconsin. Adult capture dates vary from 20 July to 1 September. Putman (1942) reported rearing this species from larvae boring in stems of *Helianthus decapetalis* L. (thinleaf sunflower) near Vineland Station, Ontario.

## 80. Unresolved specimens
(Plate GG, 80a–b; Plate 35, 80a–b)

The specimens illustrated in Plate GG:80a–b and Plate 35:80a–b are males in the USNM collected by R. W. Hodges in Chaffee County, Colorado on 15 July and by J. W. Brown in Boulder County, Colorado on 25–26 June. They are similar in forewing appearance to *S. daracana* and *S. cinerodorsana* but differ from *S. daracana* in head color (white vs. brownish gray) and from *S. cinerodorsana* in the presence of a subbasal mark that interrupts the white band on the inner margin. They are intermediate between those two species in size (mean FWL = 7.8 mm vs. 8.0 and 6.5, respectively) and forewing geometry (AR = 3.07 vs. 3.28 and 3.02). The male genitalia of these specimens exhibit no distinctive characteristics, and the female genitalia are unknown. More information is needed to ascertain the taxonomic status of these specimens.

## 81. *Suleima mendaciana* Blanchard and Knudson, 1983
(Plate GG, 81a–f; Plate 35, 81a–d)

*Suleima mendaciana* Blanchard and Knudson 1983:848; Brown 2005:576.

**Holotype** (Plate GG:81a) ♂, Texas, Brewster County, Big Bend National Park, Dugout Wells, E. C. Knudson, 28 September 1981, slide ECK 186, USNM.

*Suleima mendaciana* was described from the holotype and 4 paratypes (2 ♂, 2 ♀) collected by E. C. Knudson in west Texas (Brewster and Jeff Davis counties) and south Texas (Cameron and Hidalgo counties). Two of the paratypes, 1 ♂ from Davis Mountains State Park and 1 ♀ from South Padre Island, were deposited along with the Knudson Collection in the MGCL; the other two reside in the USNM.

*Suleima mendaciana* is the smallest member of the genus (mean FWL = 6.2 mm vs. 6.5–9.6). It is distinguished from its congeners by the combination of a dull, brownish-gray, longitudinally streaked forewing and a conspicuous whitish ocellus. The medial and lateral surfaces of the labial palpus are white and brown, respectively; the lower frons is white; the upper frons and vertex are brown; and the antenna is brown with white scaling on the ventral surface of the scape. Fasciate markings are vaguely discernable as darker shades of gray and include a thin mid-costal bar, a thin postmedian fascia, and sometimes ill-defined patches in the subbasal and pretornal positions. The white central field of the ocellus is flanked laterally by pale gray transverse bands and is crossed longitudinally by a black dash near the costal margin and a second weakly expressed dash (often reduced to a dot) on $M_3$. The costal strigulae other than number 9 are brownish gray, the termen has a narrow band of salt-and-pepper colored scales from the tornus to the apex, and the fringe is brownish gray.

In males, the socius tapers to a narrow apex, NR = 0.79, SA = 146°, and the cucullus is slightly less elongate than in most members of the genus. Females have a cylindrical sterigma. In other respects, the genitalia are typical of the genus.

We examined 10 specimens (6 ♂, 4 ♀) from Brewster, Cameron, Culberson, Hidalgo, Jeff Davis, and Uvalde counties, Texas. Adult capture dates range from mid-August to late November, with one record in early May.

## *Notocelia* Hübner [1825] 1816

*Notocelia* Hübner, [1825] 1816, Verz. bekannter Schmett. 379. Type species: *Phalaena* (*Tortrix*) *uddmanniana* Linnaeus, 1758 [subsequent designation by Westwood 1840].

*Aspis* Treitschke, 1829, *in* Ochsenheimer, Schmett. Eur. 7: 231. Type species: *Phalaena* (*Tortrix*) *solandriana* Linnaeus *sensu* Treitschke, 1829 [= *Phalaena* (*Tortrix*) *uddmanniana* Linnaeus, 1758] [preoccupied by *Aspis* Laurenti, 1768 (Reptilia)].

*Aspidia* Duponchel, 1834, in Godart, Hist. nat. Lépid. Papillons Fr. 9: 176 [replacement name for *Aspis*].

*Pardia* Guenée, 1845, Annls. Soc. Ent. Fr. (2) 3: 155. Type species: *Tortrix tripunctana* [Denis and Schiffermüller], 1775 [= *Phalaena* (*Tinea*) *cynosbatella* Linnaeus, 1758].

*Bardia*; Obraztsov, 1965, Tijdschr. Ent. 108: 371 [misspelling of *Pardia*].

*Notocelia* was thought to be a strictly Palearctic genus until Brown (1979) recognized four species in North America (*N. trimaculana*, *N. purpurissatana*, *N. illotana*, and *N. culminana*) that previously had been referred to *Eucosma* (Fernald 1903; Barnes and McDunnough 1917) or *Epiblema* (Heinrich 1923; McDunnough 1939). *Notocelia trimaculana* was believed to have been introduced into eastern North America, but Miller et al. (2000) concluded that the Nearctic specimens previously referred to that taxon are in fact *N. rosaecolana*, a similar looking Palearctic species. The other three species mentioned by Brown (1979) are indigenous to the Nearctic region. Mutuura (1980) reported the introduction in the Pacific Northwest of another Palearctic species, *N. cynosbatella*, and we discovered specimens of that species collected in eastern North America that date back to the 1960s. Here we describe a recently recognized endemic species (*N. lobana*), raising the Nearctic species count to six.

**Diagnosis.** The Nearctic species of *Notocelia* resemble those of *Epiblema* in wing venation, in the presence of a costal fold at the base of the male forewing, and in the presence of a prominent clasper near the basal excavation of the medial surface of the male valva. In both genera, the vesica has a cluster of deciduous cornuti, but *Notocelia* differs from *Epiblema* (and other *Eucosmini*) in having two non-deciduous cornuti near the distal end of the phallus. The last feature is reputed to be an autapomorphy for *Notocelia* (Razowski 2003). Females of North American *Notocelia* have two prominent scale clusters on the membrane posteriorly adjacent to the sterigma, one at each of the posterolateral corners of the lamella postvaginalis (Fig. C9). With the exception of *N. purpurissatana*, they also have numerous scales uniformly distributed on the posterior two-thirds of tergum 8. In these two respects, they differ from all Nearctic *Epiblema* except *E. tandana*.

**Description.** *Forewing* (Fig. A5). Mean FWL: 6.7–8.6 mm, average = 8.0 mm; AR: 2.51–2.69, average = 2.56; CFR: 0.40–0.51, average = 0.45; male with costal fold; costa weakly arched; termen straight to weakly convex; venation as in Fig. A1a; forewing pattern semifasciate; subbasal fascia often subsumed into prominent basal patch; pretornal mark obscure to conspicuous; ocellus obscure but usually discernable.

*Hindwing*. Color uniform in shades of grayish brown; venation as in Fig. A1b ($M_3$ and $CuA_1$ stalked); males in four of the six species with hair pencil at base of dorsal surface, resting in fold between vein $A_3$ and anal margin.

*Male genitalia*. Tegumen + appendages Type 1 (Fig. B2); uncus weakly to moderately developed; socius fingerlike; phallus stout, weakly attenuating distally, with base closely surrounded by anellus; vesica with 4–20 deciduous cornuti and two non-deciduous cornuti; valva with Type 1 clasper and (in four of the six species) small auxiliary clasper, located respectively near dorsodistal and ventrodistal corners of the basal excavation (Fig. B9); cucullus with dorsal lobe well developed, apex rounded, distal margin convex to nearly straight, ventral lobe absent to moderately developed, the last usually with basoventral margin extending ridgelike onto medial surface of neck; setae on medial surface of cucullus moderately short and stiff near ventral one-half of distal margin, grading to fine near costal and apical margins. We did not record SA or NR data for specimens of *Notocelia*. In only two of the six species does the shape of the valva lend itself to such measurements.

*Female genitalia*. Papillae anales Type 4 (Fig. C5), with pads moderately setose and densely microtrichiate; posterior two-thirds of tergum 8 sparsely scaled (except in *N. purpurissatana*); sterigma shieldlike (Fig. C9) or somewhat triangular; membrane posteriorly adjacent to sterigma with two conspicuous patches of scales, one at each posterolateral corner of lamella postvaginalis (Fig. C9); lamella antevaginalis with raised and often weakly sclerotized lip along anterior one-half of ostium; posterior margin of sternum 7 concave, medially approximate to lamella antevaginalis, weakly to moderately divergent laterally from sterigma; ductus bursae weakly sclerotized from ostium to ductus seminalis, usually with more strongly sclerotized ringlike or patchlike section near ductus seminalis (Fig. C15), always with broad band of dentate microtrichia on inner surface from constriction near ostium to ductus seminalis; corpus bursae with two signa.

**Immature stages.** MacKay (1959) described the larvae of *N. culminana* (under *E. culminana*) and *N. rosaecolana* (as *E. suffusana*) and noted that they differ from *Epiblema* larvae in several morphological characters, in feeding behavior, and in preferred host plants. Most significantly, *N. rosaecolana* has an anal comb, a structure that is absent in the other genera treated here.

*Egg*. As described for *Epiblema*.

*Larva*. Head broader than long; adfrontals tapered anteriorly, extending to vertical angle; vertical angle obtuse or right angle; stemmata convex; basal segment of the labial palpus slender; spinneret about five times as long as wide; spiracles moderately large; pinacula moderately small, and those of D1s and D2s on the dorsum well separated; seta on a pinaculum close together, especially noticeable with respect to the lateral group and D2s on segment 9; L-group on T1 trisetose, never enlarged or extending under the spiracle; abdominal L1s and L2s more or less ventral to spiracle; SV-group setal counts on A1, 2, 7, 8, 9 are 2:2:2:2:1; D1 and SD1 on the same pinaculum on A9; anal shield rounded or more rounded than tapered posteriorly; crotchets biordinal or partially triordinal; anal comb present or absent.

*Pupa*. As described for *Epiblema*.

**Biology.** Life history information is well documented for several species of Palearctic and Holarctic *Notocelia*. The summary provided here is from Heinrich (1923), Bradley et al. (1979), and Razowski (2003). Larvae of most species for which life histories have been reported feed on *Rosa* L. (rose) and other Rosaceae, including *Rubus* L. (blackberry), *Pyrus* L. (pear), *Malus* Mill. (apple), *Prunus* L. (plum), and *Cydonia* Mill. *Notocelia cynosbatella* has been reported to also feed on *Myrica* L. (sweetgale; Myricaceae), *Carpinus* L. (hornbeam; Betulaceae), and *Quercus* L. (oak; Fagaceae). Females lay 60–300 eggs in small clusters. Early instar larvae live communally and feed within flower buds or bore into young shoots, often constructing a shelter of spun leaves. Later instars disperse and feed solitarily, moving between shoots or buds. Fourth instar larvae overwinter in a flimsy cocoon constructed in a bark crevice, leaf litter, or the larval feeding site. Larvae emerge from hibernation in the spring and continue to feed on buds, shoots, or leaves. Pupation occurs in the larval feeding site, and adults emerge in approximately two weeks. Most species complete one generation per year. Larvae of *N. rosaecolana* can be a serious pest of cultivated roses, with one larva destroying up to seven buds during development.

## The *culminana* group (species 82–83)

This group is comprised of two superficially similar species, *N. culminana* and *N. lobana*, that are distinguished from one another by differences in valva shape, sterigma shape, and DNA sequence data. Group properties include:

*Forewing*. Mean FWL = 8.1, 7.8 mm; AR = 2.69, 2.55; CFR = 0.41, 0.40; interfascial areas of dorsal surface with thin gray transverse reticulations on a whitish ground, producing a pale grayish-white effect; fasciate markings dark brown; basal fascia represented by thin line from inner margin to radius; subbasal fascia variable in expression, often with black edging on lateral margins; mid-costal bar absent to vaguely discernable; discal mark reduced to a few minute dots; pretornal mark triangular and conspicuous; postmedian fascia subsumed in prominent triangular patch extending from apex to anterior margin of ocellus to termen between $CuA_1$ and $CuA_2$; patch edged by thin black line on termen and enclosing thin white bar from $M_3$ to $R_5$, the latter edged basally by narrow gray/brown bar; ocellus barely discernable, with grayish or pinkish lateral bands and white central field; costal strigulae inconspicuous, concolorous with interfascial areas; fringe pale at tornus, dark brown from $CuA_1$ to apex.

*Hindwing*. Pale brownish gray; males with hair pencil at base of dorsal surface between $A_3$ and anal margin.

*Male genitalia*. Uncus moderately developed, with apical margin rounded, straight, or weakly indented medially; socius fingerlike, tapering evenly to narrowly rounded apex; vesica with 10–18 deciduous cornuti; valva with large Type 1 clasper and small tablike auxiliary clasper; valval neck with dorsal margin weakly concave and ventral margin weakly convex or strongly lobelike; cucullus with dorsal lobe well developed, apex rounded, distal margin weakly convex, ventral lobe weakly developed, basoventral margin extending ridgelike onto medial surface of valval neck.

*Female genitalia*. Sterigma shieldlike; in other respects, typical of the genus.

## 82. *Notocelia culminana* (Walsingham, 1879)
(Plate HH, 82a–i; Plate 36, 82a–g)

*Paedisca culminana* Walsingham 1879:38; Fernald 1882:39.
*Eucosma culminana*: Fernald 1903:457; Barnes and McDunnough 1917:170.
*Epiblema culminana*: Heinrich 1923:153; McDunnough 1939:48; MacKay 1959:106.
*Notocelia culminana*: Brown 1979:27; Powell 1983:35; Miller 1987:61; Brown 2005:435; Gilligan et al. 2008:129; Powell and Opler 2009:137.

**Holotype.** ♂, California, Shasta County, Bear Creek, 27–28 July 1871, Walsingham, slide 11522, BMNH.

*Notocelia culminana* was described from a single male. We examined photographs of that specimen and its genitalia made by Obraztsov in the 1960s. Its collection data, given above, includes details regarding the type locality (Bear Creek) and the collection date (27–28) that were not mentioned by Walsingham (1879) but were recorded by Obraztsov on the plastic mounts for his 35-mm slides. The male genitalia illustrated in Heinrich (1923: fig. 282, slide 70856) as *N. culminana* are those of the then unrecognized species described below as *N. lobana*.

*Notocelia culminana* is an average sized member of the genus (mean FWL = 8.1 mm) whose forewing geometry is slightly more elongate than that of its congeners (AR = 2.69, vs. 2.51–2.55). In females, the subbasal fascia is prominent and extends in a bandlike manner from the inner margin to the costa; in males, it usually fades near the costa. The basal area of the forewing is paler than the subbasal fascia and usually somewhat darker than the interfascial spot. In other respects, the forewing appearance is typical of the group.

In males, the apical margin of the uncus varies from rounded to medially indented, the vesica has 10–18 deciduous cornuti, and the ventral margin of the neck is weakly convex. In females, the anterior margin of the sterigma is semicircular and transitions smoothly into the straight lateral margins of the lamella postvaginalis, the latter diverging slightly from one another posteriorly. The lamella postvaginalis has a weakly sinuous posterior margin and acute posterolateral corners. The sclerotization of the ductus bursae is ringlike and conspicuous near the ductus seminalis but barely discernable from there to the constriction anterior to the ostium. The signa in the corpus bursae are conelike and located near mid-bursa, with one considerably larger than the other.

We examined 120 adults (91 ♂, 29 ♀) that document a transcontinental distribution for this insect. In the West, the range extends from British Columbia and Alberta south to central California and northern New Mexico; in the Midwest, it includes Indiana, Minnesota, Michigan, and Wisconsin; and in the East, it includes Ontario, Nova Scotia, Maine, Massachusetts, and New York. Adult capture dates range from 13 May to 20 October, but most records are from July, August, and September. MacKay (1959) described the larva based on one late instar collected on *Rosa* L. (rose) at Cape George, Nova Scotia.

## 83. ***Notocelia lobana*** Wright and Gilligan, new species
(Plate HH, 83a–f; Plate 36, 83a–g)

**Diagnosis**. We found no features of the wing pattern that we consider to be reliable in separating this species from *N. culminana*. The adult illustrations (Plate HH:83a–f) suggest that the subbasal mark is less prominent in *N. lobana* than in *N. culminana*, but that might be due to the somewhat flight-worn condition of the specimens we had at our disposal. Nevertheless, consistent differences in the genitalia prompted us to propose the new name. This situation was brought to our attention by J. W. Brown, who noticed a distinctive cluster of specimens in a COI DNA barcode analysis that appeared to differ from typical *culminana*. He examined a few genitalia preparations and noticed some potentially diagnostic features but did not pursue the matter further. We examined two specimens from the distinctive cluster and found that they agree in genitalia with several *culminana*-like specimens we had collected in Ohio and Kentucky. Eventually, we assembled a sample of 24 specimens (18 ♂, 6 ♀) from Connecticut, Indiana, Kentucky, Maryland, New York, western North Carolina, Ohio, and Washington DC that share those genitalia characteristics. Males differ from *N. culminana* in the shape of the ventral margin of the valval neck (strongly lobelike vs. weakly convex). In females, the sterigma has a more "rounded" appearance. Its lateral margins are convexly curved (instead of straight), producing sharply-differentiated lobelike projections at the posterolateral corners. The medial one-third of the posterior margin of the lamella postvaginalis is semicircular (instead of weakly convex), and the sclerotization of the ductus bursae is patchlike instead of ringlike. Based on our somewhat meager sample, the range of *N. lobana* likely includes the region from northwestern Indiana to central Kentucky east to western North Carolina, Maryland, and Connecticut. *Notocelia lobana* and *N. culminana* are sympatric on Long Island, New York, but we have not seen specimens of *N. culminana* that were collected south of a line from the northwest corner of Indiana to New York City.

**Description**. *Head.* Third segment of labial palpus brown, often with whitish tip; first and second segments with lateral surfaces brown, medial surfaces mostly white; frons whitish; vertex brown; antenna brown with some black scaling on pedicel and ventral surface of scape. *Thorax.* Dorsal surface tan, with two dark brown transverse bars (on mesothorax and metathorax); fore- and mid-legs with anterior surfaces brown, posterior surfaces pale tan; hind-leg pale tan; tarsi with alternating brown and tan annulations. Forewing: ♂ FWL 7.2–8.0 mm (mean = 7.6, n = 14), AR = 2.53, CFR = 0.40; ♀ FWL 7.8–8.7 mm (mean = 8.2, n = 5), AR = 2.61; costal margin weakly arched, apical angle approximately 90°, termen straight; dorsal surface as described for group. Hindwing: Pale grayish brown. *Abdomen.* Male genitalia (n = 14): Uncus somewhat rectangular, with apical margin straight to weakly indented medially; socius fingerlike, moderately short, and sparsely setose; vesica with 10–15 deciduous cornuti; ventral margin of valval neck with conspicuous lobelike projection; cucullus with dorsal lobe gradually narrowing to rounded apex, distal margin weakly convex, basoventral margin extending ridgelike onto ventral one-half of medial surface of neck, anal angle acute with vertex narrowly rounded. Female genitalia (n = 6): Papillae anales Type 4; sterigma with lateral margins convexly curved, medial one-third of posterior margin semicircular, posterolateral corners produced into sharply-pointed lobelike projections; lip along anterior margin of ostium mostly membranous; ductus bursae with elongate sclerotized patch from ductus seminalis to three-fourths distance to ostium and with microtrichia on inner surface from ductus seminalis to ostium; corpus bursae with two signa of nearly equal size located near mid-bursa.

**Holotype** (Plate HH:83a). ♂, North Carolina, Swain County, Great Smoky Mountains National Park, J. B. Sullivan and L. Deutschman, Big Cove Road, Specimen ID: DNA-ATBI-6130, Bold ID: LGSMF 130-06, 18 August 2001, slide 150259, USNM.

**Paratypes** (16 ♂, 6 ♀). CONNECTICUT, East River, C. R. Riley, 4 September 1907 (2 ♂, slides 70856, DJW 4643). INDIANA. Lake County, Osborne, A. P. Troschel, 11 September 1904 (1 ♀, slide 70859). KENTUCKY. Boone County, Big Bone Lick State Park, L. D. Gibson, 25 August 1979 (1 ♂, slide DJW 4238); Bullitt County, Route 480, 6.9 mi E of Route 61, L. D. Gibson, 6 September 1989 (1 ♂, slide DJW 4248). MARYLAND. Baltimore County, 2 mi N of Carney, J. Glaser, 30 August 2004 (1 ♂, slide DJW 4415); Charles County, Myrtle Grove Wildlife Management Area, J. Glaser, 5 September 1996 (1 ♀, slide 150249). NEW YORK. Suffolk County, Montauk, Long Island, R. Latham, 10 June 1954 (1 ♀, slide DJW 4313); Shelter Island, R. Latham, 10 July 1946 (1 ♂, slide DJW 4314); Orient, Long Island, R. Latham, 5 September (1 ♀, slide 150260). NORTH CAROLINA. Same collection data as holotype, (2 ♂, slide 150248). OHIO. Adams County, Lynx Prairie #2, L. D. Gibson, 10 September 1988 (1 ♂, DJW 4247); Greene County, Wright-Patterson Air Force Base, E. H. Metzler, 28 August 1992 (1 ♂, DJW 4291); Hancock County, Van Buren, T. M. Gilligan, 31 August 2009, (1 ♂, DJW 4522); Cass Township, SE ¼, Section 22, T. M. Gilligan, 2 September 1991 (1 ♂, TMG 940); Lake County, Hells Hollow Park, J. Wiedmann, 24 August 2018 (2 ♂, DJW 4311), Hidden Lake Woods, J. Wiedmann, 30 August 2017 (1 ♀, DJW 4312); Wyandot County, Killdeer Plains, D. J. Wright, 16 August 2006 (1 ♀, 153230); Summit County, Liberty Metropark, J. D. Hooper, 25 August 2012 (1 ♂). WASHINGTON DC. Charles R. Ely, 13 May 1907 (1 ♂). Paratype depositories: CUIC, JW, LDG, TMG, OSUC, USNM,

**Etymology.** The specific epithet derives from the Latin *lobus*, meaning rounded projection.

## The *illotana* group (species 84–85)

This group is composed of *N. illotana* and *N. cynosbatella*, two species that are easily confused with one another due to similarities in forewing appearance and genitalia. *Notocelia illotana* was described by Walsingham (1879) from southwest Oregon. A century later, Mutuura (1980) reported the presence of the Palearctic *N. cynosbatella* (Linnaeus, 1758) in the Pacific Northwest. Mutuura examined 10 adults (3 ♂, 7 ♀) of *N. cynosbatella* that had been collected as larvae on *Rosa* L. (rose) by M. Doğanlar at Ladner, British Columbia (a suburb of Vancouver), and he provided a species-to-species comparison of 10 characters (8 maculational, 2 genitalic) that he considered to be useful in distinguishing between the two taxa. The proximity of the collection site to the Vancouver International Airport as well as to the Port of Vancouver, together with a lack of parasitoids on the larvae, led Mutuura to surmise that the introduction had been recent, possibly on imported ornamental plants.

We examined 97 North American and European specimens (67 ♂, 30 ♀) that had been determined as *N. illotana* or *N. cynosbatella*, and for the most part we were able to confirm the differences between the two species that had been noted by Mutuura (1980). For example, Mutuura reported *N. illotana* to be slightly smaller than *N. cynosbatella*, and our findings agree with that conclusion (mean FWL = 8.1 & 8.6 mm). He did not mention forewing geometry; in fact, the two taxa are identical in that respect (AR = 2.54 & 2.54). Mutuura considered the male costal fold to be slightly shorter in *N. illotana* (our data yielded CFR = 0.48 & 0.49). He described the color of the labial palpus as yellowish fuscous in *N. illotana* and bright orange yellow in *N. cynosbatella*. We found the color to be variable, ranging from pale yellow to brown to bright orange yellow, frequently with two-toned combinations of these colors. Nevertheless, in most of the specimens we determined as *N. cynosbatella*, the color of the labial palpus was bright orange yellow (Plate II:85c), and in most of the *N. illotana* specimens it was yellowish and/or brownish.

The wing pattern distinctions reported by Mutuura (1980) are less clear, but the specimens in his figures appear to be identified correctly. When we sorted spread museum specimens by forewing phenotype, we were able to identify consistent but subtle differences in wing pattern. In *N. cynosbatella*, the basal patch is uniformly blackish gray with a sharply defined distal margin, the interfascial spot is bright white or lightly suffused with pale olive gray and virtually unmarked, and the ocellus has a transverse central line of 2–4 black spots that is bordered laterally by bands of metallic-gray scales, the proximal band being most prominent. In *N. illotana*, the pattern elements are not as clearly defined (especially the discal and pretornal marks), the contrast between the basal patch and interfascial spot is less pronounced, and most of the wing is overlaid with indistinct transverse reticulations, the last being black on the basal patch, olive gray on the interfascial spot, and gray, pale brown, or olive otherwise. The obscure ocellus has 2–3 black spots but lacks metallic-gray patches on the lateral margins.

We rely upon these observations and some subtle differences in valva shape and sterigma structure to distinguish between the two species. Further diagnostic details are discussed in the following two species accounts. Group properties include:

*Forewing*. Mean FWL = 8.1, 8.6 mm; AR = 2.54, 2.54; CFR = 0.49, 0.48; basal patch prominent, blackish gray, with distal margin extending obliquely from basal one-third of inner margin to mid-costa; interfascial areas predominantly white, sometimes with pale olive-gray suffusion and/or olive-gray reticulations; interfascial spot large, extending from middle one-third of inner margin into cell; mid-costal mark usually confluent with basal patch; discal and pretornal marks olive gray to dark gray, disjunct, and sometimes vaguely expressed; ocellus often weakly defined, with white or metallic gray bands on lateral margins and 2–4 black spots on central field.

*Hindwing*. Whitish to grayish brown; males without hair pencil at base of dorsal surface.

*Male genitalia* (n = 18). Uncus weakly developed but distinguishable from dorsolateral shoulders of tegumen; socius of medium length; vesica with 4–15 deciduous cornuti; valva with Type 1 clasper but lacking auxiliary clasper; ventral margin of neck convex; medial surface of neck with conspicuous patch of moderately stiff setae; cucullus with well-developed dorsal lobe, semicircular apex, and weakly convex distal margin.

*Female genitalia* (n = 16). Sterigma somewhat V-shaped, with lateral margins folded under lamella postvaginalis in accordionlike manner (indicated in the drawings by dashed lines); lamella postvaginalis uniformly microtrichiate; lamella antevaginalis weakly indented along semicircular arc paralleling anterior margin of sterigma, with indentation transitioning posteriorly into inner ridges of accordion folds in lateral margins of lamella postvaginalis; lip on anterior margin of ostium bandlike and variable in width; ductus bursae lightly sclerotized from ostium to ductus seminalis, with inner surface microtrichiate; corpus bursae with two conelike signa, one slightly larger than the other.

## **84. *Notocelia illotana*** (Walsingham, 1879)
(Plate II, 84a–h; Plate 37, 84a–i)

*Paedisca illotana* Walsingham 1879:39; Fernald 1882:41.
*Eucosma illotana*: Fernald 1903:459; Barnes and McDunnough 1917:171.
*Epiblema illotana*: Heinrich 1923:153; McDunnough 1939:48; Mutuura 1980:511.
*Notocelia illotana*: Brown 1979:27; Powell 1983:35; Miller 1987:60; Brown 2005:435; Gilligan et al. 2008:130.

**Lectotype** (designated here). ♀, Oregon, Josephine County, Rogue River, 7 May 1872, slide 11542, BMNH.

Walsingham (1879) described this species from three syntypes (2 ♂, 1 ♀) with the same collection data. Obraztsov dissected the female and one of the males, photographed those two specimens and their genitalia, and selected the lectotype designated here.

In *N. illotana*, the fasciate markings other than the basal patch are olive gray. The median fascia is represented by a discal mark and a pretornal mark that are weakly expressed and separated from one another by small patches of whitish scales; the postmedian fascia is narrow at the costa, expands into a substantial olive-gray patch between $R_4$ and $M_2$, and attenuates into a narrow streak that invades the central field of the ocellus; and the preterminal fascia is thin, extending along the termen to approximately $CuA_1$. The interfascial spot has transverse olive-gray or grayish-brown reticulations, and the ocellus has prominent white bands on its lateral margins that converge into a conspicuous white patch at the tornus. The fringe is white at the tornus and olive gray to dark gray otherwise. Differences in forewing appearance between *N. illotana* and *N. cynosbatella* are discussed above in the introductory comments for the group.

In males, the uncus is weakly developed, with apex rounded, and the vesica has 4–15 deciduous cornuti. The convex protrusion of the ventral margin of the valval neck is a flangelike extension of the medial surface (indicated in the illustrations by a dashed line, Plate 37:84a–e). It is slightly thinner than the rest of the neck and lacks scales on its lateral surface. The cucullus has a small triangular ventral lobe whose basoventral margin projects ridgelike onto the ventral one-third of the medial surface of the neck. In females, the posterolateral corners of the sterigma are rounded, the posterior margin of the lamella postvaginalis is weakly incurved, and the signa in the corpus bursae are located opposite one another at approximately mid-bursa.

This species ranges from Nova Scotia to British Columbia and south to Massachusetts, Pennsylvania, Illinois, Missouri, Idaho, and northern California. Adults fly from late May to early August. As far as we know, no larval host has been reported.

## 85. *Notocelia cynosbatella* (Linnaeus, 1758)

(Plate II, 85a–g; Plate 37, 85a–i)

*Phalaena* (*Tinea*) *cynosbatella* Linnaeus 1758:536.
*Pyralis cynosbana* Fabricius 1775:654, unjustified emendation.
*Tortrx tripunctana* [Denis and Schiffermüller] 1775:131.
*Epiblema cynosbatella* Mutuura 1980:511.
*Notocelia cynosbatella* Razowski 2003:102; Brown 2005:435; Gilligan et al. 2020a:37.

**Types.** Syntype(s), unknown.

*Notocelia cynosbatella* is a Palearctic species whose introduction into North America near Vancouver, British Columbia was reported by Mutuura (1980). It was described from syntypes whose fates are unknown (Brown 2005), so there currently is no candidate for a name-bearing specimen.

The forewing in *N. cynosbatella* lacks the olive tones in *N. illotana*. Other differences in forewing appearance between the two species are discussed above in the introductory comments for the group. These observations compare favorably to European specimens of *N. cynosbatella* and Obraztsov's image of the lectotype of *N. illotana*.

In males, the distal margin of the cucullus is evenly curved and smoothly transitions into the ventral margin of the valval neck. Unlike *N. illotana*, the proximal extent of the cucullus is not defined by a constriction in the neck. As in *N. illotana*, the ventral margin of the valva has a flangelike protrusion, but in this case it is narrow and extends partway along the ventrodistal margin of the cucullus. In both species, the setation on the medial surface of the cucullus ceases in a conspicuous line along the distal extremity of the neck. In *N. illotana*, that line is reinforced by the ridgelike extension of the basoventral margin of the cucullus, whereas in *N. cynosbatella* the cucullus has no distinguishable ventral lobe and no such ridgelike extension. The female genitalia resemble those of *N. illotana*, but the posterolateral corners of the sterigma are more angular than rounded.

Current records indicate that in North America this species is present in the Pacific Northwest (British Columbia and Washington) and the Atlantic Northeast (Ontario, Quebec, and Connecticut), where it is sympatric with *N. illotana*. The USNM has several specimens of *N. cynosbatella* that were collected in Hamden, Connecticut in the mid-1960s by D. C. Ferguson, indicating that this species was present in North America at least a decade before it was reported by Mutuura (1980). The first specimens found in British Columbia (by M. Doğanlar) were collected in Ladner, near Vancouver, and reared from *Rosa* L. (rose) (Mutuura 1980). In its native range of Europe, Asia Minor, and the Trans-Caspian region, larvae have been recorded on *Rosa, Pyrus* L. (pear), *Malus* Mill. (apple), *Prunus* L. (plum), *Cydonia* Mill., *Myrica* L. (sweetgale), *Carpinus* L. (hornbeam), and *Quercus* L. (oak) (Bradley et al. 1979; Razowski 2003).

As part of a study to distinguish between Holarctic and introduced tortricids in North America, Gilligan et al. (2020a) examined DNA barcodes for several species of *Notocelia*, including *N. cynosbatella* and *N. illotana*, and concluded that *N. cynosbatella* is indeed an introduced species. We reexamined the COI DNA barcodes in BOLD (Barcode of Life Data System; www.boldsystems.org) for specimens identified as *N. cynosbatella* or *N. illotana* and found three clusters or BINS (Barcode Index Numbers). By examining images of specimens in those clusters, we determined that two of those BINS (BOLD:AAC3246 and BOLD:AAC9709) represent *N. cynosbatella,* with specimens from Europe, British Columbia, Washington, Quebec, and Ontario, and the third BIN (BOLD:AAC3245) represents *N. illotana*, with specimens from British Columbia, Washington, and Alberta.

## Species not assigned to a group

### 86. *Notocelia purpurissatana* (Heinrich, 1923)
(Plate JJ, 86a–f; Plate 38, 86a–c)

*Epiblema purpurissatana* Heinrich 1923:149; McDunnough 1939:48.
*Notocelia purpurissatana*: Brown 1979:27; Powell 1983:35; Brown 2005:436.
*Epiblema gratuitana* Heinrich 1923:268; McDunnough 1939:48.

**Types.** *Epiblema purpurissatana*. Holotype. ♂, British Columbia, Vernon, May 1920, slide TOR 4081, CNC. *Epiblema gratuitana*. Holotype. ♂, Washington, Eastsound, 11 July 1901, slide 72881, USNM.

Heinrich (1923) described this species from two specimens (1 ♂, 1 ♀) that he had received from J. M. McDunnough. He illustrated the genitalia of the male and referred to it as the type, thereby designating a name-bearing specimen. Brown (2005) reported that specimen as a holotype. The female paratype is in the USNM and has the same collection data as the holotype. Heinrich (1923) described *E. gratuitana* from two specimens (1 ♂, 1 ♀) in the USNM from Eastsound and Whidbey Island, respectively, in Washington State. He illustrated the genitalia of the male and referred to that specimen as the type, thereby designating a holotype. McDunnough (1939) recognized *gratuitana* as a synonym of *purpurissatana* but left the taxon in *Epiblema*. Subsequent authors (Powell 1983; Brown 2005) placed it in the synonymy under *Epiblema radicana*. Heinrich's illustration of the male genitalia (fig. 431) clearly shows two non-deciduous cornuti in the vesica, indicating that *Notocelia* is the correct generic placement.

This is the smallest Nearctic *Notocelia* (mean FWL = 6.7 vs. 7.8–8.6 mm). The forewing is brownish gray with a vaguely defined whitish interfascial spot and considerable white suffusion in the basal area.

The fasciate markings are black and include a somewhat fragmented subbasal mark, a mid-costal bar, three or four spots on the proximal margin of the ocellus (possibly remnants of a pretornal mark), and a postmedian fascia that is prominent anterior to the ocellus but barely discernable near the costa. The ocellus has a central transverse line of black spots that is flanked laterally by lustrous blue-gray bands. The costal strigulae from mid-wing to the apex are whitish, and the termen has a thin black line from tornus to apex, the last edged proximally by a narrow band of orange-brown scales from $CuA_1$ to the apex. Males have a dark gray hair pencil at the base of the hindwing (Plate JJ:86c–d, f).

In males, the apical margin of the uncus has a shallow medial indentation, the socius tapers evenly to a narrowly rounded apex, the base of the phallus is closely surrounded by the anellus, and the vesica has 17–18 deciduous cornuti (n = 2). The valva has a small Type 1 clasper, a smaller auxiliary clasper, and a neck that is well defined by a shallow ventral emargination. The cucullus has a strongly developed dorsal lobe that tapers evenly to a rounded apex, a nearly straight distal margin, and a weakly developed ventral lobe whose basoventral margin projects ridgelike onto the medial surface of the neck. In females (n = 1), *E. purpurissatana* differs from other Nearctic members of the genus in that the scaling on tergum 8 is substantially reduced (there appear to be only a few scales at the extreme ventral extremities of that tergite). The sterigma closely resembles that of *E. culminana* except that the lateral margins fold under the posterolateral corners of the lamella postvaginalis. The ductus bursae has a sclerotized ring at the juncture with the ductus seminalis, a more lightly sclerotized ring at the constriction anterior to the ostium, and a large patch of microtrichia on the inner surface between the two rings. The two signa in the corpus bursae are slightly different in size and are located a little posterior to mid-bursa.

The female paratype bears a white label with the inscription "from wild roses," which apparently is the basis for Heinrich's (1923) statement that the larval host is wild rose. We examined 22 specimens (15 ♂, 7 ♀) from British Columbia, Alberta, Ontario, Maine, New Hampshire, and Washington State. Adult capture dates range from 2 April to 20 August.

## 87. *Notocelia rosaecolana* (Doubleday, 1850)
(Plate JJ, 87a–i; Plate 38, 87a–d)

*Spilonota rosaecolana* Doubleday 1850:CVI (appendix).
*Notocelia rosaecolana*: Miller et al. 2000:1; Razowski 2003:104; Brown 2005:436; Gilligan et al. 2008:128; Gilligan et al. 2020a:36.
*Eucosma suffusana*: Kearfott 1910:541; Barnes and McDunnough 1917:170.
*Epiblema suffusana*: Heinrich 1923:152; McDunnough 1939:48; MacKay 1959:107.
*Notocelia trimaculana*: Brown 1979:27; Powell 1983:35.

North American tortricid specialists have long been aware of a rose-feeding species of *Notocelia* that is widespread in eastern United States, but there has been considerable uncertainty as to its proper name. Kearfott (1910) listed it from New Jersey as *Eucosma suffusana* (Zeller), and Heinrich (1923) treated it as *Epiblema suffusana* (Zeller 1846). *Epiblema suffusana* is now a junior synonym of *Notocelia trimaculana* (Haworth [1811]), and it is the latter name that was utilized by Brown (1979) and Powell (1983) in reference to this species.

*Notocelia trimaculana* is a Palearctic species that is remarkably similar in forewing appearance and genitalia to the Palearctic *Notocelia rosaecolana* (Doubleday 1850). Miller et al. (2000) undertook a review of prior attempts to distinguish between the two species. They performed an allometric analysis of seven characters that various authors had used for that purpose and concluded that none of those character states is truly diagnostic. In the process, they discovered a previously overlooked character (melanic sex scaling at the base of the male hindwing) that is present in *N. rosaecolana* and absent in *N. trimaculana*. The North American males examined by Miller et al. (2000) were found to have the melanic sex scaling and thus were referred to *N. rosaecolana*. To date, no one has reported a reliable means of separating the females of these two taxa. Having encountered no correctly determined males of *N. trimaculana* in North America, Miller et al. (2000) suggested that, at least tentatively, the females they examined should be interpreted as *N. rosaecolana*. Gilligan et al. (2020a) examined DNA sequence data for these species and confirmed that *N. rosaecolana* is present in Europe and North America, and that *N. trimaculana* is a separate taxon. Consequently, all historical North American records of *trimaculana* and *suffusana* likely refer to *rosaecolana*.

*Notocelia rosaecolana* is a moderately large member of the genus (mean FWL = 8.4 mm) with slightly broader than average forewing geometry (AR = 2.51 vs. 2.54–2.69). The forewing has mottled grayish–brown fasciate markings. The basal patch is well defined and has inconspicuous black reticulations; the interfascial spot is large and white, with the distal one-third overlaid with gray; the mid-costal bar is moderately well defined in females (Plate JJ:87b, d, f–h) but tends to merge with the basal patch in males; the weakly defined pretornal mark is mostly brown, includes three black spots along the proximal margin of the ocellus, and is separated from the mid-costal bar by a vaguely defined white streak along the radius. The ocellus has gray bands along its lateral margins and up to three black dots on its white central field. The costal strigulae are whitish and sharply defined, and the fringe is white at the tornus and brownish gray from about $CuA_1$ to the apex. Males of *N. rosaecolana* have a gray hair pencil at the base of the hindwing (Plate JJ:87a, c, e, i) and melanic scales in the fold beneath the hair pencil. Miller et al. (2000: figs. 3–5) provide scanning electron micrographs illustrating those scales in *N. rosaecolana* and their absence in *N. trimaculana*.

Males have a small semicircular uncus that is weakly differentiated from the dorsolateral shoulders of the tegumen. The socius is long and narrow (length about six times width), with hairlike setae on the proximal four-fifths and scalelike setae at the apex (Plate 38:87b). The latter setae do not occur in any other Nearctic member of the genus but are present in *N. trimaculana*. The valva has a Type 1 clasper and a tablike auxiliary clasper, both offset somewhat from the distal margin of the basal excavation. The dorsal margin of the valva is strongly concave, the neck narrows evenly from the sacculus to the base of the cucullus, and the saccular corner is broadly rounded. The cucullus is elongate (length about three times width). Its dorsal lobe is strongly developed, with lateral margins nearly parallel, and its triangular ventral lobe is moderately developed, with the basoventral margin extending ridgelike onto the medial surface of the neck. In females, the sterigma resembles that of *N. purpurissatana* except that its anterolateral corners are laterally produced into rounded convex projections. The sclerotization of the ductus bursae is ringlike near the ductus seminalis and has a lobelike extension to about two-thirds the distance to the ostium.

Our sample, along with distributional data reported by Miller et al. (2000), indicates that this species ranges over eastern North America from Newfoundland, Quebec, and Ontario south to North Carolina, Alabama, and Missouri. We also examined several specimens from Seattle, Washington, one of which (a male in the EME) was collected as a larva on *Rosa* L. (rose). The larva feeds in buds, twigs, and webbed leaves of *Rosa* (Razowski 2003) and can be a serious pest of cultivated roses.

# Literature Cited


Baixeras, J. 2002. An overview of the genus level taxonomic problems surrounding *Argyroploce* Hübner (Lepidoptera: Tortricidae), with description of a new species. Annals of the Entomological Society of America 95: 422–431.

Barnes, W. and J. McDunnough. 1917. Check list of the Lepidoptera of Boreal America. Decatur, Illinois. 200 pp.

Blanchard, A. 1979. New status for *Epiblema minutana* (Kearfott) and new species of *Epiblema* Hübner and *Sonia* Heinrich (Tortricidae). Journal of the Lepidopterists' Society 33: 179–188.

Blanchard, A. 1984. *Epiblema luctuosana* A. Blanchard, a homonym, is changed to *Epiblema luctuosissima*, new name. Journal of the Lepidopterists' Society 38: 245.

Blanchard, A. and E. Knudson. 1983. New North American species of Eucosmini (Lepidoptera: Tortricidae). Proceedings of the Entomological Society of Washington 85: 845–852.

Bottimer, I. J. 1926. Notes on some Lepidoptera from eastern Texas. Journal of Agricultural Research 33: 797–819.

Bradley, J. D., W. G. Tremewan and A. Smith. 1979. British Tortricoid Moths, Tortricidae: Olethreutinae. The Ray Society, London. 336 pp.

Brown, J. W. 2005. World catalogue of insects. Volume 5: Tortricidae (Lepidoptera). Apollo books, Stenstrup, Denmark. 741 pp.

Brown, J. W. and J. Lewis. 2000. Catalogue of the type specimens of Tortricidae (Lepidoptera) in the collection of the National Museum of Natural History, Smithsonian Institution, Washington, D.C. Proceedings of the Entomological Society of Washington 102: 1014–1069.

Brown, R. L. 1973. Phylogenetic systematics: Its application to the genus *Epiblema* (Lepidoptera). Unpublished Master's thesis, University of Arkansas. 179 pp.

Brown, R. L. 1979. Nomenclatorial changes in Eucosmini (Tortricidae). Journal of the Lepidopterists' Society 33: 21–28.

Brown, R. L. 1980. A revision of the genus *Epinotia* (Hübner) (Tortricidae: Eucosmini), Part 1. The North-American species of the *stroemiana* group. Ph. D. Thesis, Cornell University, 460 pp.

Brown, R. L. 1983. Taxonomic and morphological investigations of Olethreutinae: *Rhopobota*, *Griselda*, *Melissopus*, and *Cydia* (Lepidoptera: Tortricidae). Entomography 2: 97–120.

Brown, R. L. 1987. Tortricidae (Tortricoidea), pp. 419-433. *In*: Stehr, F. W. (ed.), Immature Insects. Kendall/Hunt, Dubuque, Iowa.

Brown, R. L., J. F. G. Clarke and D. H. Habeck. 1983. New host records for Olethreutineae (Tortricidae). Journal of the Lepidopterists' Society 37: 224–227.

Brown, R. L. and J. A. Powell. 1991. Description of a new species of *Epiblema* (Lepidoptera: Tortricidae: Olethreutinae) from coastal redwood forests in California with an analysis of the forewing pattern. Pan-Pacific Entomologist 67: 104–114.

Busck, A. 1907. New genera and species of American microlepidoptera. Journal of the New York Entomological Society 15: 134–140.

Ĉapek, M. 1971. The possibility of biological control of imported weeds of the genus *Solidago* L. in Europe. Acta Instituti Forestalis Zvolenens. pp. 429–441.

Clarke, J. F. G. 1952. Two new species of Olethreutidae from California (Lepidoptera). Bulletin of the Southern California Academy of Sciences 51: 60–62.

Clarke, J. F. G. 1953. New species of Olethreutidae from Illinois (Lepidoptera). Journal of the Washington Academy of Sciences 43:226–231.

Clemens, B. 1860. Contributions to American Lepidopterology - No. 6. Proceedings of the Academy of Natural Sciences (Philadelphia) 1860: 345–362.

Clemens, B. 1865. North American micro-lepidoptera. Proceedings of the Entomological Society of Philadelphia 5: 133–147.

Darlington, E. P. 1947. Notes on certain types of Lepidoptera described by Brackenridge Clemens. Transactions of the American Entomological Society 73: 85–104.

Decker, G. C. 1932. Biology of the bidens borer, *Epiblema otiosana* (Clemens) (Lepidoptera, Olethretidae). Journal of the New York Entomological Society 30: 503–509.

Doubleday, H. 1850. Descriptions of lepidopterous insects of the genera *Hypenodes, Eupithecia*, and *Spilonota* recently discovered in Britain. Zoologist 8 (Append.): cv–cvi.

Duponchel P.-A.-J. 1834. *In*: Godart, J. B. and P.-A.-J. Duponchel, Histoire naturelle des Lépidoptères ou Papillons de France. Nocturnes. Tome sixième. Méquignon-Marvis, Libraire-Editeur, Paris. 9, 627 pp.

Fernald, C. H. 1882. A synonymical catalogue of the described Tortricidae of North America north of Mexico. Transactions of the American Entomological Society 10: 1–64.

Fernald, C. H. 1901. New Pyralidae and Tortricidae from Palm Beach, Florida. New York Entomological Society 9: 51.

Fernald, C. H. 1903 [1902]. Family Tortricidae, pp. 448–489. *In*: Dyar, H. G. (ed.), A list of North American Lepidoptera. Bulletin of the United States National Museum 52.

Fernald, C. H. 1908. The genera of Tortricidae and their types. Carpenter & Morehouse, Amherst, Massachusetts. 68 pp.

Gilligan, T. M., D. J. Wright and L. D. Gibson. 2008. Olethreutine moths of the midwestern United States, an identification guide. Ohio Biological Survey Bulletin New Series, Vol. 16. 334 pp.

Gilligan, T. M. and D. J. Wright. 2013. Revised world catalogue of *Eucopina*, *Eucosma*, *Pelochrista*, and *Phaneta* (Lepidoptera: Tortricidae: Eucosmini). Zootaxa 3746: 302–337.

Gilligan, T. M., D. J. Wright, J. Munz, K. Yakobson and M. P. Simmons. 2014. Molecular phylogeny and revised classification of *Eucosma* Hübner and related genera (Lepidoptera: Tortricidae: Eucosmini). Systematic Entomology 39: 49–67.

Gilligan, T. M., Baixeras, J., and Brown, J. W. 2018. T@RTS: Online World Catalogue of the Tortricidae (Ver. 4.0). Available from: http://www.tortricid.net/catalogue.asp.

Gilligan, T. M., J. W. Brown, and J. Baixeras. 2020a. Immigrant Tortricidae: Holarctic versus introduced species in North America. Insects 11, 594. doi:10.3390/insects11090594.

Gilligan, T. M., D. J. Wright, R. L. Brown, B. A. Augustinus, and U. Schaffner. 2020b. Taxonomic issues related to biological control prospects for the ragweed borer, *Epiblema strenuana* (Lepidoptera: Tortricidae). Zootaxa 4729: 347–358.

Godfrey, G. L., E. D. Cashatt and M. O. Glenn. 1987. Microlepidoptera from the Sandy Creek and Illinois River region: an annotated checklist of the suborders Dacnonypha, Monotrysia, and Ditrysia (in part) (Insecta). Illinois Natural History Survey Special Publication 7. 44 pp.

Hancock, E. F., K. P. Bland and J. Razowski. 2014. *In*: Bland, K. P. (ed.), The Moths and Butterflies of Great Britain and Ireland, Volume 5, Part 2. Tortricidae, Olethreutinae. Brill, Leiden, Boston. 377 pp.

Heinemann, H. 1863. Die Schmetterlinge Deutschlands und der Schweiz. Zweite Abtheilung. Kleinschmetterlinge. Band I. Heft I. Die Wickler. Braunschweig (C. A. Schwetschke und Sohn). 248 pp.

Heinrich, C. 1921. Some Lepidoptera likely to be confused with the pink bollworm. Journal of Agricultural Research 20: 807–836, Plates 93–109.

Heinrich, C. 1923. Revision of the North American moths of the subfamily Eucosminae of the family Olethreutidae. Bulletin of the United States National Museum 123: 1–128.

Horak, M. 1984. Assessment of taxonomically significant structures in Tortricinae (Lep., Tortricidae). Mitteilungen der Schweizerischen Entomologischen Gesellschaft 57: 3–64.

Horak, M. 1998. The Tortricoidea, pp. 199-215. *In*: Kristensen, N. P. (ed.), Handbook of Zoology, Lepidoptera, Moths and Butterflies, Vol 1: Evolution, Systematics, and Biogeography. Walter de Gruyter, Berlin, New York.

Horak, M. 2008. Olethreutine moths of Australia (Lepidoptera: Tortricidae). Monographs on Australian Lepidoptera, Volume 10, CSIRO Publishing, Collingwood, Victoria, Australia. 522 pp.

Hübner, J. [1825] 1816. Verzeichniss bekannter Schmettlinge [sic]. Augsburg [no publisher listed]. 431 pp.

Kearfott, W. D. 1905. Descriptions of new species of tortricid moths from North Carolina, with notes. Proceedings of the United States National Museum 28: 349–364.

Kearfott, W. D. 1907. New North American Tortricidae. Transactions of the American Entomological Society 33: 1–98.

Kearfott, W. D. 1908. New North American Tortricidae and Tineina. Journal of the New York Entomological Society 16: 167–188.

Kearfott, W. D. 1910. Family Tortricidae, pp. 537–552. *In* Smith, J. B. (ed.), Annual report of the New Jersey State Museum including a report of the insects of New Jersey, 1909. Trenton, New Jersey.

Kellicott, D. S. 1882. Is *Paedisca scudderiana* a gall-maker? The Canadian Entomologist 14: 161–163.

Kimball, C. P. 1965. Arthropods of Florida and neighboring land areas. Vol. 1. Florida Department of Agriculture. 363 pp.

Klots, A. B. 1942. Type material of North American Microlepidoptera other than Aegeriidae in the American Museum of Natural History. Bulletin of the American Museum of Natural History 79: 391–424.

Knodel, J. J., L. D. Charlet and J. Gavloski. 2010. Integrated pest management of sunflower insect pests in the northern Great Plains. North Dakota State University Extension Service, E-1457. 20 pp.

Kuznetsov, V. I. 1987. Family Tortricidae, *In*: Medvedev, G. S. (ed.), Keys to the insects of the European part of the USSR, Part 4, Lepidoptera, Part 1, Nauka Publishers, Leningrad Division, 1978. English translation published for The United States Department of Agriculture and the National Science Foundation, Washington, D.C. by Amerind Publishing Co., Pvt. Ltd., New Delhi, pp. 279–991.

Looney, C. and M. Murray. 2022. Collections of *Epiblema rudei* Powell, 1975, (Tortricidae) in Utah. Journal of the Lepidopterists' Society 76: 152–154.

MacKay, M. R. 1959. Larvae of the North American Olethreutidae (Lepidoptera). The Canadian Entomologist 91, Supplement 10. 338 pp.

McClay, A. S. 1987. Observations on the biology and host specificity of *Epiblema strenuana* [Lepidoptera, Tortricidae], a potential biocontrol agent for *Parthenium hysterophorus* [Compositae]. Entomophaga 32, 23–34.

McDunnough, J. H. 1925a. New Canadian Lepidoptera with notes. The Canadian Entomologist 57: 11–23.

McDunnough, J. H. 1925b. New Canadian Eucosminae (Lepidoptera). The Canadian Entomologist 57: 115–116.

McDunnough, J. H. 1935. New Canadian eucosmids with notes (Lepidoptera). The Canadian Entomologist 67: 140–149.

McDunnough, J. H. 1939. Check List of the Lepidoptera of Canada and the United States of America. Part II. Microlepidoptera. Memoirs of the Southern California Academy of Sciences 2: 3–171.

McDunnough, J. H. 1959. On some changes in nomenclature of microlepidoptera, with description of a new species. American Museum Novitates 1954, New York, American Museum of Natural History, 9 pp.

Meyrick E. 1895. A handbook of British Lepidoptera. MacMillan & Co., London, New York. 843 pp.

Meyrick, E. 1912. On some impossible specific names in micro-lepidoptera. Entomologist's Monthly Magazine, Series 2, 48: 32–36.

Meyrick, E. 1920. Exotic microlepidoptera. 2(11). 321–352.

Miller, S. E. and R. W. Hodges. 1990. Primary types of microlepidoptera in the Museum of Comparative Zoology (with a discursion on V. T. Chambers' work). Bulletin of the Museum of Comparative Zoology 152: 45–87.

Miller, W. E. 1973. Clemens types of Olethreutinae (Lepidoptera: Tortricidae). Transactions of the American Entomological Society 99: 205–234.

Miller, W. E. 1976. Biology and taxonomy of three gall forming species of *Epiblema* (Olethreutidae). Journal of the Lepidopterists' Society 30: 50–58.

Miller, W. E. 1977. Wing measure as a size index in Lepidoptera: the family Olethreutidae. Annals of the Entomological Society of America 70: 235–256.

Miller, W. E. 1985. Nearctic *Epiblema*: a new synonymy, a revised identity, and two new species (Lepidoptera: Tortricidae). The Great Lakes Entomologist 18: 33–38.

Miller, W. E. 1987. Guide to the olethreutine moths of Midland North America (Tortricidae). United States Department of Agriculture. Forest Service Agriculture Handbook 660. 104 pp.

Miller, W. E. 1990. A new species of *Sonia* from eastern North America (Tortricidae). Journal of the Lepidopterists' Society 44: 88–90.

Miller, W. E. 1995. Beringian Tortricidae: a new synonymy in *Epiblema* and a recount of Holarctic species. Journal of the Lepidopterists' Society 49: 250–252.

Miller, W. E. and M. G. Pogue. 1984. Ragweed borer (Lepidoptera: Tortricidae: Eucosmini): taxonomic implications of an allometric analysis of adult characters. Annals of the Entomological Society of America 77: 227–231.

Miller, W. E., R. L. Brown, and K. R. Tuck. 2000. Taxonomic clarification of *Notocelia rosaecolana* (Doubleday) and *N. trimaculana* (Haworth) (Tortricidae). Journal of the Lepidopterists' Society 54: 111–118.

Mosher, E. 1916. A classification of the Lepidoptera based on characters of the pupa. Illinois Natural History Survey Bulletin 12: 17–159.

Muise A. M. and K. B. Storey. 2001. Regulation of hexokinase in a freeze avoiding insect: role in the winter production of glycerol. Archives of Insect Biochemistry and Physiology 47: 29–34.

Mutuura, A. 1980. *Epiblema* (*Notocelia*) *cynosbatella* (Lepidoptera: Tortricidae), a recently introduced Old World species potentially injurious to *Rosaceae*. The Canadian Entomologist 112: 511–514.

Obraztsov, N. S. 1946. Versuch einer systematischen Übersicht der europäischen Eucosmini-Gattungen (Lepidoptera, Tortricidae). Zeitschrift der wiener entomologischen Gesellschaft 30: 20–46.

Obraztsov, N. S. 1964. *Griselda radicana* Heinrich, 1923 (Insecta, Lepidoptera): Proposed validation under the plenary powers. Z.N. (S.) 1612. Bulletin of Zoological Nomenclature 21: 144–145.

Obraztsov, N. S. 1965a. Die Gattungen der Palaearktischen Tortricidae. II. Die Unterfamilie Olethreutinae. 6. Teil. Tijdschrift voor Entomologie 108: 365–387.

Obraztsov, N. S. 1965b. On *Paedisca radicana* and the type species of the genus *Griselda* (Lepidoptera, Tortricidae, Olethreutinae). American Museum Novitates No. 2213: 1–16.

Obraztsov, N. S. 1967. Die Gattungen der Palaearktischen Tortricidae. II. Die Unterfamilie Olethreutinae. 7 Teil. Tribus Eucosmini (Heinri. 1923) – fortsetzung. Tijdschrift voor Entomologie 110: 65–88.

Obraztsov, N. S. 1968. Die Gattungen der Palaearktischen Tortricidae. II. Die Unterfamilie Olethreutinae. 8 Teil und Schluss. Tijdschrift voor Entomologie 111: 1–48.

Passoa, S. 2008. Part III: Immature stages, pp. 295–314. *In*: Gilligan, T. M. et al., Olethreutine moths of the midwestern United States, an identification guide. Ohio Biological Survey Bulletin New Series, Vol. 16.

Peterson, A. 1965. Some eggs of moths among the Olethreutidae and Tortricidae (Lepidoptera). Florida Entomologist 48: 1–8.

Pierce, F. N. and J. W. Metcalfe. 1922. The genitalia of the group Tortricidae of the Lepidoptera of the British Islands. Oundle, Liverpool, England. 101 pp.

Powell, J. A. 1964. A review of *Griselda*, with descriptions of a related new genus and two species (Lepidoptera: Tortricidae). Pan-Pacific Entomologist 40: 85–97.

Powell, J. A. 1975. Biological records and descriptions of some little known *Epiblema* in the southwestern United States. The Pan-Pacific Entomologist 51: 99–112.

Powell, J. A. 1983. Tortricidae, pp. 31-41. *In*: Hodges, R. W. (ed.), Check List of the Lepidoptera of America North of Mexico. E. W. Classey & Wedge Entomological Research Foundation, London.

Powell, J. A. and P. A. Opler. 2006. Larval host plant records of *Asteraceae* root-feeding Eucosmini in California and adjacent states (Tortricidae). Journal of the Lepidopterists' Society 60: 189–193.

Powell, J. A. and P. A. Opler. 2009. Moths of Western North America. University of California Press. Berkeley, Los Angeles, London. 369 pp.

Putman, W. L. 1942. Hostplants and parasites of some lepidopterous larvae. The Canadian Entomologist 74: 219–224.

Razowski, J. 1989. The genera of Tortricidae (Lepidoptera). Part II: Palaearctic Olethreutinae. Acta Zoologica Cracoviensia 32: 107–328.

Razowski, J. 2003. The Tortricidae of Europe, Vol. 2, Olethreutinae. František Slamka. Bratislava, Slovakia. 301 pp.

Rebel, H. 1901. Catalog der Lepidopteren des palaearctischen Faunengebietes. II. Theil: Famil. Pyralidae-Micropterigidae. R. Friedländer & Sohn, Berlin. 368 pp.

Riley, C. V. 1881. Descriptions of some new Tortricidae (leaf-rollers). Transactions of the St. Louis Academy of Sciences 4: 316–324.

Riley, C. V. 1888. Proceedings of the Entomological Society of Washington 1: 33.

Satterthwait, A. F. 1948. Important sunflower insects and their insect enemies. Journal of Economic Entomology 41: 725–731.

Stegmaier, C. E. Jr. 1971. Lepidoptera, Diptera, and Hymenoptera associated with *Ambrosia artemisiifolia* (Compositae) in Florida. The Florida Entomologist 54: 259–272.

Stephens, J. F. 1834. Illustrations of British entomology or, a synopsis of indigenous insects: containing their generic and specific distinctions; with an account of their metamorphoses, times of appearance, localities, food, and economy, as far as practicable. Insecta Haustellata, Vol. IV, London. 436 pp.

Stephens, J. F. 1852. List of the specimens of British animals in the collection of the British Museum. Part X.–Lepidoptera (continued). The Trustees of the British Museum, London. 120 pp.

Swatschek, B. 1958. Die larvalsystematik der Wickler (Tortricidae und Carposinidae). Abhandlungen Zur Larvalsystematik der Insekten nr. 3. Akademie-Verlag, Berlin. 269 pp.

Thompson, R. W. and O. C. D. Guelph. 1928. The golden-glow borer (*Epiblema carolinana*, Walsingham). Annual Report of the Entomological Society of Ontario (1927) 58: 73–75.

Treitschke F. 1829. Die Schmetterlinge von Europa. (Fortsetzung des Ochsenheimer'schen Werkes). Band 7. Gerhard Fleischer, Leipzig. 252 pp.

Walker, F. 1863. List of the specimens of Lepidopterous insects in the Collection of the British Museum. Tortricites & Tineites. XXVIII. British Museum (Natural History), London, 287–561.

Walsingham, Lord T. de Grey. 1879. Illustrations of typical specimens of Lepidoptera Heterocera in the collection of The British Museum. Part IV. North American Tortricidae. Department of Zoology, British Museum, London. 84 pp., 17 pls.

Walsingham, Lord T. de Grey. 1884. North American Tortricidae. The Transactions of the Entomological Society of London, Part I: 121–148.

Walsingham, Lord T. de Grey. 1895. New Species of North American Tortricidae. The Transactions of the Entomological Society of London, Part IV: 495–520.

Walsingham, Lord T. de Grey. 1914 [1909–1915]. Insecta. Lepidoptera-Heterocera, Vol. IV. Tineina, Pterophorina, Orneodina, and Pyralidina and Hepialina (part), pp. 1–482. *In*: Godman, F. D. and O. Salvin (eds.), Biologia Centrali-Americana. R. H. Porter, London.

Westwood, 1840. An introduction to the modern classification of insects; Founded on the natural habits and corresponding organization of the different families, Vol. II (Synopsis Genera Br. Insects): 108.

Williamson S. M., W. A. Ingelson-Filpula, H. Hadj-Moussa and K. B. Storey. 2021. Epigenetic underpinnings of freeze avoidance in the goldenrod gall moth, *Epiblema scudderiana*. Journal of Insect Physiology 134: 104298. doi: 10.1016/j.jinsphys.2021.104298.

Wocke, M. 1871. II. Microlepidoptera. *In*: Staudinger, O. and M. Wocke, Catalog der Lepidopteren des europäischen Faunengebiets. Dresden. 426 pp.

Wright, D. J. 2002. A new species of *Epiblema* previously confused with *E. tripartitana* (Zeller) and *E. infelix* Heinrich (Tortricidae). Journal of the Lepidopterists' Society 56: 277–285.

Wright, D. J. and C. V. Covell Jr. 2003. A new species of *Epiblema* (Tortricidae) formerly misidentified as *E. walsinghami* (Kearfott) and *E. infelix* Heinrich. Journal of the Lepidopterists' Society 57: 230–234.

Wright, D. J. and T. M. Gilligan. 2015. *Eucosma* Hübner of the contiguous United States and Canada (Lepidoptera: Tortricidae: Eucosmini). Wedge Entomological Research Foundation, Alamogordo, New Mexico. 256 pp.

Wright, D. J. and T. M. Gilligan. 2017. *Pelochrista* Lederer of the contiguous United States and Canada (Lepidoptera: Tortricidae: Eucosmini). Wedge Entomological Research Foundation, Alamogordo, New Mexico. 375 pp.

Wright, D. J. and T. M. Gilligan. 2021. Three Nearctic Eucosmini (Tortricidae: Olethreutinae) with discordant genus level characteristics: two new species and a new combination. Journal of the Lepidopterists' Society 75: 128–134.

Zeller, P. C. 1875. Beiträge zur Kenntniss der nordamericanischen Nachtfalter, besonders der Microlepidopteren. Verhandlungen der Zoologisch-Botanischen Gesellschaft in Wien 25: 205–360.

Zlatkov, B. 2016. On the vesica of Eucosmini and Grapholitini (Insecta: Lepidoptera: Tortricidae). Zootaxa 4168: 297–312.

# PLATES

Plates A–JJ

# Adult Photos

# PLATE A

1. ***Epiblema strenuana*** (Walker). FWL: 4.0–9.0 mm (mean = 7.1).
   a. ♂. LTP. North America, Carter Collection, BMNH.
   b. ♂. North America, Carter Collection, BMNH. [LTP. *G. exvagana*]
   c. ♂. Kentucky, Gallatin County, Markland Dam, 27 May 1990, L. D. Gibson, LDG.
   d. ♀. Kentucky, Gallatin County, Markland Dam, 21 June 1990, L. D. Gibson, LDG.
   e. ♂. Ohio, Hamilton County, Shawnee Lookout Park, 26 August 1988, D. J. Wright, USNM.
   f. ♀. Ohio, Wyandot County, Killdeer Plains WLA, 16 June 1995, D. J. Wright, USNM.
   g. ♀. Alabama, Baldwin County, Weeks Bay NERS, 22 June 2008, D. J. Wright, USNM.
   h. ♀. Alabama, Baldwin County, Weeks Bay NERS, 22 June 2008, D. J. Wright, USNM.
   i. ♂. Louisiana, Bossier Parish, Barksdale AFB, 9 May 1996, D. M. Pollock, MEM.
   j. ♀. Mississippi, Lowndes County, Black Belt Prairie, 10 June 1993, D. M. Pollock, MEM.
   k. ♂. Texas, Tarrant County, Benbrook, 20 May 1991, J. A. Powell, EME.
   l. ♀. Florida, Brevard County, Cocoa Beach, 17 July 1967, R. H. Leuschner, EME.
   m. ♀. California, Los Angeles County, San Clemente Island, 11 April 1990, J. Powell and D. Faulkner, EME.
   n. ♀. California, Kern County, Palm Avenue and Blankenship at Poso Creek, 31 May 2012, K. Richers and J. Forrest, KR.
   o. ♂. California, San Diego County, La Puerta Valley, July 1911, G. H. Field, EME.

1a
1b
1c
1d
1e
1f
1g
1h
1i
1j
1k
1l
1m
1n
1o

# PLATE B

**2.** ***Epiblema minutana*** (Kearfott). FWL: 4.3–7.9 mm (mean = 6.0).

a. ♂. LTP. New Jersey, Essex County, Montclair, July 1908, W. D. Kearfott, USNM.
b. ♂. Ohio, Erie County, Resthaven WLA, 24 May 1991, D. J. Wright, USNM.
c. ♂. Ohio, Wyandot County, Killdeer Plains, 16 August 2006, D. J. Wright, USNM.
d. ♀. Kentucky, McCracken County, Paducah Zip Track, 12 August 2008, D. J. Wright, USNM.
e. ♂. Mississippi, Oktibbeha County, Osborn Prairie, 26 August 2003, D. J. Wright, USNM.
f. ♂. Florida, Volusia County, Blue Springs State Park, 12 June 1986, Powell and Heppner, EME.
g. ♂. California, Contra Costa County, Richmond Field Station, 18 March 1992, J. A. Powell, EME.
h. ♀. California, Contra Costa County, Pointe Molate, 12 December 1989, J. A. Powell, EME. [reared from *Ambrosia chamissonis*]

**Pale Phenotype**. FWL: 5.1–8.7 mm (mean = 6.8).

i. ♀. Alabama, Baldwin County, Bon Secour NWR, 8–9 August 1994, R. Brown and D. Pollock, MEM.
j. ♂. California, Los Angeles County, El Segundo Dunes, 14 June 1988, R. Leuschner, EME.
k. ♀. California, Ventura County, San Nicolas Island, 7 May 1978, J. A. Powell, EME.
l. ♂. California, Los Angeles County, San Clemente Island, 23 March 1972, J. A. Powell, EME.

**Mixed Phenotypes.**

m. ♂. Rhode Island, Washington County, Block Island, 6 August 2020, A. S. Hunt, USNM.
n. ♀. Rhode Island, Washington County, Block Island, 6 August 2020, A. S. Hunt, USNM.
o. ♂. California, San Bernardino County, 6 mi W of Rice, 1 April 1978, Powell and Dietz, EME.

2a
2b
2c
2d
2e
2f
2g
2h
2i
2j
2k
2l
2m
2n
2o

# PLATE C

**3.** ***Epiblema luctuosissima*** Blanchard. FWL: 4.7–6.8 mm (mean = 6.1).

a. ♂. Colorado, Baca County, Picture Canyon Picnic Area, 6 August 2005, D. J. Wright, USNM.
b. ♀. Colorado, Baca County, Picture Canyon Picnic Area, 6 August 2005, D. J. Wright, USNM.
c. ♂. Kansas, Morton County, 7.5 mi N of Elkhart, 2 August 1999, D. J. Wright, USNM.
d. ♀. Colorado, Yuma County, Bonny Reservoir SRA, 5 August 1996, D. J. Wright, USNM.
e. ♂. Texas, Hemphill County, Gene Howe WMA, 31 July 2005, L. D. Gibson, LDG.
f. ♀. Texas, Nueces County, North Padre Island, 9 June 1978, A. and M. E. Blanchard, USNM.
g. ♀. Texas, Padre Island National Seashore, 18 May 1979, A. and M. E. Blanchard, USNM.
h. ♂. Louisiana, Bossier Parish, Barksdale AFB, 3 April 1996, D. M. Pollock, MEM.

**4.** ***Epiblema adusta*** Wright and Gilligan. FWL: 4.4–6.3 mm (mean = 5.3).

a. ♀. HTP. Mississippi, Jackson County, Grand Bay Savanna, 30 April 2004, Brown, Lee and Martinez, USNM.
b. ♂. PTP. Mississippi, Jackson County, Grand Bay Savanna, 30 April 2004, Brown, Lee and Martinez, MEM.
c. ♀. PTP. Alabama, Baldwin County, Plash Island, 16–19 June 1984, R. and B. Brown, MEM.
d. ♀. PTP. Alabama, Baldwin County, Plash Island, 16–19 June 1984, R. and B. Brown, MEM.

**5.** **Morphospecies**. FWL: 5.0–6.3 mm (mean = 5.6).

a. ♂. Mexico, Baja California, 1 mi E Santo Domingo, 18 March 1972, J. A. Powell, EME.

**6.** ***Epiblema ochraceana*** Fernald. FWL: 5.5 mm.

a. ♂. HTP. Florida, Palm Beach County, Palm Beach, USNM.
b. ♂. Florida, Palm Beach County, Palm Beach, USNM.

3a
3b
3c
3d
3e
3f
3g
3h
4a
4b
4c
4d
5
6a
6b

# PLATE D

**7.** ***Epiblema sosana*** (Kearfott). FWL: 5.5–9.1 mm (mean = 7.3).

a. ♂. CoTP. Utah, Tooele County, Stockton, 1 June 1904, T. Spalding, USNM.
b. ♀. CoTP. Utah, Tooele County, Stockton, 1 August 1904, T. Spalding, USNM.
c. ♂. California, San Bernardino County, Cucamonga, 16 January 1969, R. Goeden and D. Ricker, USNM. [reared on *Ambrosia acanthicarpa*]
d. ♀. California, San Bernardino County, Cucamonga, 16 January 1969, R. Goeden and D. Ricker, USNM. [reared on *Ambrosia acanthicarpa*]
e. ♂. California, Imperial County, 17 mi NW of Glamis, 27 June 1978, D. and J. Powell, EME.
f. ♂. Arizona, Yuma County, 1.27 mi SE I-8, 8 February 2014, S. Lee, ASUHIC.

**8.** ***Epiblema cimarronensis*** Wright and Gilligan. FWL: 5.2–8.4 mm (mean = 7.0).

a. ♀. HTP. Kansas, Morton County, 7.5 mi N of Elkhart, 2 August 1999, D. J. Wright, USNM.
b. ♀. PTP. Kansas, Morton County, 7.5 mi N of Elkhart, 25 August 2000, D. J. Wright, USNM.
c. ♂. PTP. Texas, Coleman County, Santa Anna, 19 May 1985, E. C. Knudson, USNM.
d. ♀. PTP. Texas, Coleman County, Santa Anna, 19 May 1985, E. C. Knudson, MGCL.
e. ♀. PTP. Texas, Cottle County, Paducah, 17 April 1968, A. and M. E. Blanchard, USNM.
f. ♀. PTP. Texas, Briscoe County, Caprock Canyon State Park, 29–30 August 1994, E. C. Knudson, MGCL.
g. ♂. PTP. California, San Diego County, Banner, 10 August 1970, R. Goeden and D. Ricker, USNM. [reared on *Ambrosia confertiflora*]
h. ♀. PTP. California, San Diego County, San Diego, 16–23 August, USNM.
i. ♂. PTP. California, Riverside County, Banning, 24 September 1970, Goeden and D. Ricker, USNM. [reared on *Ambrosia confertiflora*]

7a
7b
7c
7d
7e
7f

8a
8b
8c
8d
8e
8f
8g
8h
8i

# PLATE E

**9.** ***Epiblema wielgusi*** Wright and Gilligan. FWL: 7.5–9.2 mm (mean = 8.5).

a. ♀. HTP. Arizona, Maricopa County, Thunderbird Regional Mountain Park, 6 March 1977, R. S. Wielgus, USNM.

b. ♂. PTP. Arizona, Maricopa County, Thunderbird Regional Mountain Park, 20 April 1975, R. and J. Wielgus, ASUHIC.

c. ♂. PTP. Arizona, Maricopa County, Thunderbird Regional Mountain Park, 20 April 1975, R. and J. Wielgus, ASUHIC.

**10.** ***Epiblema carolinana*** (Walsingham). FWL: 6.9–10.9 mm (mean = 9.2).

a. ♀. Iowa, Pocahontas County, Kalsow Prairie, 18 July 2003, D. J. Wright, USNM.

b. ♂. Iowa, Pocahontas County, Kalsow Prairie, 18 July 2003, D. J. Wright, USNM.

c. ♀. Iowa, Howard County, Hayden Prairie, 18 July 1998, D. J. Wright, USNM.

d. ♂. North Carolina, Macon County, Highlands, 17 July 1958, J. G. Franclemont, USNM.

e. ♂. Kentucky, Menifee County, Red River Gorge, 29 June 1984, L. D. Gibson, LDG.

f. ♂. Kentucky, Bullitt County, Route 480, 6.9 mi E Route 61, 8 July 1988, L. D. Gibson, LDG.

**11.** ***Epiblema arizonana*** Powell. FWL: 8.9–11.9 mm (mean = 11.0).

a. ♂. PTP. Arizona, Coconino County, Oak Creek Canyon, 28 June 1936, G. P. Englehardt, USNM. [reared from *Rudbeckia laciniata*]

b. ♀. PTP. Arizona, Coconino County, Oak Creek Canyon, G. P. Englehardt, USNM. [reared from *Rudbeckia laciniata*]

c. ♀. PTP. Arizona, Coconino County, Oak Creek Canyon, 26 June 1936, G. P. Englehardt, USNM. [reared from *Rudbeckia laciniata*]

d. ♀. PTP. Arizona, Coconino County, Oak Creek Canyon, 26 June 1936, G. P. Englehardt, USNM. [reared from *Rudbeckia laciniata*]

e. ♀. PTP. Arizona, Coconino County, Oak Creek Canyon, 2 July 1936, G. P. Englehardt, USNM. [reared from *Rudbeckia laciniata*]

f. ♀. Colorado, Larimer County, Fort Collins, Fossil Creek, P. A. Opler, 2 August 1988, EME.

9a
9b
9c
10a
10b
10c
10d
10e
10f
11a
11b
11c
11d
11e
11f

# PLATE F

**12.** ***Epiblema tripartitana*** (Zeller). FWL: 3.7–9.8 mm (mean = 6.3).

a. ♀. Texas, Shelby County, 3 mi E of Liberty, 8 April 1924, I. J. Bottimer, USNM. [reared from *Rudbeckia maxima*]

b. ♂. Illinois, Putnam County, 19 June 1967, M. O. Glenn, USNM.

c. ♂. New Mexico, Guadalupe County, 5 mi S of Santa Rosa, 27 May 1995, R. Watkins, USNM.

d. ♀. Kansas, Morton County, Cimarron National Grassland, 2 August 1999, D. J. Wright, USNM.

e. ♀. Kansas, Morton County, Cimarron National Grassland, 26 August 2000, D. J. Wright, USNM.

f. ♀. Kansas, Morton County, Cimarron National Grassland, 26 August 2000, D. J. Wright, USNM.

g. ♂. New York, Thompkins County, Ithaca, 14 June 1957, J. G. Franclemont, USNM.

h. ♂. Ohio, Erie County, Resthaven WLA, 13 July 1991, L. D. Gibson, LDG.

i. ♂. Mississippi, Oktibbeha County, Starkville, R. L. Brown, 5 July 1980, MEM.

**13.** ***Epiblema glenni*** Wright. FWL: 5.0–8.4 mm (mean = 6.1).

a. ♀. PTP. Ohio, Adams County, 1 mi SE of Lynx, 12 August 1998, D. J. Wright, USNM.

b. ♂. PTP. Illinois, Putnam County, 8 July 1965, M. O. Glenn, USNM.

c. ♂. PTP. Kentucky, Laurel County, Marsh Branch Road, 25 July 1997, L. D. Gibson, LDG.

d. ♀. PTP. Ohio, Adams County, 1 mi SE of Lynx, 12 August 1998, D. J. Wright, USNM.

e. ♂. PTP. Ohio, Adams County, 1 mi SE of Lynx, 12 August 1998, D. J. Wright, USNM.

f. ♂. Ohio, Adams County, Lynx Prairie, 10 September 1988, L. D. Gibson, LDG.

12a
12b
12c
12d
12e
12f
12g
12h
12i

13a
13b
13c
13d
13e
13f

# PLATE G

**14.** ***Epiblema scudderiana*** (Clemens). FWL: 5.2–11.3 mm (mean = 8.2).

a. ♂. Iowa, Pocahontas County, Kalsow Prairie, 22 June 2000, D. J. Wright, USNM.
b. ♂. Ohio, Wyandot County, Kildeer Plains, 8 June 1991, L. D. Gibson, LDG.
c. ♂. New Hampshire, Coos County, Whitefield, 8 June 1996, W. Kiel, MGCL.
d. ♂. Texas, Bexar County, D. M. Pollock, 6 May 1993, MEM.
e. ♀. Kansas, Morton County, Cimarron National Grassland, 2 August 1999, D. J. Wright, USNM.
f. ♀. Canada, Ontario, Central Experimental Farm, 24-25 May 1986, J. A. Powell, EME. [reared from *Solidago*]
g. ♂. Oklahoma, Roger Mills County, Black Kettle National Grassland, 22 August 2015, J. Vargo, USNM.
h. ♂. Florida, Manatee County, Oneco, 24 March 1957, J. G. Franclemont, USNM.
i. ♀. Florida, Manatee County, Oneco, 23 March 1954, J. G. Franclemont, USNM.

**15.** ***Epiblema benignata*** McDunnough. FWL: 5.7–9.0 mm (mean = 7.6).

a. ♂. HTP. Canada, Manitoba, Aweme, 13 June 1923, N. Criddle, CNC.
b. ♂. Iowa, Pocahontas County, Kalsow Prairie, 22 June 2000, D. J. Wright, USNM.
c. ♀. California, Siskiyou County, Shasta Retreat, 16–23 June, USNM.
d. ♂. Washington, Whitman County, Snake River, 1 March 1931, J. F. G. Clarke, USNM.
e. ♂. Washington, Whitman County, Wilma, 6 April 1934, J. F. G. Clarke, USNM. [reared from *Artemisia vulgaris*]
f. ♀. Washington, Whitman County, Snake River, 3 March 1931, J. F. G. Clarke, USNM. [reared from *Artemisia dracunculus*]

14a
14b
14c
14d
14e
14f
14g
14h
14i
15a
15b
15c
15d
15e
15f

# PLATE H

**16.** ***Epiblema infelix*** Heinrich. FWL: 5.3–10.2 mm (mean = 7.9).
- a. ♂. Kentucky, Laurel County, Bolton Branch, 18 May 1996, D. J. Wright, USNM.
- b. ♀. Kentucky, Laurel County, Forest Service Road 615a, 4 May 1996, D. J. Wright, USNM.
- c. ♂. Kentucky, Laurel County, Bolton Branch, 18 May 1996, D. J. Wright, USNM.
- d. ♀. Kentucky, Owsley County, 3 mi NE of Booneville, 14 June 1980, L. D. Gibson, LDG.
- e. ♀. South Carolina, McClellanville, Wedge Plantation, 22 April 1974, D. C. Ferguson, USNM.

**17.** ***Epiblema gibsoni*** Wright and Covell. FWL: 5.9–9.8 mm (mean = 7.8).
- a. ♂. PTP. Ohio, Adams County, 1 mi SE of Lynx, 3 August 2000, D. J. Wright, USNM.
- b. ♀. PTP. Ohio, Adams County, 1 mi SE of Lynx, 3 August 2000, D. J. Wright, USNM.
- c. ♂. Ohio, Adams County, 1 mi SE of Lynx, 5 July 2002, D. J. Wright, USNM.
- d. ♀. PTP. Kentucky, Bullitt County, Pine Creek Forest, 22 July 1989, L. D. Gibson, LDG.
- e. ♂. PTP. Illinois, Cook County, Gensburg-Markham Prairie, 7 June 1999, R. Panzer, USNM.

**18.** ***Epiblema desertana*** (Zeller). FWL: 6.7–9.6 mm (mean = 8.0).
- a. ♂. Ohio, Summit County, Springfield Bog Park, 17 June 2011, J. D. Hooper, USNM.
- b. ♀. Texas, Montgomery County, Conroe, 9 March 1971, A. and M. E. Blanchard, USNM.
- c. ♀. Connecticut, New Haven County, Hamden, 9 June 1967, D. C. Ferguson, USNM.
- d. ♂. Kentucky, Boone County, Big Bone Lick State Park, 24 June 1981, L. D. Gibson, LDG.
- e. ♂. Alabama, Baldwin County, Weeks Bay NER Reserve, 5–6 April 2003, R. L. Brown, MEM.

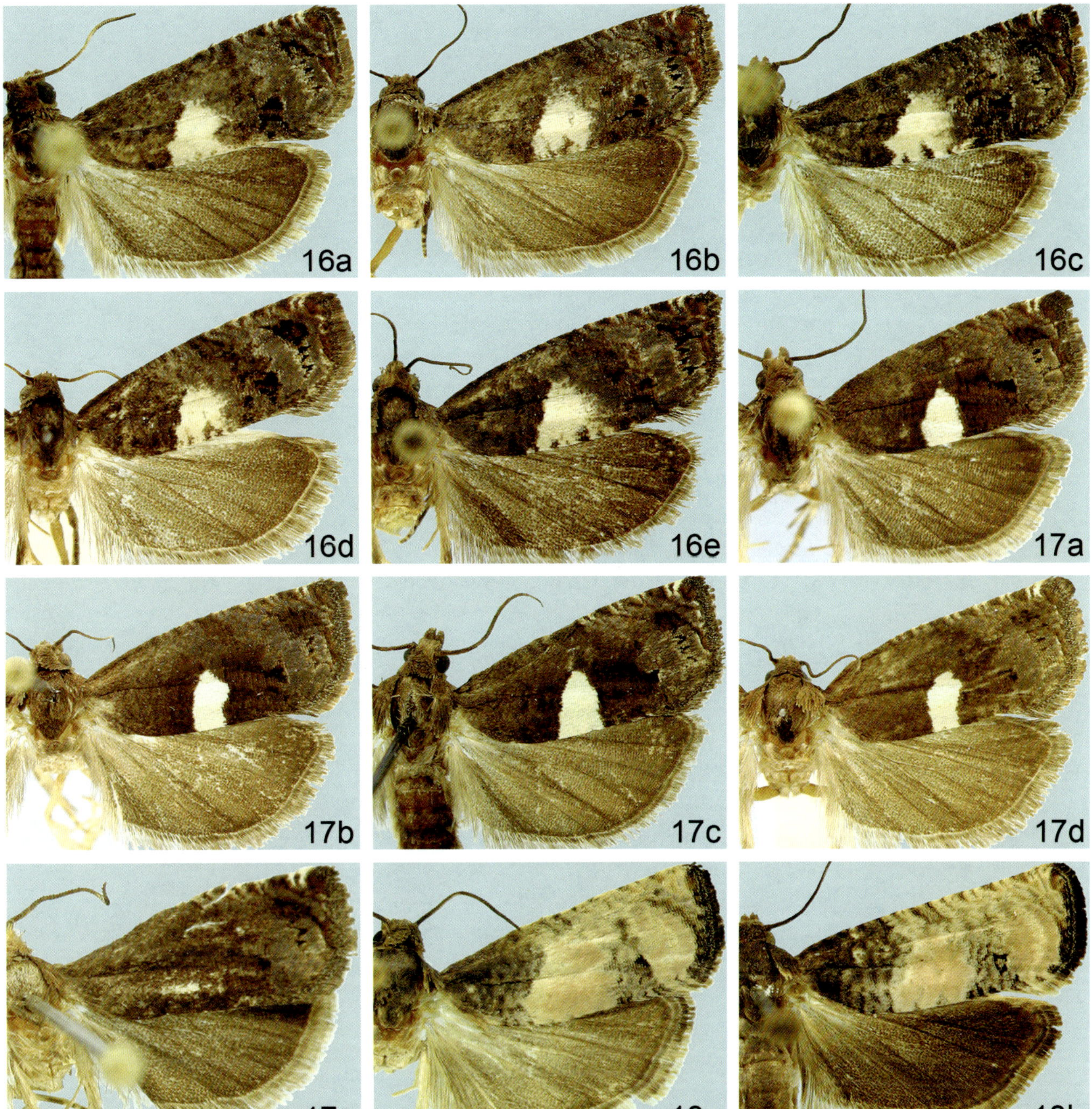
16a
16b
16c
16d
16e
17a
17b
17c
17d
17e
18a
18b
18c
18d
18e

# PLATE I

**19.** ***Epiblema rudei*** Powell. FWL: 7.6–11.0 mm (mean = 9.4).

a. ♂. PTP. California, San Benito/Fresno County line, Big Panoche Creek, 21 April 1967, P. A. Rude, USNM. [reared from *Gutierrezia*]

b. ♀. California, Kings County, 2 mi S of Kettleman, 17 March 1975, J. A. Powell, EME. [reared from *Gutierrezia*]

c. ♂. Texas, Cameron County, Laguna Atascosa, 3 March 1973, A. and M. E. Blanchard, USNM.

d. ♂. Texas, Hemphill County, Canadian, 13 March 1969, A. and M. E. Blanchard, USNM.

e. ♀. PTP. California, San Benito/Fresno County line, Big Panoche Creek, 4 March 1968, J. A. Powell, EME. [reared from *Gutierrezia*]

**20.** ***Epiblema discretivana*** Heinrich. FWL: 5.8–8.4 mm (mean = 6.9).

a. ♂. Louisiana, Bossier Parish, Barksdale AFB, 9 May 1996, D. M. Pollock, USNM.

b. ♀. Mississippi, Oktibbeha County, 6 mi SE of Starkville, 7 May 1987, R. and B. Brown, MEM.

c. ♀. Florida, Manatee County, Oneco, 28 March 1957, J. G. Franclemont, USNM.

d. ♂. North Carolina, Hyde County, Ocracoke, 9 June 2005, L. D. Gibson, LDG.

e. ♀. Florida, Manatee County, Oneco, 22 March 1955, J. G. Franclemont, USNM.

**21.** ***Epiblema obfuscana*** (Dyar). FWL: 6.9–9.3 mm (mean = 8.0).

a. ♂. Iowa, Johnson County, Williams Prairie, 15 June 1992, D. J. Wright, USNM.

b. ♀. Washington, Whitman County, Almota, 18 April 1934, J. F. G. Clarke, USNM. [reared from *Solidago* stalk]

c. ♀. Kentucky, Gallatin County, Markland Dam, 7 June 1990, L. D. Gibson, LDG.

d. ♂. Indiana, Dearborn County, Sparta Township, 13 May 1992, D. J. Wright, USNM.

e. ♂. Kentucky, Boone County, Big Bone Lick State Park, 18 May 1980, L. D. Gibson, LDG.

19a
19b
19c
19d
19e
20a
20b
20c
20d
20e
21a
21b
21c
21d
21e

# PLATE J

**22.** ***Epiblema radicana*** (Walsingham). FWL: 6.7–8.4 mm (mean = 7.3).

a. ♂. LTP. Oregon, Jackson County, “to Rogue River,” 4–6 May 1872, Walsingham, BMNH.
b. ♀. California, San Mateo County, San Bruno Mountains, 6 April 1962, J. A. Powell, EME.
c. ♀. California, San Mateo County, San Bruno Mountains, 7 April 1962, J. A. Powell, EME.
d. ♂. California, San Mateo County, San Bruno Mountains, 7 April 1962, J. A. Powell, EME.
e. ♀. California, San Mateo County, San Bruno Mountains, 7 April 1962, J. A. Powell, EME.
f. ♂. California, Placer County, Cisco, 1 June 1905, A. H. Vachell, USNM. [CoTP: *Eucosma vomonana* Kearfott]

**23.** ***Epiblema hirsutana*** (Walsingham). FWL: 5.8–7.2 mm (mean = 6.4).

a. ♂. LTP. California, Sonoma County, 23 May 1871, Walsingham, BMNH.
b. ♂. California, Alameda County, Mills College, 26 April 1908, G. R. Pilate, USNM.
c. ♂. California, Walsingham, USNM. [Fernald Collection].
d. ♂. California, Alameda County, Berkeley Hills, 17 April 1962, J. A. Powell, EME.
e. ♂. California, Alameda County, Berkeley Hills, 17 April 1962, J. A. Powell, EME.
f. ♂. California, Alameda County, Berkeley Hills, 17 April 1962, J. A. Powell, EME.

**24.** ***Epiblema lyallana*** McDunnough. FWL: 6.8–7.3 mm (mean = 7.0).

a. ♂. HTP. Canada, Quebec, Gaspe County, Mont Lyall, 25 July 1933, W. J. Brown, CNC.
b. ♂. PTP. Canada, Quebec, Gaspe County, Mont Lyall, 2 August 1933, W. J. Brown, USNM.
c. ♀. ALTP. Canada, Quebec, Gaspe County, 27 July 1933, E. B. Watson, CNC.

22a
22b
22c
22d
22e
22f
23a
23b
23c
23d
23e
23f
24a
24b
24c

# PLATE K

**25.** ***Epiblema deverrae*** Brown. FWL: 5.0–7.1 mm (mean = 6.2).

a. ♀. PTP. California, Sonoma County, Cazadero, 6 June 1979, J. A. Powell, EME.
b. ♀. California, Monterey County, Big Creek Reserve, 7–9 June 1993, Hsu and Powell, EME.
c. ♂. PTP. California, Monterey County, Big Creek Reserve, 5–8 June 1989, J. A. Powell, MEM.
d. ♂. PTP. California, Monterey County, Big Creek Reserve, 4–6 June 1982, J. A. Powell, MEM.

**26.** ***Epiblema periculosana*** Heinrich. FWL: 6.6–9.5 mm (mean = 8.0).

a. ♂. Oregon, Crook County, Ochoco National Forest, 3 July 1967, J. H. Baker, USNM.
b. ♂. Idaho, Bonner County, Priest River Experimental Forest, 21 June 1979, D. C. Ferguson, USNM.
c. ♀. Canada, Alberta, Waterton Lakes, 26 June 1923, J. McDunnough, CNC.
d. ♂. Colorado, Chaffee County, Cottonwood Pass, 16 July 1982, R. W. Hodges, USNM.
e. ♂. Wyoming, Albany County, Upper Blair Picnic Area, 7 July 2004, J. S. Nordin, USNM.
f. ♂. Wyoming, Albany County, Upper Blair Picnic Area, 15 July 2003, J. S. Nordin, USNM.

**27.** ***Epiblema arctica*** Miller. FWL: 7.7–9.5 mm (mean = 8.7).

a. ♂. PTP. Alaska, Schrader Lake, 4 July 1973, J. F. G. Clarke, USNM.
b. ♂. PTP. Alaska, Schrader Lake, 2 July 1973, K. W. Phillip, USNM.
c. ♂. PTP. Alaska, Schrader Lake, 14 July 1973, K. W. Phillip, USNM.
d. ♀. PTP. Alaska, Eagle Summit, 2 July 1979, Opler and Powell, USNM.
e. ♂. PTP. Alaska, Schrader Lake, 24 July 1973, J. F. G. Clarke, USNM.

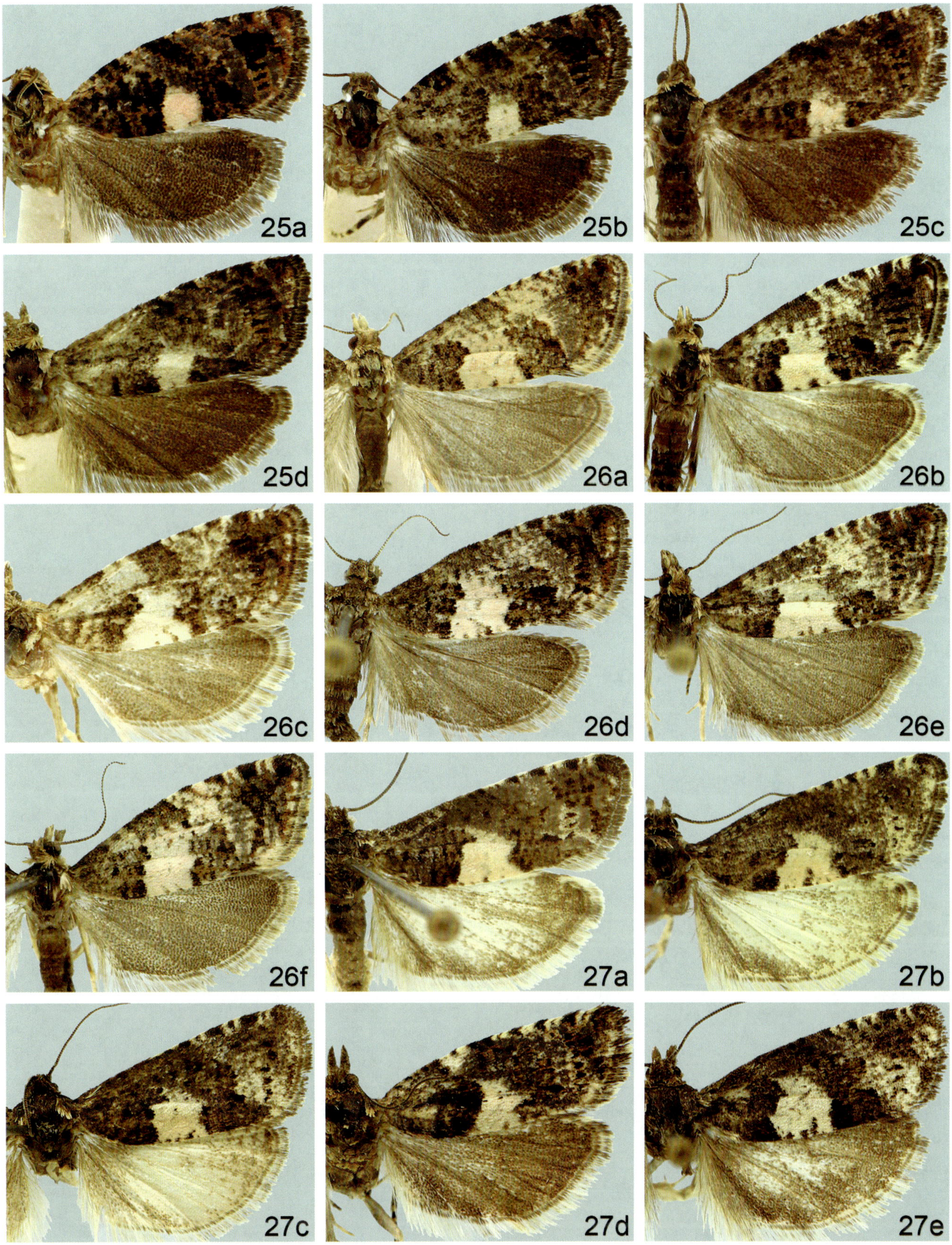
25a
25b
25c
25d
26a
26b
26c
26d
26e
26f
27a
27b
27c
27d
27e

# PLATE L

**28.** ***Epiblema walsinghami*** (Kearfott). FWL: 6.0–7.9 mm (mean = 7.1)

a. ♀. New Jersey, Passaic County, Great Notch, 10 May 1914, USNM.
b. ♀. New Jersey, Palisades, 25 April 1915, USNM.
c. ♀. Ohio, Montgomery County, Germantown, 21 April 1987, V. Albu, LDG.

**29.** ***Epiblema iowana*** McDunnough. FWL: 6.0–9.0 mm (mean = 7.3).

a. ♂. HTP. Iowa, Humboldt County, Renwick, 19 May 1928, G. S. Walley, CNC.
b. ♂. Illinois, Putnam County, 22 May 1967, M. O. Glenn, INHS.
c. ♀. llinois, Putnam County, 29 May 1950, M. O. Glenn, USNM. [PTP: *E. naoma*, reared from *Ratibida pinnata*]
d. ♀. llinois, Putnam County, 22 May 1960, M. O. Glenn, INHS.
e. ♂. Kentucky, Boone County, Big Bone Lick State Park, 1 May 1982, L. D. Gibson, LDG.
f. ♀. Illinois, Putnam County, 13 June 1950, M. O. Glenn, INHS. [PTP: *E. naoma*, reared from *Ratibida pinnata*]
g. ♂. Ohio, Erie County, Resthaven WLA, 24 May 1991, D. J. Wright, USNM.
h. ♀. Illinois, Putnam County, 28 April 1948, M. O. Glenn, INHS. [PTP: *E. naoma*]
i. ♂. Missouri, Benton County, 3 mi NW of Warsaw, 5 April 1976, J. R. Heitzman, USNM.

**30.** ***Epiblema coracinana*** Wright and Gilligan. FWL: 5.6–7.4 mm (mean = 6.6).

a. ♀. HTP. Pennsylvania, Allegheny County, Oak Station, 15 June 1908, F. Marloff, CMNH.
b. ♀. PTP. Pennsylvania, Allegheny County, Oak Station, 14 June 1907, F. Marloff, USNM.
c. ♀. PTP. Pennsylvania, Huntington County, Stone Valley Recreation Area, 26 May 2008, F. D. Fee, CMNH.

28a
28b
28c
29a
29b
29c
29d
29e
29f
29g
29h
29i
30a
30b
30c

# PLATE M

**31.** ***Epiblema otiosana*** (Clemens). FWL: 5.4–9.2 mm (mean = 7.6).

- a. ♂. Iowa, Pocahontas County, Kalsow Prairie, 22 June 2000, D. J. Wright, USNM.
- b. ♀. Kentucky, Henderson County, Sauerhaber WLA, 22 August 1992, L. D. Gibson, LDG.
- c. ♂. Ohio, Wyandot County, Killdeer Plains, 15 June 1995, D. J. Wright, USNM.
- d. ♀. Maryland, Talbot County, Pickering Creek Audubon Center, 14 August 2008, M. G. Pogue, USNM.
- e. ♀. Kentucky, Boone County, Florence, 17 June 1998, L. D. Gibson, LDG.
- f. ♂. Kentucky, Boone County, Middle Creek Road, 8 June 1984, L. D. Gibson, LDG.

**32.** ***Epiblema dorsisuffusana*** (Kearfott) FWL: 6.9–9.0 mm (mean = 8.1).

- a. ♂. Illinois, Putnam County, 3 July 1965, M. O. Glenn, INHS.
- b. ♀. Illinois, Putnam County, 11 June 1966, M. O. Glenn, INHS.
- c. ♂. North Carolina, Jackson County, Balsam, 17 July 1974, D. C. Ferguson, USNM.
- d. ♂. Arkansas, Johnson County, 8 mi N of Clarksville, 10–12 May 1987, R. L. Brown, MEM.

**33.** ***Epiblema boxcana*** (Kearfott). FWL: 5.7–8.0 mm (mean = 7.0).

- a. ♂. Indiana, Dearborn County, SE corner Sparta Township, 13 May 1992, D. J. Wright, USNM.
- b. ♂. Ohio, Erie County, Resthaven WLA, 24 May 1991, D. J. Wright, USNM.
- c. ♂. Kentucky, Bracken County, Meldahl Dam, 9 May 1993, L. D. Gibson, LDG.
- d. ♂. Texas, San Patricio County, Sinton, 30 June 1975, A. and M. E. Blanchard, USNM.
- e. ♂. Washington, Whitman County, Snake River opposite Clarkston, 12 April 1934, J. F. G. Clarke, USNM. [reared from *Xanthium spaciosum*]

31a
31b
31c
31d
31e
31f
32a
32b
32c
32d
33a
33b
33c
33d
33e

# PLATE N

**34.** ***Epiblema resumptana*** (Walker). FWL: 4.4–6.7 mm (mean = 5.6).

a. ♂. Connecticut, Windham County, 1.6 mi S of Hampton, 16 May 1994, D. L Wagner, USNM.
b. ♂. Illinois, Putnam County, 13 May 1946, M. O. Glenn, USNM.
c. ♀. Ohio, Hamilton County, Cincinnati, 2 June 1906, A. Braun, USNM.
d. ♀. Ohio, Hamilton County, Cincinnati, 2 June 1906, A. Braun, USNM.
e. ♂. Washington, Whitman County, Union Flatt, 15 May1929, J. F. G. Clarke, USNM.
f. ♂. Oregon, Baker County, 6.4 mi N of Richland, 7 June 1970, J. F. G. Clarke, USNM.

**35.** ***Epiblema macneilli*** Powell. FWL: 10.8–13.2 mm (mean = 11.9).

a. ♀. PTP. California, Inyo County, near Mono Pass, 30 August 1965, C. D. MacNeill, EME.
b. ♀. PTP. California, Inyo County, near Mono Pass, 3 September 1965, J. A. Powell, EME.
c. ♂. California, Inyo County, Mt. Whitney Trail, E side Trail Crest, el. 13,700', 13 July 2005, T. M. Gilligan, TMG.

**36.** ***Epiblema brightonana*** (Kearfott). FWL: 5.2–7.9 mm (mean = 6.5).

a. ♂. Ohio, Adams County, 1 mi SE of Lynx, 3 August 1998, D. J. Wright, USNM.
b. ♂. North Carolina, Macon County, Highlands, 25 July 1958, J. G. Franclemont, USNM.
c. ♂. Maryland, Dukes County, Martha's Vineyard, 6 August 1949, F. M. Jones, USNM.
d. ♂. Illinois, Putnam County, 21 July 1965, M. O. Glenn, USNM.
e. ♂. Illinois, Putnam County, 23 July 1965, M. O. Glenn, USNM.
f. ♂. Arkansas, Polk County, Queen Wilhelmina State Park, 23 June 2017, K. Austin, L. and J. Santiago, MEM.

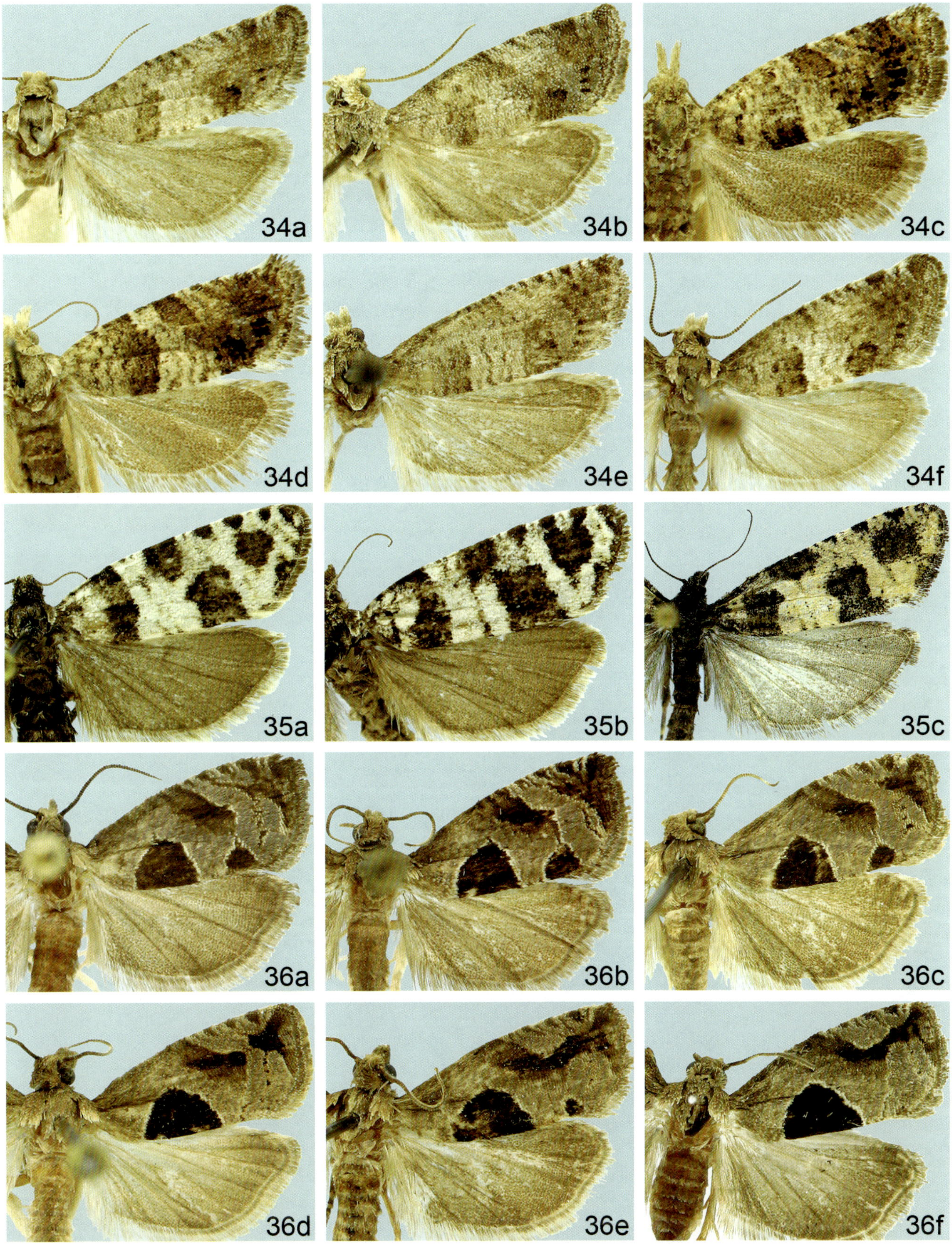
34a
34b
34c
34d
34e
34f
35a
35b
35c
36a
36b
36c
36d
36e
36f

# PLATE O

**37.** ***Epiblema browni*** Wright and Gilligan. FWL: 5.0–7.0 mm (mean = 5.9).

a. ♂. HTP. Mississippi, Oktibbeha County, Osborn Prairie, 30 August 1997, R. Brown and L. Koehn, USNM.

b. ♀. PTP. Arkansas, Arkansas County, 3 mi E of Ethel, 9 September 1970, R. L. Brown, MEM.

c. ♂. PTP. Mississippi, Oktibbeha County, 7 mi S of Starkville, 31 August 1995, D. M Pollock, MEM.

d. ♀. PTP. Louisiana, Bossier Parish, Bodcau WMA, 13 September 1996, R. L. Brown, MEM.

e. ♀. Texas, Harris County, 18 September 2020, J. Schneider, USNM.

**38.** ***Epiblema fulminana*** (Walsingham). FWL: 7.3–14.0 mm (mean = 11.3).

a. ♂. Ohio, Adams County, 1 mi SE of Lynx, 16 July 1990, D. J. Wright, USNM.

b. ♀. Ohio, Wyandot County, Killdeer Plains, 21 July 1995, E. H. Metzler, USNM.

c. ♂. Illinois, Macon County, Decatur, 18 August 1890, Barnes, USNM.

b. ♂. Kansas, Riley County, Konza Prairie, 23 July 1995, D. J. Wright, USNM.

**39.** ***Epiblema alba*** Wright and Gilligan. FWL: 4.2–5.7 mm (mean = 5.1).

a. ♂. HTP. Florida, Putnam County, Palatka, 23 August 1992, H. D. Baggett, USNM.

b. ♀. PTP. Alabama, Baldwin County, Bon Secour NWR, 8–9 August 1994, R. Brown and D. Pollock, MEM.

c. ♀. PTP. Florida, Hernando County, Withlacoochee State Forest, 18 September 1990, W. L. Adair Jr., MGCL.

**40.** ***Epiblema tandana*** (Kearfott) FWL: 6.6–10.2 mm (mean = 8.7).

a. ♂. Kentucky, Gallatin County, Markland Dam, 24 August 2006, L. D. Gibson, LDG.

b. ♀. Kentucky, Owsley County, 3 mi NE of Booneville, 25 June 1994, L. D. Gibson, LDG.

c. ♀. Pennsylvania, York County, near Dillsburg, 19 June 1969, G. Werschkow, USNM.

37a
37b
37c
37d
37e
38a
38b
38c
38d
39a
39b
39c
40a
40b
40c

# PLATE P

**41.** ***Epiblema numerosana*** (Zeller). FWL: 5.2–7.9 mm (mean = 6.5).

a. ♂. HTP. Texas, Dallas, J. Boll, MCZ.
b. ♂. Texas, Anderson County, Tennessee Colony, 23 August 1978, A. and M. E. Blanchard, USNM.
c. ♀. Texas, Anderson County, Tennessee Colony, 23 August 1978, A. and M. E. Blanchard, USNM.
d. ♂. Nebraska, Cherry County, Valentine NWR, 25 June 1988, Hackberry Lake, R. H. Hodges, USNM.
e. ♂. Illinois, Putnam County, 17 July 1970, M. O. Glenn, INHS.
f. ♀. Illinois, Putnam County, 17 July 1969, M. O. Glenn, INHS.

**42.** ***Epiblema deflexana*** Heinrich. FWL: 5.8–7.2 mm (mean = 6.3).

a. ♂. HTP. Texas, Cameron County, Brownsville, 27 May 1917, A. Busck, USNM.
b. ♀. Texas, Bexar County, 6 May 1993, D. M. Pollock, MEM.
c. ♂. Mexico, Nuevo Leon, 4 mi W of Iturbide, 13–14 September 1976, Powell and Chemsak, EME.

**43.** ***Epiblema abruptana*** (Walsingham). FWL: 5.2–8.8 mm (mean = 6.4).

a. ♂. Texas, Harris County, Houston, 7 September 1965, A. and M. E. Blanchard, USNM.
b. ♂. Florida, Putnam County, Palatka, 25 September 1990, H. D. Baggett, MEM.
c. ♀. Texas, Comal County, near Spring Branch, 27 June 1992, D. M. Pollock, MEM.
d. ♂. Ohio, Adams County, 1 mi SE of Lynx, 25 July 1998, D. J. Wright, USNM.
e. ♂. Texas, Kimble County, Junction, 26 October 1973, A. and M. E. Blanchard, USNM.
f. ♂. Tennessee, Wilson County, Vista Cedar Glade NA, 21 July 2009, Sangmi Lee, MEM.

41a
41b
41c
41d
41e
41f
42a
42b
42c
43a
43b
43c
43d
43e
43f

# PLATE Q

**44.** ***Epiblema mandana*** Wright and Gilligan. FWL: 5.3–8.4 mm (mean = 6.9).

a. ♂. Maryland, Worcester County, Vaughn WMA, 20 June 2000, J. Glaser, USNM.
b. ♂. Louisiana, Bossier Parish, Barksdale AFB, 13 September 1996, R. L. Brown, MEM.
c. ♀. Alabama, Baldwin County, Bon Secour NWR, 8–9 August 1994, R. Brown and D. Pollock, MEM.
d. ♂. Ohio, Adams County, 1 mi SE of Lynx, 25 July 1997, D. J. Wright, USNM.
e. ♂. Kansas, Riley County, Konza Prairie, 12 June 1999, J. MacGown, MEM.

**45.** ***Epiblema exacerbatricana*** Heinrich. FWL: 5.0–6.9 mm (mean = 6.1).

a. ♂. South Carolina, Chesterfield County, Patrick, 15 September 2012, J. Glaser, USNM.
b. ♂. Florida, Highlands County, Highlands Hammock State Park, 15 November 1987, G. J. Balogh, USNM.
c. ♂. Alabama, Baldwin County, 1 mi E of Oyster Bay, 13 October 1990, R. L Brown, USNM.
d. ♂. Alabama, Baldwin County, 1 mi E of Oyster Bay, 13 October 1990, R. L Brown, MEM.
e. ♂. Massachusetts, Plymouth County, Myles Standish State Forest, 14 August 1990, D. L. Wagner, MEM.

**46.** ***Epiblema glaseri*** Wright and Gilligan. FWL: 5.2–7.9 mm (mean = 6.1).

a. ♂. HTP. Georgia, Manuel County, Ohoopee Dunes, 16 September 2012, J. Glaser, USNM.
b. ♂. PTP. Alabama, Bibb County, Bibb County Glades Preserve, 6 August 2003, R. Brown and J. MacGown, MEM.
c. ♂. PTP. Alabama, Mobile County, Saraland, 5 September 1982, E. C. Knudson, MGCL.
d. ♂. PTP. Kentucky, Laurel County, ridgetop SW of Ned Branch, 28 July 1992, L. D. Gibson, LDG.
e. ♂. Florida, Collier County, Big Cypress Preserve, 28 January 2010, J. Vargo, CMNH.

44a
44b
44c
44d
44e
45a
45b
45c
45d
45e
46a
46b
46c
46d
46e

# PLATE R

**47.** ***Epiblema praesumptiosa*** Heinrich. FWL: 6.2–9.5 mm (mean = 7.9).

a. ♀. PTP. Texas, Cameron County, Brownsville, 27 May 1917, A. Busck, USNM.
b. ♂. Texas, Cameron County, Laguna Atascosa, 1 April 1978, A. and M. E. Blanchard, USNM.
c. ♀. Texas, Cameron County, Laguna Atascosa, 3 March 1973, A. and M. E. Blanchard, USNM.
d. ♀. Texas, Cameron County, Laguna Atascosa, 3 March 1973, A. and M. E. Blanchard, USNM.
e. ♀. Texas, Jefferson County, Sabine Pass, 11 June 1994, E. C. Knudson, MGCL.

**Louisiana, Mississippi, Alabama, Georgia**

f. ♂. Mississippi, Jackson County, Grand Bay Savanna, 30 April 2004, Brown and Martinez, MEM.
g. ♀. Alabama, Baldwin County, Oyster Bay, 4–8 September 1988, R. Brown and J. MacGown, MEM.
h. ♂. Alabama, Baldwin County, Bon Secour NWR, 12–16 October 1991, R. Brown and D. Pollock, MEM.
i. ♂. Alabama, Baldwin County, Plash Island, 16–19 June 1984, R. L. and B. B. Brown, MEM.
j. ♂. Georgia, McIntosh County, Sapelo Island, 18 April 2013, B. G. Scholtens, BGS.

**North Carolina**

k. ♂. Carteret County, Fort Macon State Park, 17 May 1996, J. B. Sullivan, MEM.
l. ♂. Carteret County, Fort Macon State Park, 24 August 2016, J. B. Sullivan, JBS.
m. ♂. Carteret County, Fort Macon State Park, 16 July 1996, J. B. Sullivan, MEM.
n. ♂. Carteret County, Fort Macon State Park, 3 August 1997, J. B. Sullivan, MEM.
o. ♂. Carteret County, Fort Macon State Park, 11 August 1996, J. B. Sullivan, MEM.

47a
47b
47c
47d
47e
47f
47g
47h
47i
47j
47k
47l
47m
47n
47o

# PLATE S

**48.** ***Epiblema grossbecki*** Heinrich. FWL: 7.2–9.1 mm (mean = 8.3).
- a. ♂. HTP. Florida, Everglade, 10 April 1912, AMNH.
- b. ♂. Florida, Everglade, 8–15 April, USNM.
- c. ♂. Florida, Everglade, 8–15 April, USNM.
- d. ♀. PTP. Florida, Everglade, 14 April 1912, USNM.

**49.** ***Epiblema chromata*** Miller. FWL: 6.9–9.7 mm (mean = 8.3).
- a. ♂. Texas, San Patricio County, Welder WLR, 19 March 1969, A. and M. E. Blanchard, USNM.
- b. ♂. Texas, Comal County, Guadalupe River, 12 March 1972, A. and M. E. Blanchard, USNM.
- c. ♂. Texas, Cameron County, Laguna Atascosa, 24 February 1973, A. and M. E. Blanchard, USNM.
- d. ♂. Texas, Cameron County, Brownsville, 11 March 1977, A. and M. E. Blanchard, USNM.
- e. ♂. Texas, San Patricio County, Welder WLR, 30 June 1975, A. and M. E. Blanchard, USNM.

**50.** ***Epiblema insidiosana*** Heinrich. FWL: 4.9–6.1 mm (mean = 5.5).
- a. ♂. PTP. North Carolina, Moore County, Southern Pines, 1–7 July, USNM.
- b. ♀. Florida, Putnam County, Palatka, 20 May 1990, H. D. Baggett, MEM.
- c. ♂. Florida, Levy County, Goethe State Forest, 15 June 2006, J. W. Brown, USNM.
- d. ♂. Florida, Marion County, Marion Oaks, 17 May 1990, J. S. Kutis, MEM.
- e. ♂. Florida, Suwanee County, Suwannee River State Park, 17–20 May 1987, H. D. Baggett, MEM.
- f. ♂. Florida, Levy County, Goethe State Forest, 15 June 2006, J. W. Brown, USNM.

48a
48b
48c
48d
49a
49b
49c
49d
49e
50a
50b
50c
50d
50e
50f

# PLATE T

**51.** ***Epiblema separationis*** Heinrich. FWL: 3.7–6.5 mm (mean = 5.1).

a. ♂. PTP. Texas, Cameron County, Brownsville, 27 May 1917, A. Busck, USNM.
b. ♀. Texas, Cameron County, Laguna Atascosa, 16 May 1974, A. and M. E. Blanchard, USNM.
c. ♂. Texas, Jefferson County, Sabine Pass, 11 June 1994, E. C. Knudson, MGCL.
d. ♂. Texas, Jefferson County, Sabine Pass, 11 June 1994, E. C. Knudson, MGCL.
e. ♂. Texas, Jefferson County, Sabine Pass, 11 June 1994, E. C. Knudson, MGCL.
f. ♂. Mississippi, Jackson County, Grand Bay Savanna, 30 April 2004, Brown, Lee and Martinez, MEM.
g. ♀. Mississippi, Jackson County, Grand Bay Savanna, 30 April 2004, Brown, Lee and Martinez, MEM.
h. ♂. Mississippi, Jackson County, Grand Bay Savanna, 30 April 2004, Brown, Lee and Martinez, MEM.
i. ♀. Mississippi, Jackson County, Grand Bay Savanna, 30 April 2004, Brown, Lee and Martinez, MEM.
j. ♀. Florida, Monroe County, Sugarloaf Key, March 1945, USNM.
k. ♂. North Carolina, Carteret County, Fort Macon State Park, 3 August 1997, J. B. Sullivan, USNM. [determination uncertain]
l. ♂. Georgia, McIntosh County, Sapelo Island, 8 September 2013, B. G. Scholtens, BGS. [determination uncertain]

**52.** ***Epiblema colfaxana*** Wright and Gilligan. FWL: 5.6–7.6 mm (mean = 6.6).

a. ♂. HTP. New Mexico, Colfax County, Exit 419 off I-25, 5 August 2008, T. M. Gilligan, USNM.
b. ♀. PTP. Colorado, Cheyenne County, Wild Horse Post Office, 15 September 2007, T. M. Gilligan and C. E. Harp, USNM.
c. ♂. California, Lassen County, Ravendale, 20 June 1959, G. I. Stage, EME. [determination uncertain]

51a
51b
51c
51d
51e
51f
51g
51h
51i
51j
51k
51l
52a
52b
52c

# PLATE U

**53.** ***Sonia constrictana*** (Zeller). FWL: 6.0–8.3 mm (mean = 7.2).

a. ♂. Texas, Hidalgo County, Santa Ana Refuge, 15 September 1974, A. and M. E. Blanchard, USNM.

b. ♂. Texas, Cameron County, Sabal Palm Sanctuary, 4–5 August 1994, E. C. Knudson, MGCL.

c. ♂. Texas, San Patricio County, Welder WLR, 30 June 1975, A. and M. E. Blanchard, USNM.

d. ♀. Texas, Hidalgo County, Bentsen-Rio Grande Valley State Park, 11 October 1980, E. C. Knudson, MGCL.

e. ♂. Texas, San Patricio County, Welder WLR, 5 July 1975, A. and M. E. Blanchard, USNM.

f. ♀. Texas, San Patricio County, Welder WLR, 30 June 1975, A. and M. E. Blanchard, USNM.

**54.** ***Sonia paraplesiana*** Blanchard. FWL: 4.2–8.5 mm, (mean = 6.3).

a. ♂. PTP. Texas, Harris County, Houston, 6 June 1968, A. and M. E. Blanchard, USNM.

b. ♂. PTP. Texas, Harris County, Houston, 26 August 1966, A. and M. E. Blanchard, USNM.

c. ♂. Kentucky, McCracken County, Paducah Zip Track, 12 August 2008, D. J. Wright, USNM.

d. ♂. Maryland, Kent County, Millington WMA, 12 July 2001, J. Glaser, USNM.

e. ♂. Alabama, Baldwin County, Weeks Bay NERS, 22 June 2008, D. J. Wright, USNM.

f. ♂. Florida, Highlands County, Lake Placid, 30 April 1964, R. W. Hodges, USNM.

**55.** ***Sonia roseomaculana*** Wright and Gilligan. FWL: 7.2–8.6 mm, (mean = 7.9).

a. ♂. HTP. Florida, Franklin County, Apalachicola National Forest, 1 September 1991, H. D. Baggett, USNM.

b. ♂. PTP. Florida, Franklin County, Apalachicola National Forest, 1 September 1991, H. D. Baggett, MEM.

c. ♀. PTP. Florida, Franklin County, Apalachicola National Forest, 1 September 1991, H. D. Baggett, MEM.

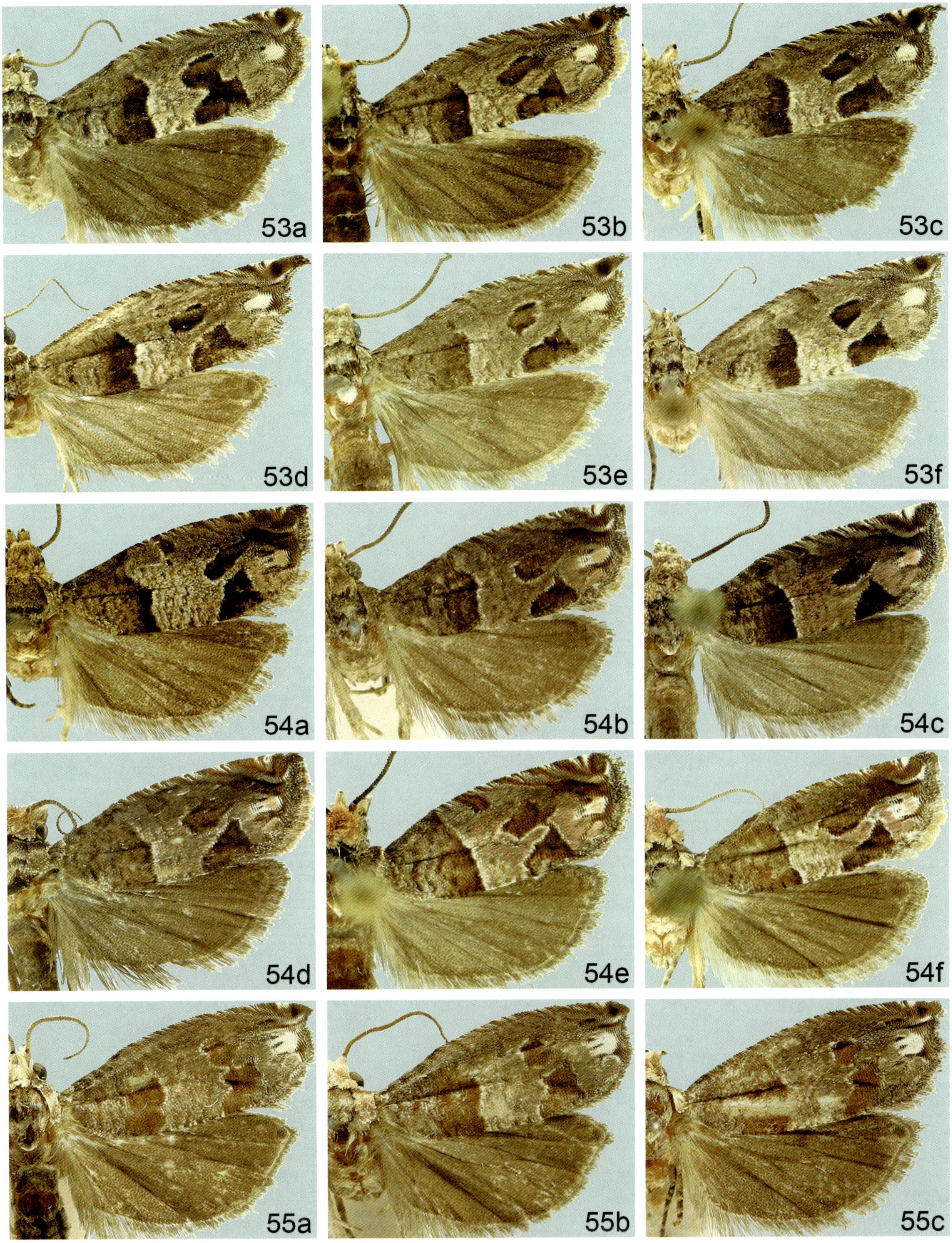
53a
53b
53c
53d
53e
53f
54a
54b
54c
54d
54e
54f
55a
55b
55c

# PLATE V

**56.** ***Sonia ferruginana*** Wright and Gilligan. FWL: 4.0–9.1 mm (mean = 7.0).

a. ♂. HTP. Kentucky, Bracken County, Meldahl Dam, 21 June 2007, L. D. Gibson, USNM.
b. ♂. PTP. Ohio, Adams County, 1 mi SE of Lynx, 27 June 1998, D. J. Wright, USNM.
c. ♀. PTP. Ohio, Adams County, 1 mi SE of Lynx, 27 June 1998, D. J. Wright, USNM.
d. ♂. PTP. Iowa, Johnson County, Williams Prairie, 15 June 1992, D. J. Wright, USNM.
e. ♀. PTP. Iowa, Johnson County, Williams Prairie, 15 June 1992, D. J. Wright, USNM.
f. ♂. PTP. Mississippi, Oktibbeha County, 6 mi SW of Starkville, 30 July 1984, R. L. and B. B. Brown, MEM.
g. ♂. PTP. Ohio, Wyandot County, County Road T-108, 30 June 1990, D. J. Wright, USNM.
h. ♂. PTP. New York, Tompkins County, Ithaca, Snyder Heights, 6 August 1976, J. G. Franclemont, MEM.
i. ♂. PTP. Ohio, Delaware County, Delaware Reservoir WA, 29 June 1990, L. D. Gibson, LDG.

**57.** ***Sonia griseolana*** Wright and Gilligan. FWL: 5.5–8.2 mm, (mean = 6.8).

a. ♂. HTP. Mississippi, Oktibbeha County, 6 mi SW of Starkville, 23 July 1984, R. L. and B. B. Brown, USNM.
b. ♂. PTP. Kentucky, McCracken County, Paducah Zip Track, 12 August 2008. L. D. Gibson, LDG.
c. ♂. PTP. Texas, Newton County, Deweyville, 18 August 1979, E. C. Knudson, MGCL.
d. ♂. PTP. Louisiana, Bossier Parrish, Barksdale AFB, 21 August 1996, D. M. Pollock, MEM.
e. ♂. PTP. Kentucky, McCracken County, Paducah Zip Track, 12 August 2008. L. D. Gibson, LDG.
f. ♀. PTP. Mississippi, Oktibbeha County, 6 mi SW of Starkville, 21 August 1984, R. L. and B. B. Brown, MEM.

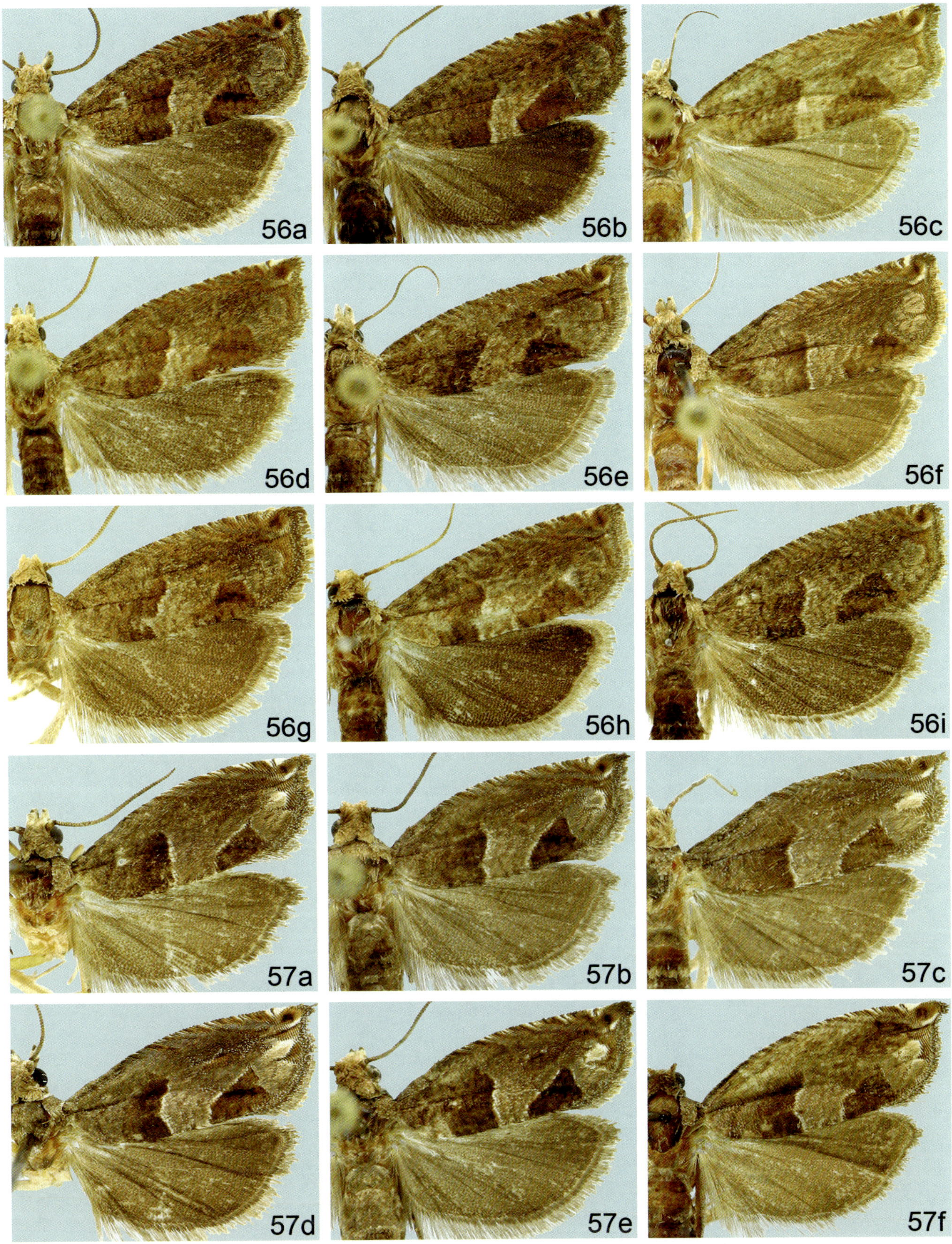
56a
56b
56c
56d
56e
56f
56g
56h
56i
57a
57b
57c
57d
57e
57f

# PLATE W

**58.** ***Sonia ferruginana* complex, grayish phenotype.** FWL: 5.8–9.2 mm (mean = 7.4).

a. ♂. Alabama, Baldwin County, Weeks Bay NER Reserve, 5 August 2000, R. L. Brown, MEM.
b. ♂. Florida, Collier County, Fakahatchee Strand, 30 June 1984, H. D. Baggett, MEM.
c. ♀. Florida, Collier County, Fakahatchee Strand, 30 June 1984, H. D. Baggett, MEM.
d. ♂. New Jersey, Cumberland County, Hansey Creek Road, 29 June 2019, S. Johnson, SJ.
e. ♀. New Jersey, Cumberland County, Hansey Creek Road, 18 June 2020, S. Johnson, USNM.
f. ♂. New Jersey, Cumberland County, Hansey Creek Road, 29 June 2019, S. Johnson, USNM.
g. ♂. Kentucky, Boone County, Boone Cliffs Nature Preserve, 8 June 1984, L. D. Gibson, LDG.
h. ♂. Ohio, Erie County, Resthaven WLA, 12 July 1991, L. D. Gibson, LDG.
i. ♀. Illinois, Putnam County, 6 August 1964, M. O. Glenn, INHS.

***Sonia ferruginana* complex, reddish phenotype.** FWL: 5.8–8.1 mm (mean = 6.9).

j. ♂. Maryland, Dorchester County, Fishing Bay WMA, 7 August 1999, J. Glaser, USNM.
k. ♂. Maryland, Worcester County, Pocomoke State Forest, 28 July 1998, J. Glaser, USNM.
l. ♀. North Carolina, Jones County, Croatan National Forest, 10 august 2016, J. B. Sullivan, JBS.
m. ♂. Florida, Putnam County, Palatka, 1 September 1990, H. D. Baggett, MEM.
n. ♀. Florida, Highlands County, Archbold Biological Station, 29 March 1959, R. W. Hodges, USNM.
o. ♂. Florida, Marion County, Lynne, 27 September 1990, J. S. Kutis, MEM.

58a
58b
58c
58d
58e
58f
58g
58h
58i
58j
58k
58l
58m
58n
58o

# PLATE X

**59.** ***Sonia canadana*** McDunnough. FWL: 6.1–8.6 mm, (mean = 7.2).
- a. ♂. New York, Tompkins County, Ithaca, Snyder Heights, 12 August 1972, J. G. Franclemont, CUIC.
- b. ♀. New York, Tompkins County, Ithaca, Snyder Heights, 26 August 1983, J. G. Franclemont, CUIC.
- c. ♀. New York, Tompkins County, Ithaca, Snyder Heights, 13 August 1972, J. G. Franclemont, CUIC.
- d. ♂. Ohio, Adams County, 1 mi SE of Lynx, 12 August 1998, D. J. Wright, USNM.
- e. ♂. Ohio, Adams County, 1 mi SE of Lynx, 21 August 1993, D. J. Wright, USNM.
- f. ♂. Ohio, Adams County, 1 mi SE of Lynx, 31 August 1989, L. D. Gibson, LDG.

**60.** ***Sonia nigrana*** Wright and Gilligan. FWL: 5.0–8.0 mm (mean = 6.4).
- a. ♂. HTP. Ohio, Adams County, 1 mi SE of Lynx, 5 July 2002, D. J. Wright, USNM.
- b. ♀. PTP. Ohio, Adams County, 1 mi SE of Lynx, 25 July 1998, D. J. Wright, USNM.
- c. ♂. PTP. Maryland, Washington County, Sideling Hill WMA, 24 August 2000, J. Glaser, USNM.
- d. ♂. PTP. Iowa, Pocahontas County, Kalsow Prairie, 18 July 2003, D. J. Wright, USNM.
- e. ♀. PTP. Iowa, Pocahontas County, Kalsow Prairie, 19 July 1998, D. J. Wright, USNM.
- f. ♂. PTP. Kansas, Riley County, Konza Prairie, 12 June 1999, D. M. Pollock, MEM.

**61.** ***Sonia albimacula*** Wright and Gilligan. FWL: 5.4–7.1 mm (mean = 6.4).
- a. ♂. HTP. Ohio, Adams County, 1 mi SE of Lynx, 27 June 1998, D. J. Wright, USNM.
- b. ♂. PTP. Massachusetts, Barnstable County, Barnstable, 3 July 1960, C. P. Kimball, USNM.
- c. ♂. PTP. Vermont, Caledonia County, Peacham, 5 July 2007, M. Sabourin, USNM.

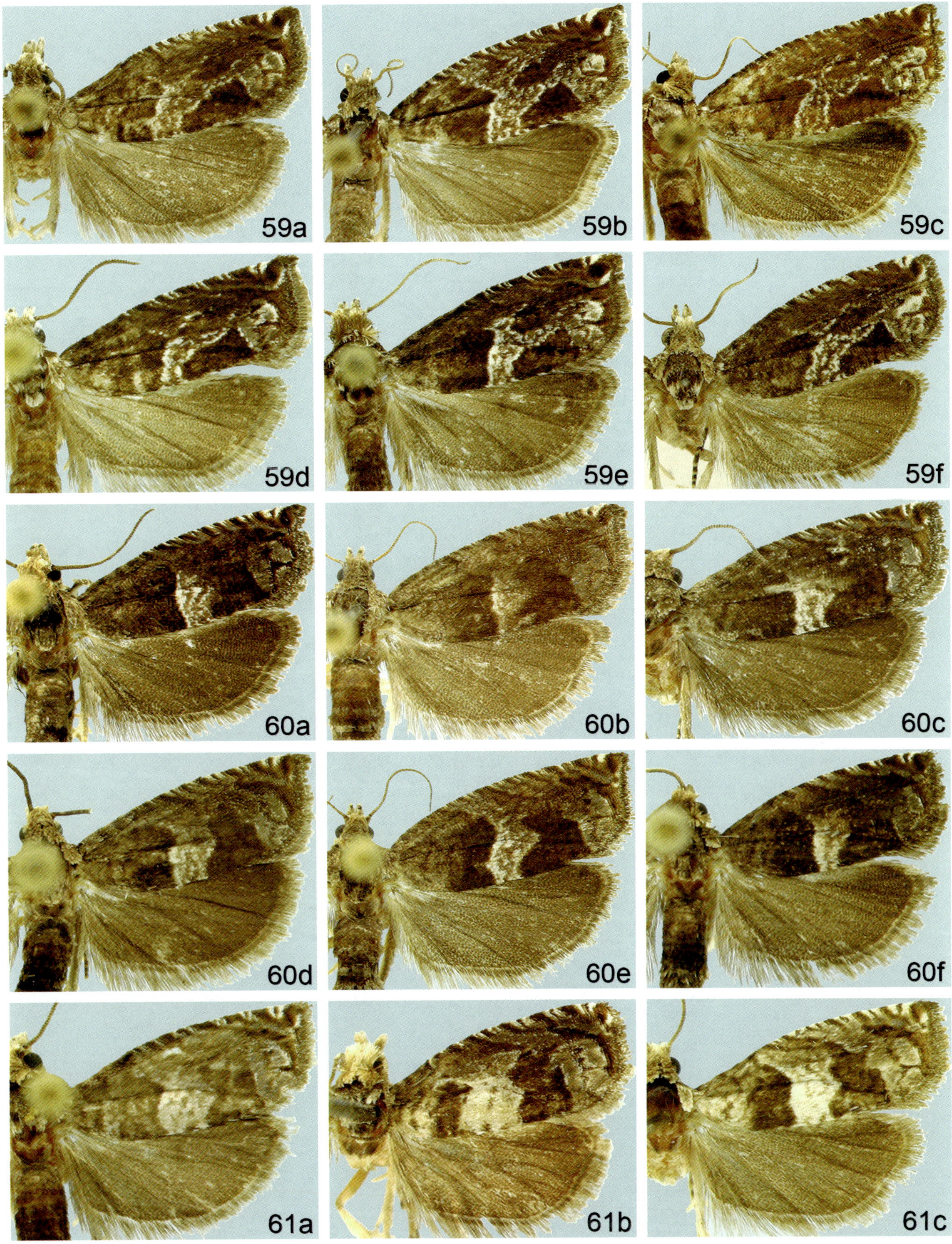
59a
59b
59c
59d
59e
59f
60a
60b
60c
60d
60e
60f
61a
61b
61c

# PLATE Y

**62.** ***Sonia parva*** Wright and Gilligan. FWL: 5.0–6.3 mm (mean = 5.7).

a. ♂. HTP. Iowa, Pocahontas County, Kalsow Prairie, 18 July 2003, D. J. Wright, USNM.
b. ♂. PTP. Ohio, Erie County, Resthaven WLA, 20 July 1990, D. J. Wright, USNM.
c. ♂. PTP. Kentucky, Menifee County, Indian Creek Road, 6 July 1991, L. D. Gibson, LDG.
d. ♀. PTP. Kentucky, Menifee County, Indian Creek Road, 6 July 1991, L. D. Gibson, LDG.
e. ♂. PTP. Missouri, Mississippi County, Big Oak Tree State Park, 28 May 1985, G. J. Balogh, USNM.

**63.** ***Sonia canadana* complex.** FWL: 6.1–7.2 mm (mean = 6.7).

a. ♂. Wisconsin, Sauk County, Spring Green NCP, 26 June 1995, J.-F. Landry and B. Landry, CNC.
b. ♂. Canada, Manitoba, Manitoba Tall Grass Prairie Preserve, 17 July 2007, J.-F. Landry and V. Nazari, CNC.

**64.** ***Sonia divaricata*** Miller. FWL: 5.9–9.2 mm (mean = 8.0).

a. ♀. PTP. Kentucky, Powell County, Red River Gorge, 14 May 1988, L. D. Gibson, LDG.
b. ♂. Kentucky, Laurel County, 0.5 mi E of Marsh Branch, 21 May 2009, L. D. Gibson, LDG.
c. ♂. Louisiana, Bossier Parish, Barksdale AFB, 31 March 1996, D. M Pollock, MEM.
d. ♂. Arkansas, Logan County, Cove Lake, 14–20 May 1989, J. MacGown, MEM.
e. ♂. Alabama, DeKalb County, DeSoto State Park, 18 May 1990, R. L. Brown, MEM.

**65.** ***Sonia comstocki*** Clarke. FWL: 9.2–11.3 mm (mean = 10.1).

a. ♂. PTP. California, San Diego County, San Felipe Wash, 11 September 1938, J. A. Comstock, USNM.
b. ♂. Texas, Brewster County, Big Bend National Park, 20 August 1995, E. C. Knudson, MGCL.
c. ♀. Texas, Brewster County, Big Bend National Park, 20–21 August 1995, E. C. Knudson, MGCL.

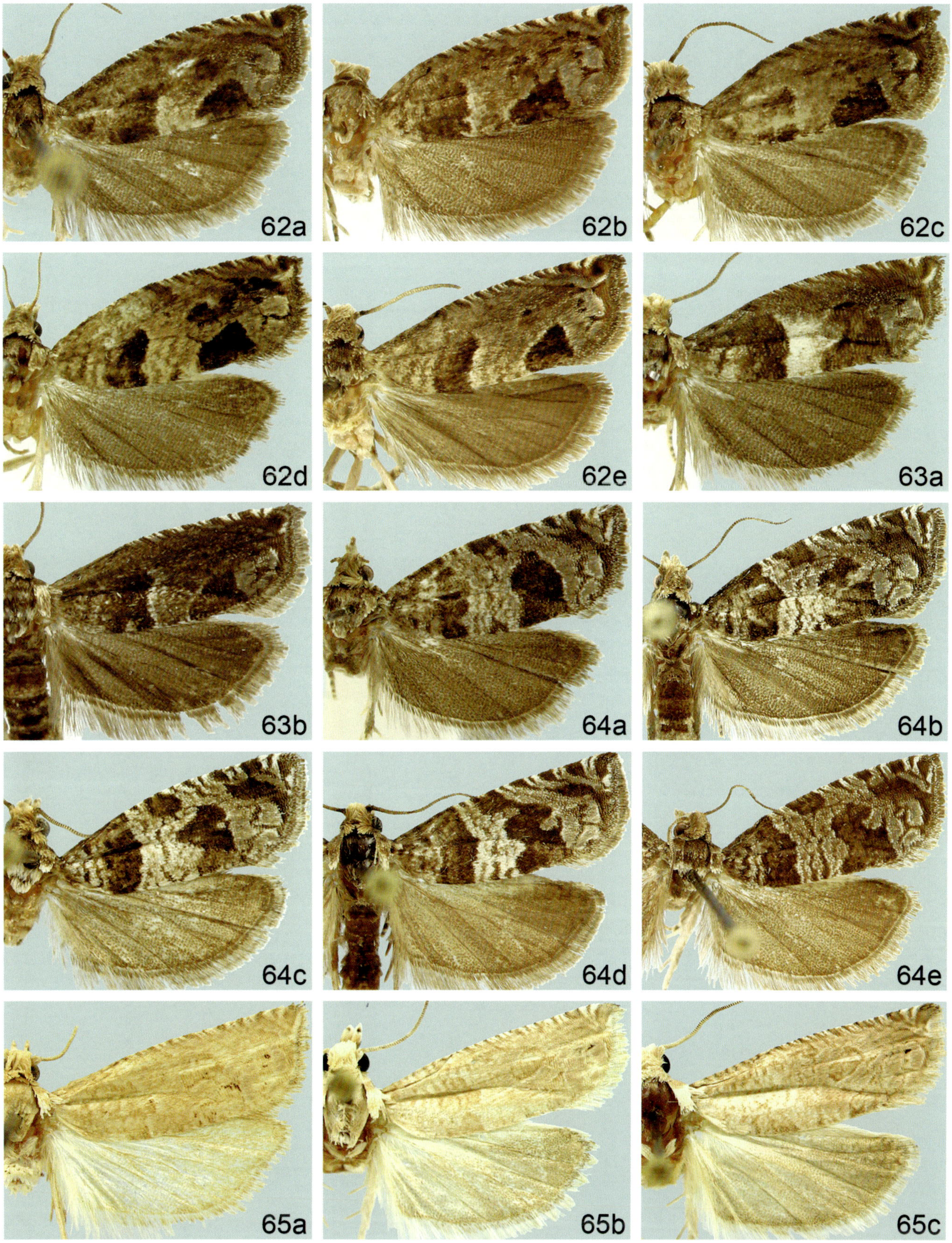
62a
62b
62c
62d
62e
63a
63b
64a
64b
64c
64d
64e
65a
65b
65c

# PLATE Z

**66.** ***Sonia filiana*** (Busck). FWL: 10.1–12.2 mm (mean = 11.1).

a. ♂. California, San Diego County, San Diego, 28 October 1908, W. S. Wright, USNM.
b. ♀. California, San Diego County, San Diego, 12 October, G. H. Field, USNM.
c. ♂. California, San Diego County, Torrey Pines State Preserve, 18–25 September 2005, N. Bloomfield, USNM.

**67.** ***Sonia vovana*** (Kearfott). FWL: 6.9–11.3 mm (mean = 9.2).

a. ♂. Arizona, Coconino County, Walnut Canyon, 12 August 1964, J. G. Franclemont, USNM.
b. ♀. Colorado, Otero County, 15 mi S of La Junta, 18 August 1997, D. J. Wright, USNM.
c. ♂. California, San Bernardino, New York Mountains, 31 August 1941, C. Henne, USNM.
d. ♀. Arizona, Mohave County, Beaver Dam Canyon, 26 September 2000, G. J. Balogh, USNM.
e. ♂. New Mexico, Hidalgo County, 1 mi N of Rodeo, 16 August 2013, B. Walsh, MEM.
f. ♀. Colorado, Otero County, 15 mi S of La Junta, 18 August 1997, D. J. Wright, USNM.

**68.** ***Sonia hodgesi*** Wright and Gilligan. FWL: 4.7–7.0 mm (mean = 5.8).

a. ♂. HTP. Arizona, Santa Rita Mountains, Madera Canyon, 5 August 1959, R. W. Hodges, USNM.
b. ♂. PTP. Arizona, Cochise County, Coronado National Forest, 14 August 2011, R. Brown and S. Lee, USNM.
c. ♀. PTP. Arizona, Cochise County, SW Research Station, 14 August 2011, R. Brown and S. Lee, MEM.
d. ♂. PTP. Arizona, Cochise, Coronado National Forest, 14 August 2011, R. Brown and S. Lee, MEM.
e. ♀. PTP. Arizona, Santa Cruz County, Pena Blanca Canyon, 11 August 1959, R. W. Hodges, CUIC.

**69.** ***Sonia gibsoni*** Wright and Gilligan. FWL: 8.1 mm.

♂. HTP. New Mexico, Grant County,14 mi SW of Silver City, 2 August 2005, L. D. Gibson, USNM.

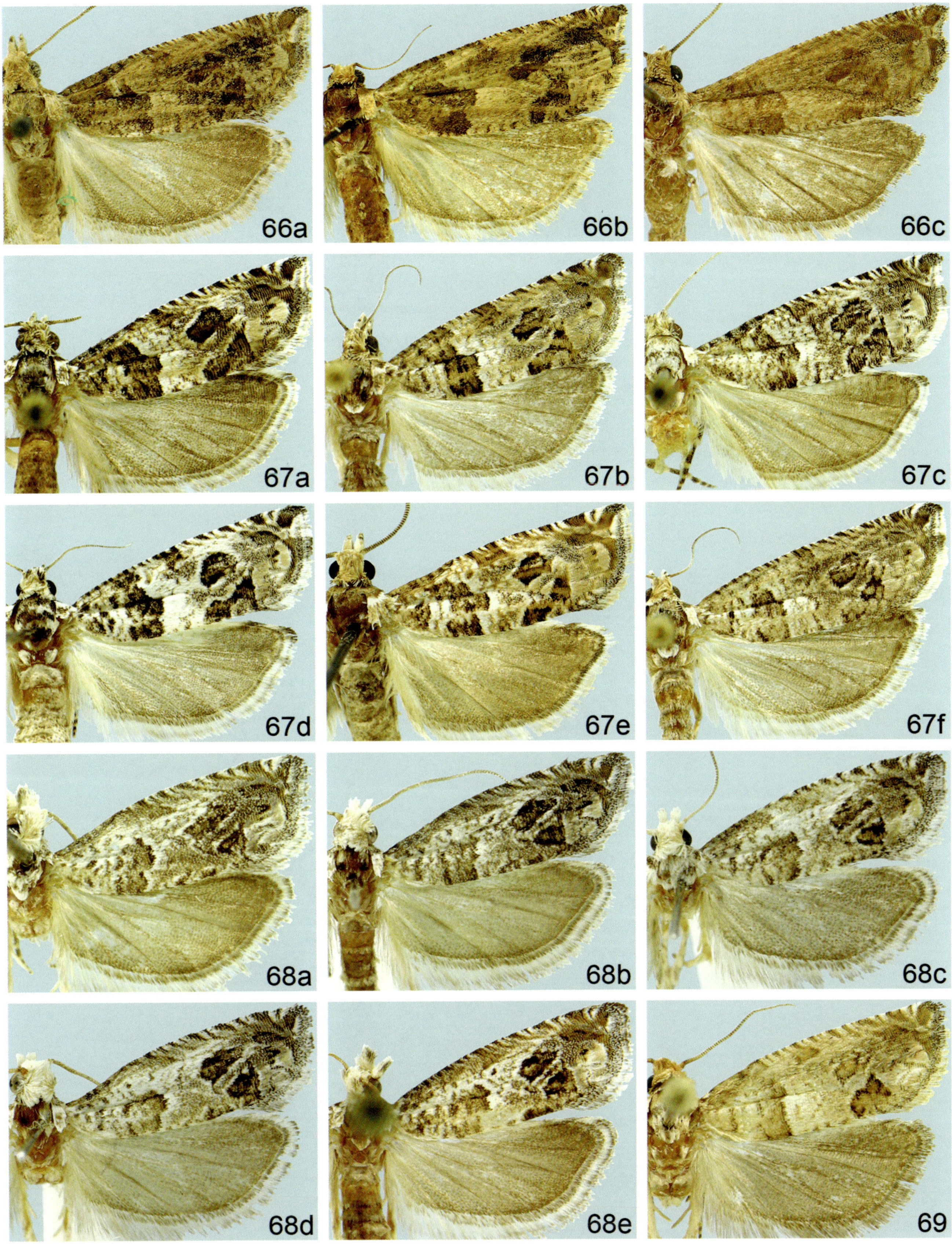
66a
66b
66c
67a
67b
67c
67d
67e
67f
68a
68b
68c
68d
68e
69

# PLATE AA

**70.** ***Suleima helianthana*** (Riley). FWL: 5.3–10.8 mm (mean = 7.7).

a. ♂. LTP. Texas, 21 August 1873, C. F. Riley, USNM.
b. ♂. Texas, Padre Island National Seashore, 17 May 1976, A. and M. E. Blanchard, USNM.
c. ♂. Texas, Briscoe County, Caprock Canyon State Park, 29 March 2015, J. D. Hooper, USNM.
d. ♂. Arizona, Cochise County, Huachuca Mountains, 26 March 1986, R. S. Wielgus, USNM.
e. ♂. New Mexico, Eddy County, Dark Canyon Road, 13 April 2005, G. J. Balogh, USNM.
f. ♀. Colorado, Arapahoe County, Deer Trail, 19 September 2007, T. Gilligan and C. Harp, TMG.
g. ♂. Colorado, Alamosa County, Great Sand Dunes, 28 June 1982, R. W. Hodges, USNM.
h. ♀. Kansas, Morton County, 7.5 mi N of Elkhart, 2 August 1999, D. J. Wright, USNM.
i. ♂. Ohio, Wyandot County, Killdeer Plains WA, 16 August 2006, D. J. Wright, USNM.
j. ♀. Ohio, Wyandot County, Killdeer Plains WA, 16 August 2006, D. J. Wright, USNM.
k. ♀. Kentucky, McCracken County, Paducah Zip Track, 12 August 2008, L. D. Gibson, LDG.
l. ♂. Kentucky, McCracken County, Paducah Zip Track, 12 August 2008, L. D. Gibson, LDG.
m. ♂. Illinois, Putnam County, 17 July 1959, M. O. Glenn, INHS.
n. ♂. Illinois, Putnam County, 12 July 1965, M. O. Glenn, INHS.
o. ♂. Kentucky, Campbell County, A. J. Jolly County Park, 15 July 1982, L. D. Gibson, LDG.

71a
71b
71c
71d
71e
71f
72a
72b
72c
72d
72e
72f
72g
72h
72i

# PLATE CC

**73.** ***Suleima baracana*** (Kearfott). FWL: 6.2–10.9 mm (mean = 8.5).

a. ♂. LTP. Utah, [Tooele County], Stockton, 5 September 1904, T. Spalding, AMNH.
b. ♂. Colorado, Denver, 3 August 1904, A. J. Oslar, AMNH. [LTP. *S. caracana*]
c. ♂. Washington, Whitman County, Wilma, 12 April 1934, J. F. G. Clarke, USNM. [reared from *Helianthus annuus*]
d. ♀. Washington, Whitman County, Wilma, 6 April 1934, J. F. G. Clarke, USNM. [reared from *Helianthus annuus*]
e. ♂. California, Riverside County, Temecula, 13 May 1994, R. Leuschner, EME.
f. ♀. California, Riverside County, Temecula, 12 May 1994, R. Leuschner, EME.
g. ♂. Colorado, Logan County, 1.2 mi W of County Road 61, 8 August 2007, L. D. Gibson, LDG.
h. ♀. Colorado, Weld County, 0.25 mi N of I-76 at Exit 57, 7 August 1996, L. D. Gibson, LDG.
i. ♀. Nebraska, Cherry County, Valentine NWR, 8 June 1988, R. W. Hodges, USNM.
j. ♀. Arizona, Cochise County, Pueblo Del Sol, 22 June 1986, R. S. Wielgus, USNM.
k. ♂. Texas, Brown County, Lake Brownwood State Park, 21 April 1966, A. and M. E. Blanchard, USNM.
l. ♂. Texas, Culberson County, Guadalupe Mountains, Cherry Canyon, 22 May 1973, R. W. Hodges, USNM.
m. ♀. Texas, Culberson County, Guadalupe Mountains, Nickel Canyon, 22 May 1973, R. W. Hodges, USNM.
n. ♂. Texas, Culberson County, Guadalupe Mountains, Frijoles, 24 May 1973, A. and M. E. Blanchard, USNM.
o. ♂. New Mexico, Dona Ana County, Las Cruces, 8 April 1979, R. Leuschner, EME.

73a
73b
73c
73d
73e
73f
73g
73h
73i
73j
73k
73l
73m
73n
73o

# PLATE DD

**74.** ***Suleima leeana*** Wright and Gilligan. FWL: 6.2–8.6 mm (mean = 7.6).

a. ♂. HTP. Arizona, Cochise County, Paradise Road, 17 August 2015, R. L. Brown, USNM.
b. ♀. PTP. Arizona, Cochise County, Chiricahua Mountains, 9 August 2013, J. W. Brown, USNM.
c. ♀. PTP. Arizona, Cochise County, John Hands Campground, 20 August 2015, Sangmi Lee, ASUHIC.
d. ♂. PTP. Arizona, Cochise County, Paradise Road, 17 August 2015, R. L. Brown, ASUHIC.
e. ♂. PTP. Arizona, Cochise County, Paradise Road, 17 August 2015, R. L. Brown, ASUHIC.
f. ♂. PTP. New Mexico, Socorro County, Cibola National Forest, 15 August 2015, S. Johnson and J. Vargo, SJ.
g. ♀. PTP. Idaho, Oneida County, Curlew National Grassland, 1 August 2001, D. J. Wright, USNM.
h. ♂. PTP. Texas, Davis Mountains, Mount Locke, 30 August 1969, A. and M. E. Blanchard, USNM.
i. ♀. PTP. Utah, Sanpete County, Ephraim Canyon Road, 18 July 2006, D. J. Wright, USNM.

**75.** ***Suleima blanchardi*** Wright and Gilligan. FWL: 5.8–7.7 mm (mean = 6.7).

a. ♂. HTP. Texas, Padre Island National Seashore, 29 September 1975, A. and M. E. Blanchard, USNM.
b. ♀. PTP. Texas, Padre Island National Seashore, 17 May 1976, A. and M. E. Blanchard, USNM.
c. ♂. PTP. Texas, Padre Island National Seashore, 17 May 1976, A. and M. E. Blanchard, USNM.
d. ♀. PTP. Texas, Padre Island National Seashore, 17 May 1976, A. and M. E. Blanchard, USNM.
e. ♂. PTP. Texas, Padre Island National Seashore, 5 June 1978, A. and M. E. Blanchard, USNM.
f. ♂. PTP. Texas, Padre Island National Seashore, 29 September 1975, A. and M. E. Blanchard, USNM.

74a
74b
74c
74d
74e
74f
74g
74h
74i
75a
75b
75c
75d
75e
75f

# PLATE EE

**76.** ***Suleima skinnerana*** Heinrich. FWL: 8.0–9.6 mm (mean = 9.0).

a. ♂. HTP. Arizona, Cochise County, Huachuca Mountains, Carr Canyon, August 1905, H. Skinner, USNM.

b. ♂. PTP. Arizona, Cochise County, Huachuca Mountains, Carr Canyon, August 1905, H. Skinner, USNM.

c. ♂. New Mexico, Grant County, Emory Pass, Gila National Forest, 16 August 2015, S. Johnson and J. Vargo, SJ.

d. ♀. PTP. Arizona, Cochise County, Huachuca Mountains, Carr Canyon, August 1905, H. Skinner, USNM.

e. ♂. Texas, Davis Mountains, Mount Locke, 30 August 1969, A. and M. E. Blanchard, USNM.

f. ♂. Texas, Davis Mountains, Mount Locke, 30 August 1969, A. and M. E. Blanchard, USNM.

**77.** ***Suleima flavidana*** Wright and Gilligan. FWL: 5.7–8.6 mm (mean = 7.5).

a. ♂. HTP. Arizona, Cochise County, Chiricahua Mountains, 31 July–2 August 1964, D. R. Davis, USNM.

b. ♀. PTP. Arizona, Cochise County, Chiricahua Mountains, 3–6 August, 1964, D. R. Davis, USNM.

c. ♀. PTP. Arizona, Cochise County, Chiricahua Mountains, 3–6 August, 1964, D. R. Davis, USNM.

d. ♀. PTP. Texas, Culberson County, Guadalupe Mountains National Park, 1 August 1992, E. C. Knudson, MGCL.

e. ♀. PTP. Texas, Culberson County, Guadalupe Mountains, Nickel Canyon, 10 July 1968, A. and M. E. Blanchard, USNM.

f. ♂. PTP. Texas, Culberson County, Sierra Diablo WMA, 14 July 1969, A. and M. E. Blanchard, USNM.

g. ♂. PTP. Texas, Culberson County, Sierra Diablo WMA, 18 August 1984, E. C. Knudson, MGCL.

h. ♂. PTP. Texas, Brewster County, BIBE Green Gulch, 16–18 May 1999, E. C. Knudson, MGCL.

i. ♂. PTP. Arizona, Santa Rita Mountains, Madera Canyon, 3 September 1959, R. W. Hodges, USNM.

76a
76b
76c
76d
76e
76f
77a
77b
77c
77d
77e
77f
77g
77h
77i

# PLATE FF

**78.** ***Suleima daracana*** (Kearfott). FWL: 6.3–9.2 mm (mean = 8.0).

a. ♂. LTP. California, Placer County, 1 May, A. H. Vachell, USNM.
b. ♀. SNTP. Californian, Placer County, 20 April, A. H. Vachell, USNM.
c. ♂. SNTP. California, Placer County, 1 May, A. H. Vachell, USNM.
d. ♂. Washington, Whitman County, Pullman, 8 June 1930, J. F. G. Clarke, USNM.
e. ♀. Washington, Whitman County, Pullman, 12 May 1930, J. F. G. Clarke, USNM.
f. ♀. Washington, Whitman County, Union Flat, 15 May 1929, J. F. G. Clarke, USNM.
g. ♂. Washington, Whitman County, Pullman, 12 May 1930, J. F. G. Clarke, USNM.
h. ♀. Washington, Whitman County, Pullman, 17 April 1934, J. F. G. Clarke, USNM. [reared from *Helianthella douglasii*]
i. ♂. Washington, Whitman County, Union Flat, 15 May 1929, J. F. G. Clarke, USNM.
j. ♂. Oregon, Baker County, 6.4 mi N of Richland, 7 June 1970, J. F. G. Clarke, USNM.
k. ♀. Oregon, Baker County, 9 mi E of Halfway, 7 June 1970, J. F. G. Clarke, USNM.
l. ♀. Oregon, 28 mi SE of Joseph, 19 June 1970, J. F. G. Clarke, USNM.
m. ♂. Oregon, 28 mi SE of Joseph, 21 June 1970, J. F. G. Clarke, USNM.
n. ♂. Oregon, 28 mi SE of Joseph, 21 June 1970, J. F. G. Clarke, USNM.
o. ♂. Oregon, 28 mi SE of Joseph, 19 June 1970, J. F. G. Clarke, USNM.

78a
78b
78c
78d
78e
78f
78g
78h
78i
78j
78k
78l
78m
78n
78o

# PLATE GG

**79.** ***Suleima cinerodorsana*** Heinrich. FWL: 5.0–7.4 mm (mean = 6.5).

- a. ♂. HTP. Pennsylvania, [Montgomery County], Oak Station, 11 August 1911, F. Marloff, AMNH.
- b. ♀. Ohio, Wyandot County, Killdeer Plains, 16 August 2006, D. J. Wright, USNM.
- c. ♂. Illinois, Putnam County, 14 August 1967, M. O. Glenn, INHS.
- d. ♀. Illinois, Putnam County, 27 July 1965, M. O. Glenn, INHS.
- e. ♂. Canada, Ontario, Vineland Station, 5 August 1936, W. L. Putnam, CNC. [reared from *Helianthus decapetalis*]
- f. ♀. Canada, Ontario, Vineland Station, 23 August 1936, W. L. Putnam, CNC. [reared from *Helianthus decapetalis*]
- g. ♂. Kentucky, Owsley County, E side County Road 1411, 20 July 1980, L. D. Gibson, LDG.

**80.** ***Suleima* species.** FWL: 7.7–7.8 mm (mean = 7.8).

- a. ♂. Colorado, Chaffee County, 4 mi SW of Buena Vista, 15 July 1982, R. W. Hodges, USNM.
- b. ♂. Colorado, Boulder County, Silver Saddle Motel, 25–26 June 2015, J. W. Brown, USNM.

**81.** ***Suleima mendaciana*** Blanchard and Knudson. FWL: 5.6–6.9 mm (mean = 6.2).

- a. ♂. HTP. Texas, Brewster County, Big Bend National Park, 28 September 1981, E. C. Knudson, USNM.
- b. ♂. Texas, Cameron County, Southmost, 27 October 1986, E. C. Knudson, MGCL.
- c. ♀. Texas, Brewster County, Big Bend National Park, 5–7 May 1997, E. C. Knudson, MGCL.
- d. ♀. PTP. Texas, Cameron County, South Padre Island, 24 October 1982, E. C. Knudson, MGCL.
- e. ♂. Texas, Uvalde County, Garner State Park, 6 October 1984, E. C. Knudson, MGCL.
- f. ♀. Texas, Culberson County, Sierra Diablo WMA, 18 August 1984, E. C. Knudson, MGCL.

79a
79b
79c
79d
79e
79f
79g
80a
80b
81a
81b
81c
81d
81e
81f

# PLATE HH

**82.** ***Notocelia culminana*** (Walsingham). FWL: 7.0–9.3 mm (mean = 8.1).

- a. ♀. Canada, British Columbia, Tranquille Ecological Preserve, 15 August 2010, J. J. Dombroskie, CUIC.
- b. ♂. Canada, Alberta, Porcupine Hills, 31 July 2007, J. J. Dombroskie, CUIC.
- c. ♂. Washington, Whatcom County, Bellingham, 13 August 1932, J. F. G. Clarke, CUIC.
- d. ♀. Montana, Sweet Grass County, 7.75 mi N of Big Timber, 27 August 1969, J. G. Franclemont, USNM.
- e. ♀. Colorado, Moffat County, 1 mi S of Craig, 4 August 2007, L. D. Gibson, LDG.
- f. ♂. Nevada, Elko County, Pine Creek Campground, 5 August 1981, J. Powell and E. Randall, EME.
- g. ♂. Michigan, Otsego County, Old State Road at Turtle Creek, 19 August 2007, L. D. Gibson, LDG.
- h. ♀. Michigan, Cheboygan County, Ocqueoc Lake, 22 August 1985, R. W. Hodges, USNM.
- i. ♀. Massachusetts, Barnstable County, S. Chatham, 1982, E. Jäckh, USNM.

**83.** ***Notocelia lobana*** Wright and Gilligan. FWL: 7.2–8.7 mm (mean = 7.8).

- a. ♂. HTP. North Carolina, Swain County, Great Smoky Mountains National Park, 18 August 2001, J. B. Sullivan and L. Deutschman, USNM.
- b. ♂. PTP. Kentucky, Boone County, Big Bone Lick State Park, 25 August 1979, L. D. Gibson, LDG.
- c. ♀. PTP. Ohio, Wyandot County, Killdeer Plains, 16 August 2006, D. J. Wright, USNM.
- d. ♂. PTP. Ohio, Hancock County, Van Buren, 31 August 2009, T. M. Gilligan, TMG.
- e. ♀. PTP. Maryland, Charles County, Myrtle Grove WMA, 5 September 1996, J. Glaser, USNM.
- f. ♀. PTP. Ohio, Lake County, Hell Hollow Park, 24 August 2018, J. Wiedmann, JW.

82a
82b
82c
82d
82e
82f
82g
82h
82i
83a
83b
83c
83d
83e
83f

# PLATE II

**84.** ***Notocelia illotana*** (Walsingham). FWL: 6.5–8.8 mm (mean = 8.1).

a. ♂. Canada, British Columbia, 4.5 km NW of Kelowna, 1 June 2006, J.-F. Landry and L. Humble, CNC.
b. ♀. Washington, Chelan County, Leavenworth, 9 July 2010, D. G. Holden, WSDA.
c. ♀. Idaho, Latah County, 6 mi NE of Moscow, 29 June 1975, W. Turner and J. Powell, EME.
d. ♂. Pennsylvania, Allegheny County, Oak Station, 25 May 1912, F. Marloff, CUIC.
e. ♀. Pennsylvania, Allegheny County, Oak Station, 24 May 1912, F. Marloff, EME.
f. ♂. New York, Clinton County, Peru, 12 June 1916, CUIC.
g. ♀. Canada, Manitoba, Aweme, 20 June 1920, N. Criddle, CNC.
h. ♂. Canada, Nova Scotia, Parrsboro, Ottawa House, 30 June 1944, J. McDunnough, CNC.

**85.** ***Notocelia cynosbatella*** (Linnaeus). FWL: 6.1–9.8 mm (mean = 8.6).

a. ♂. France, St. Nazaire en Royens, 8 August 2009, F. Groenen, TMG.
b. ♂. France, Ain, Le Poizat, 6 June 2014, B. Landry, USNM.
c. ♀. Washington, King County, Seattle, 18 May 2018, T. M. Mustelin, TMM.
d. ♂. Washington, King County, Seattle, 25 May 2020, T. M. Mustelin, TMM.
e. ♂. Washington, Thurston County, 3 mi NW of Tenino, 6 March 2005, E. LaGasa, WSDA.
f. ♀. Washington, Thurston County, 1 mi N of Olympia, 10 April 2005, E. LaGasa, WSDA.
g. ♂. Connecticut, New Haven County, Hamden, 9 June 1967, D. C. Ferguson, USNM.

84a
84b
84c
84d
84e
84f
84g
84h
85a
85b
85c
85d
85e
85f
85g

# PLATE JJ

**86.** ***Notocelia purpurissatana*** Heinrich. FWL: 5.6–7.9 mm (mean = 6.7).

a. ♀. PTP. Canada, British Columbia, Vernon, May 1920, USNM.
b. ♀. Washington, Okanogan County, Lost Lake, J. F. G. Clarke, 25 July 1962, USNM.
c. ♂. Canada, Ontario, Renfrew County, 20 August 2016, J. and L. Dombroskie, CUIC.
d. ♂. Canada, British Columbia, Sheridan Lake, 26 July 2009, D. G. Holden, WSDA.
e. ♀. Canada, Alberta, Kananaskis, 9 August 2010, J. R. Dupris, CUIC.
f. ♂. Washington, Kittitas County, Lake Kachess, 9–15 July 2017, T. M. Mustelin, TMM.

**87.** ***Notocelia rosaecolana*** (Doubleday). FWL: 7.4–9.6 mm (mean = 8.5).

a. ♂. New York, Chautauqua County, Jamestown, 25 June 2017, J. J. Dombroskie, CUIC.
b. ♀. New York, Tompkins County, Ithaca, 6 June 2012, J. J. Dombroskie, CUIC.
c. ♂. New York, Tompkins County, Ithaca, 16 June 2018, J. J. Dombroskie, CUIC.
d. ♀. Kentucky, Gallatin County, Markland Dam, 15 June 1989, L. D. Gibson, LDG.
e. ♂. Kentucky, Larue County, Thompson Creek Glade SNA, 3 May 2006, L. D. Gibson, LDG.
f. ♀. Kentucky, Meade County, Otter Creek Park, 10 May 2007, L. D. Gibson, LDG.
g. ♀. Kentucky, Gallatin County, Markland Dam, 20 May 2007, L. D. Gibson, LDG.
h. ♀. Kentucky, Pendleton County, 2.3 mi NW of Foster, 23 May 2007, L. D. Gibson, LDG.
i. ♂. Canada, Nova Scotia, Kings County, Sheffield Mills, 26 June 2001, K. Neil, EME.

86a
86b
86c
86d
86e
86f
87a
87b
87c
87d
87e
87f
87g
87h
87i

Plates 1–38

# Genitalia Drawings

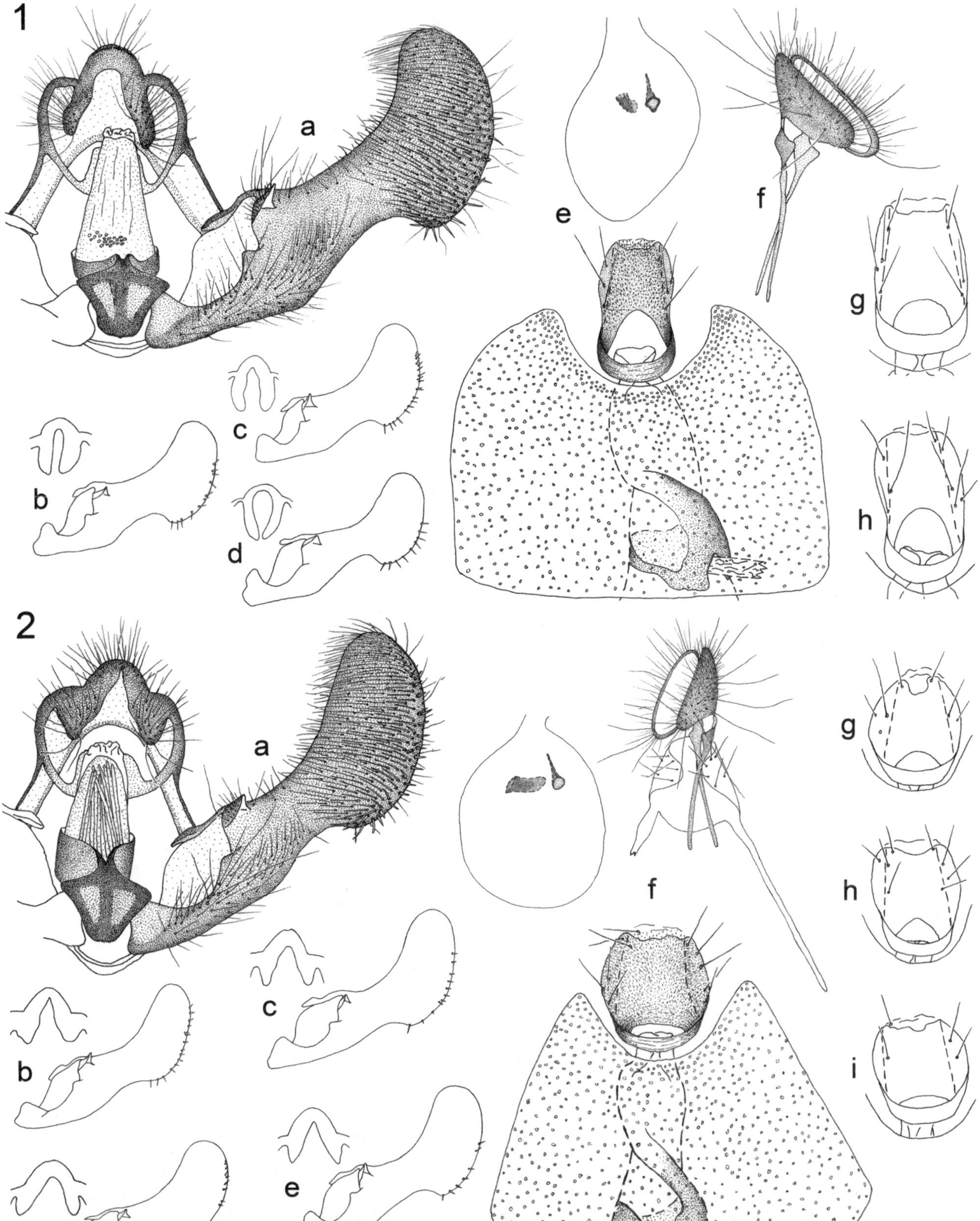


PLATE 1. **1. *Epiblema strenuana*:** a, male; b–d, valva/uncus/socii; e–f, female; g–h, sterigma.
**2. *Epiblema minutana*:** a, male; b–e, valva/uncus/socii; f, female; g–i, sterigma.

PLATE 2. **3. *Epiblema luctuosissima*:** a, male; b–c, valva/uncus/socii; d, female; e–f, sterigma. **4. *Epiblema adusta*:** a, male; b–c, valva/uncus/socii; d–e, female. **5. Unnamed morphospecies.** a, valva/uncus/socii; b, female. **6. *Epiblema ochraceana*:** a, male; b, valva/uncus/socii.

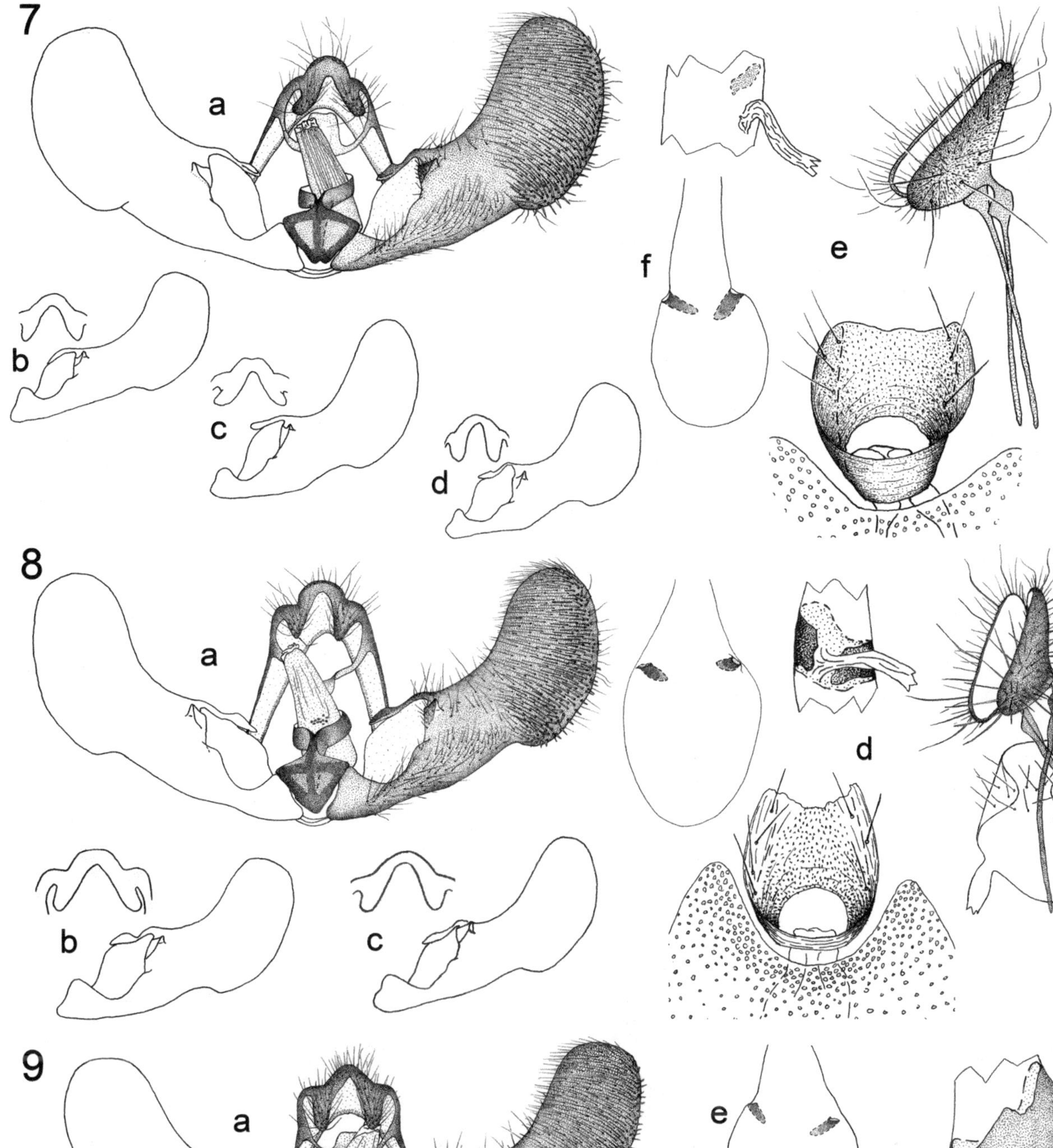


PLATE 3. **7.** ***Epiblema sosana*:** a, male; b–d, valva/uncus/socii; e–f, female. **8.** ***Epiblema cimarronensis*:** a, male; b–c, valva/uncus/socii; d, female. **9.** ***Epiblema wielgusi*:** a, male; b–c, valva/uncus/socii; d–e, female.

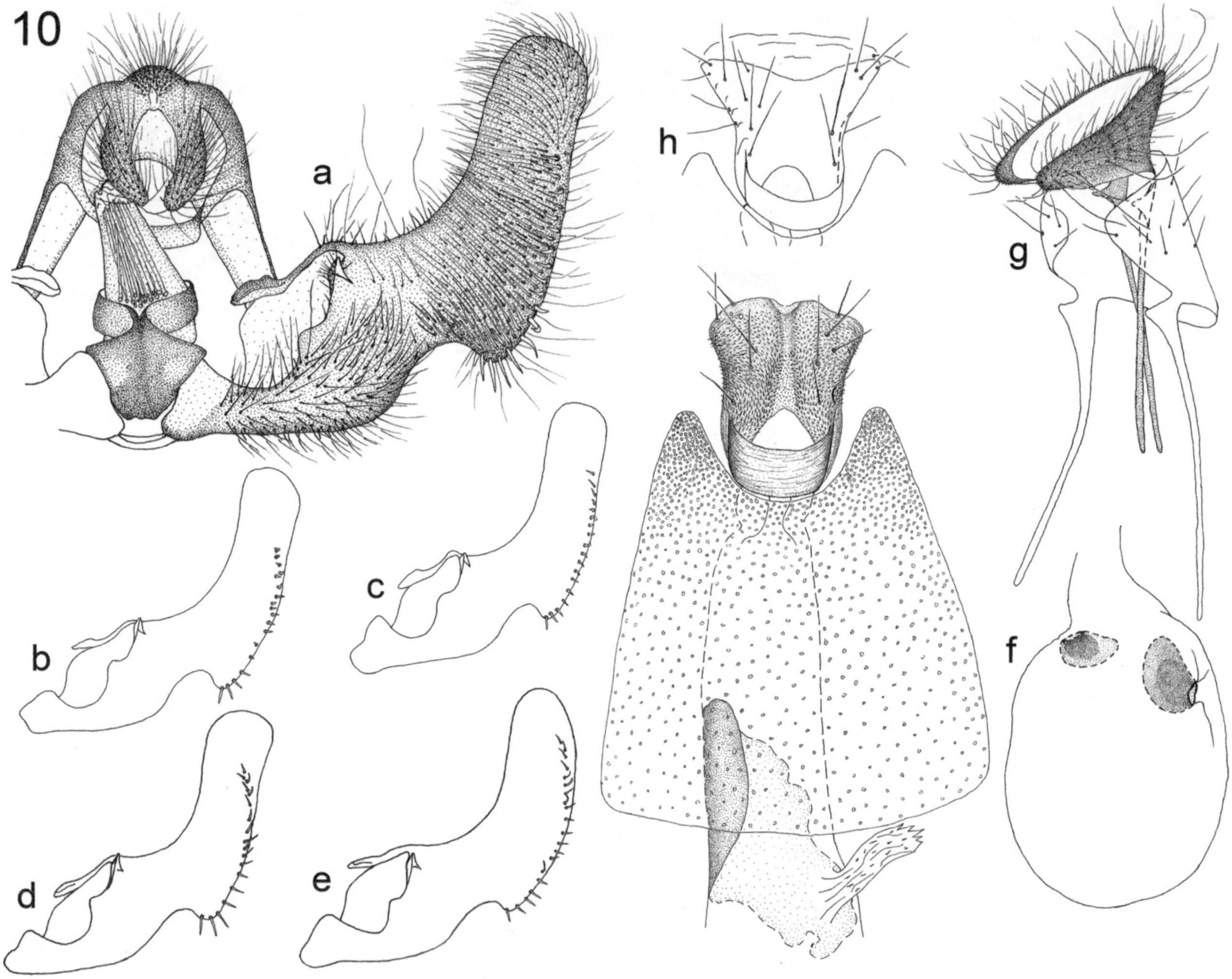


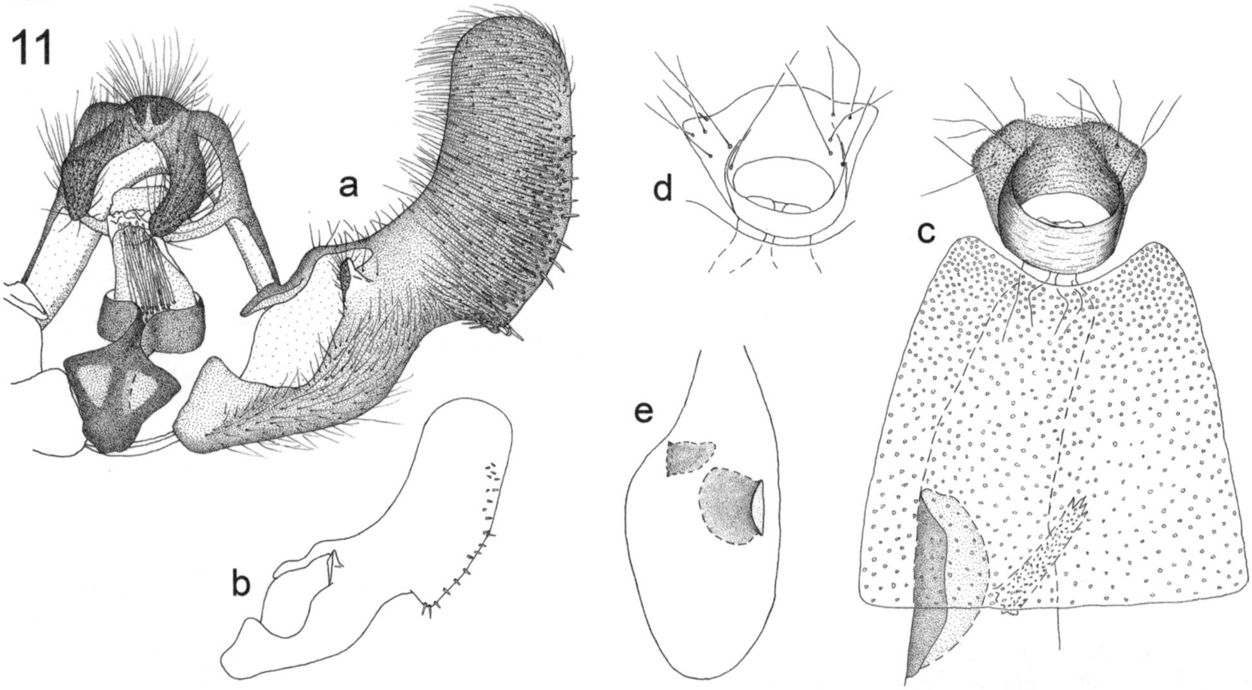


PLATE 4. **10.** ***Epiblema carolinana*:** a, male; b–e, valva; f–g, female; h, sterigma. **11.** ***Epiblema arizonana*:** a, male; b, valva; c, sterigma/sternum 7/ductus bursae; d, sterigma; e, corpus bursae.

12

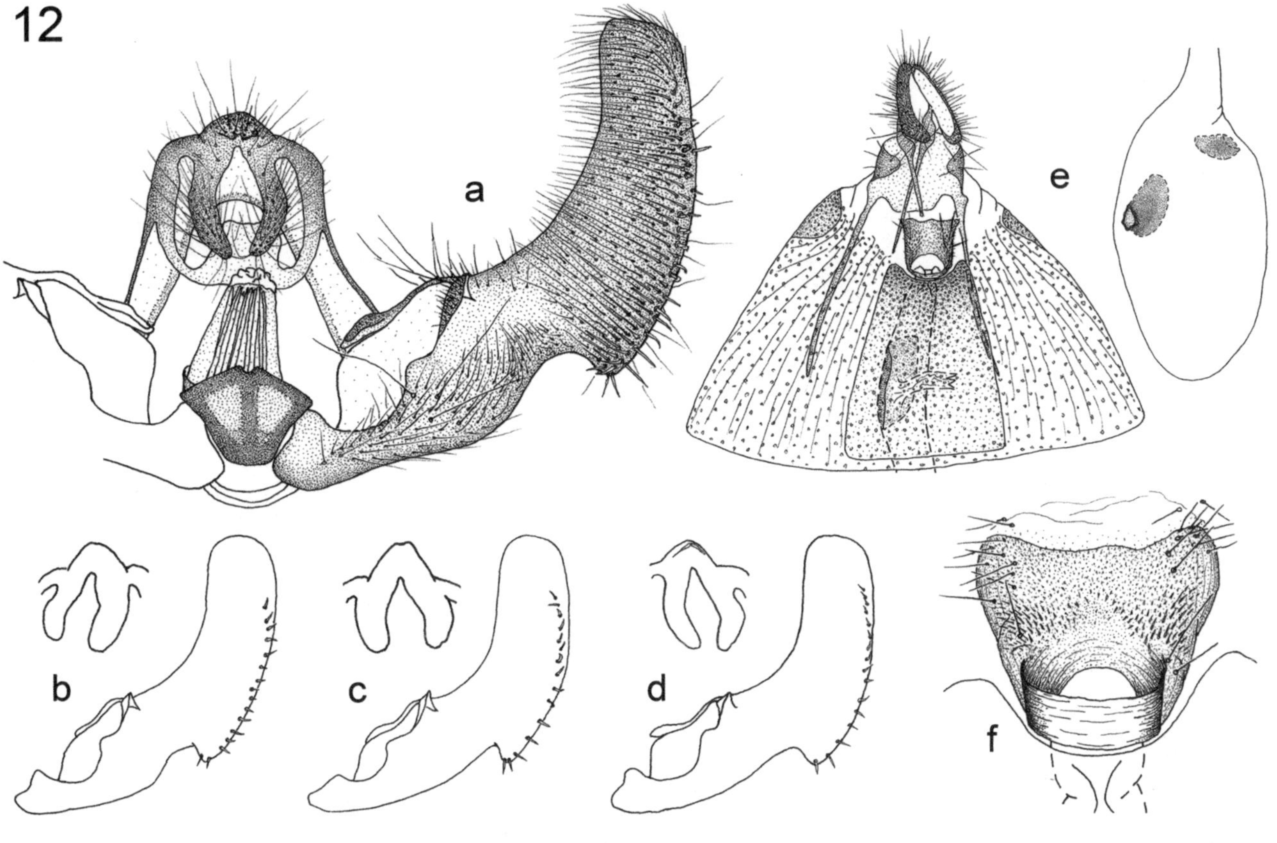


13

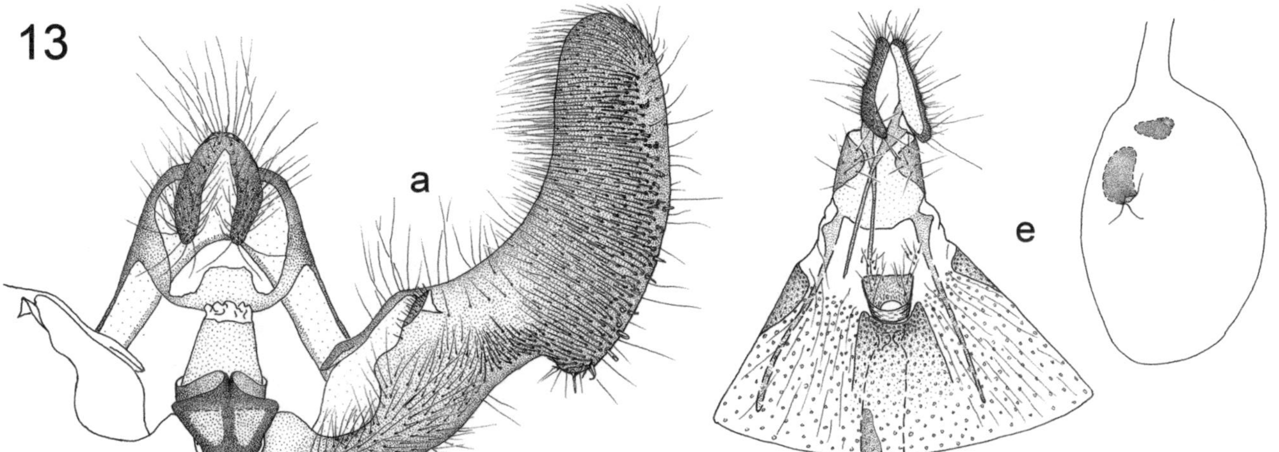


PLATE 5. **12.** ***Epiblema tripartitana*****:** a, male; b–d, valva/uncus/socii; e, female; f, sterigma; **13.** ***Epiblema glenni*****:** a, male; b–d, valva/uncus/socii; e, female; f, sterigma.

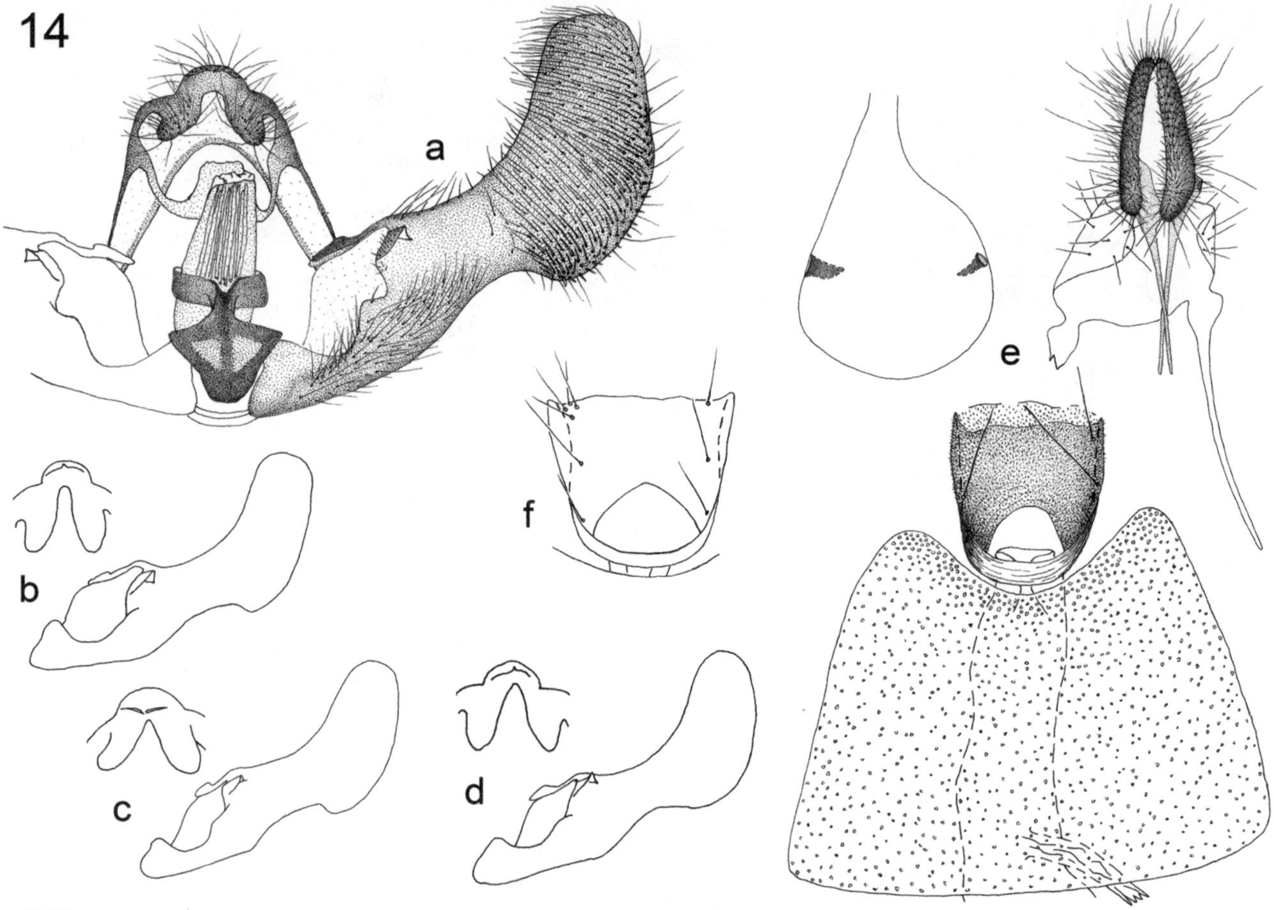


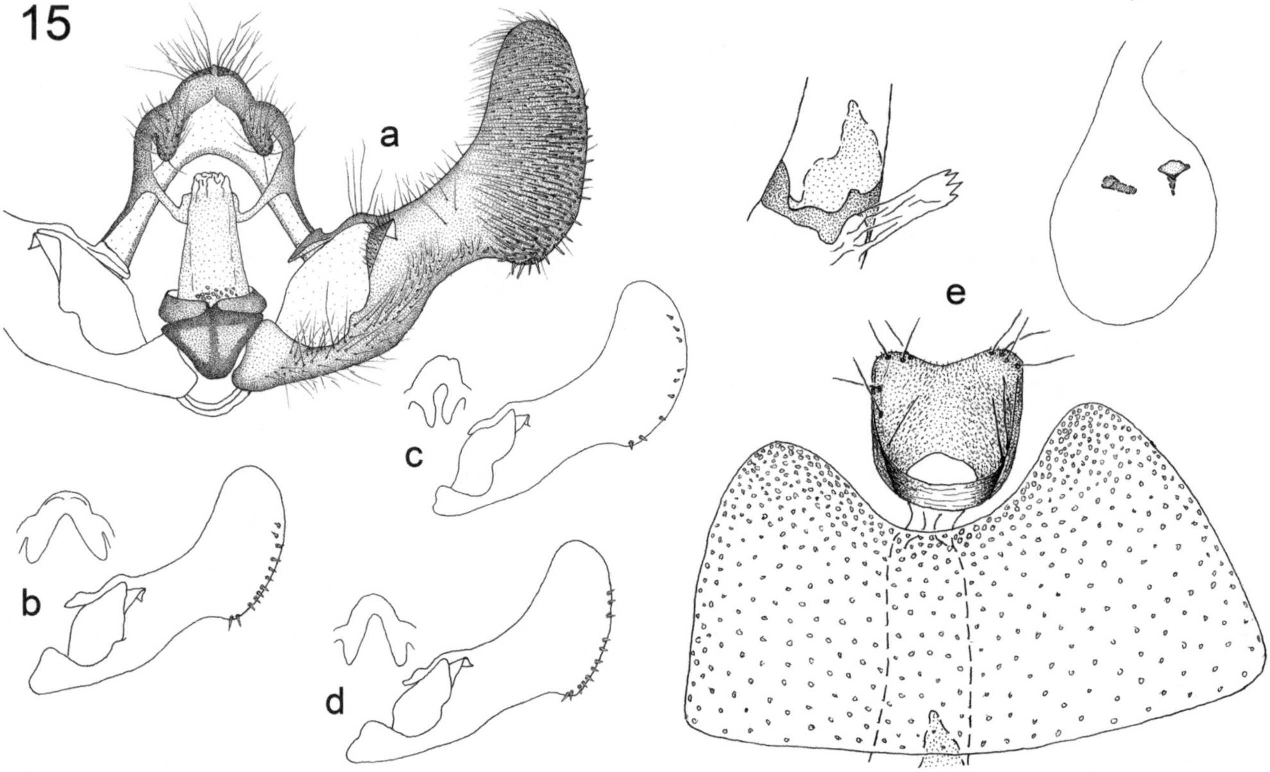


PLATE 6. **14.** ***Epiblema scudderiana*:** a, male; b–d, valva/uncus/socii; e, female; f, sterigma; **15.** ***Epiblema benignata*:** a, male; b–d, valva/uncus/socii; e, female.

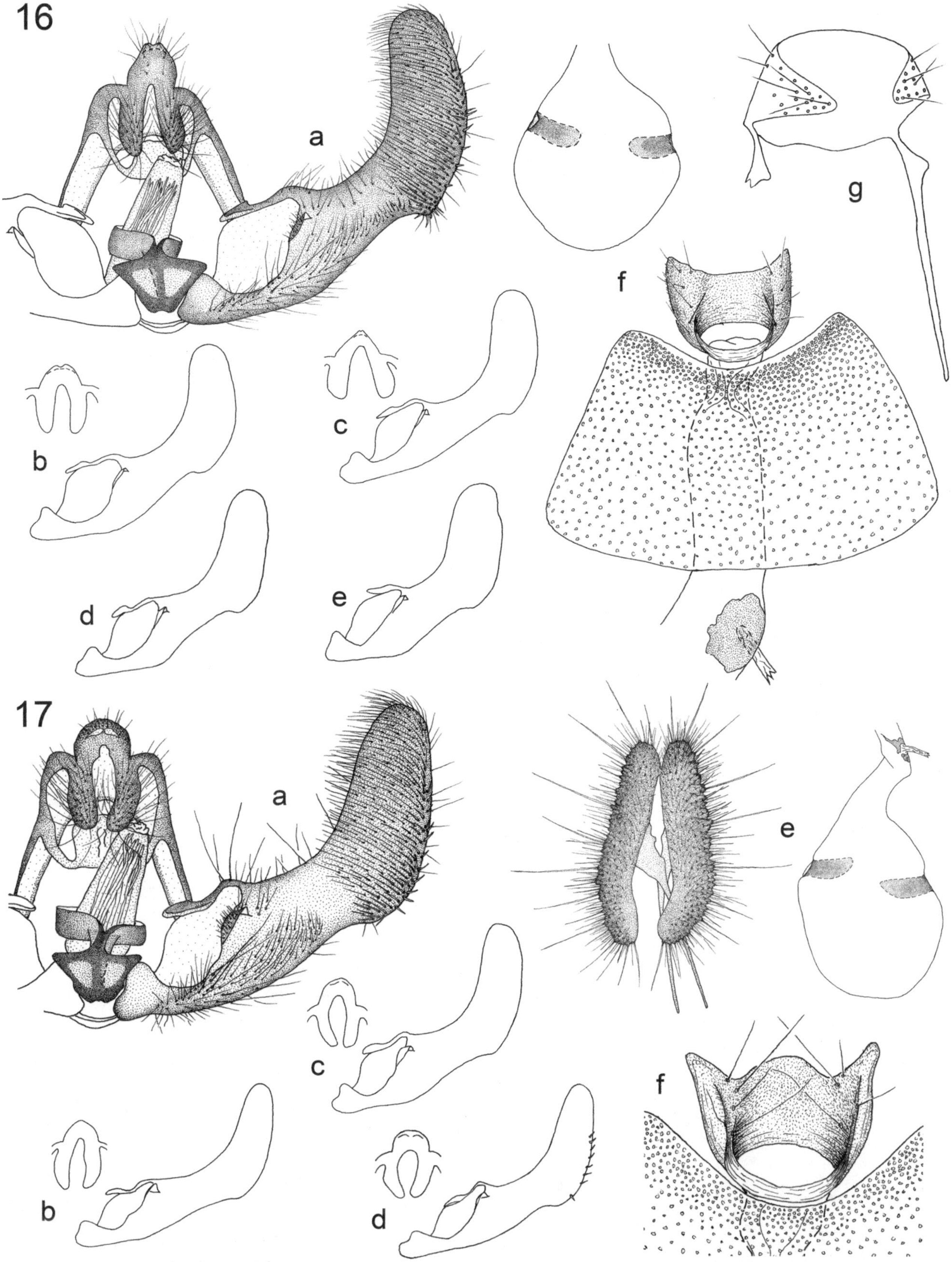


PLATE 7. **16.** ***Epiblema infelix*:** a, male; b–c, valva/uncus/socii; d–e, valva; f–g, female.
**17.** ***Epiblema gibsoni*:** a, male; b–d, valva/uncus/socii; e–f, female.

18

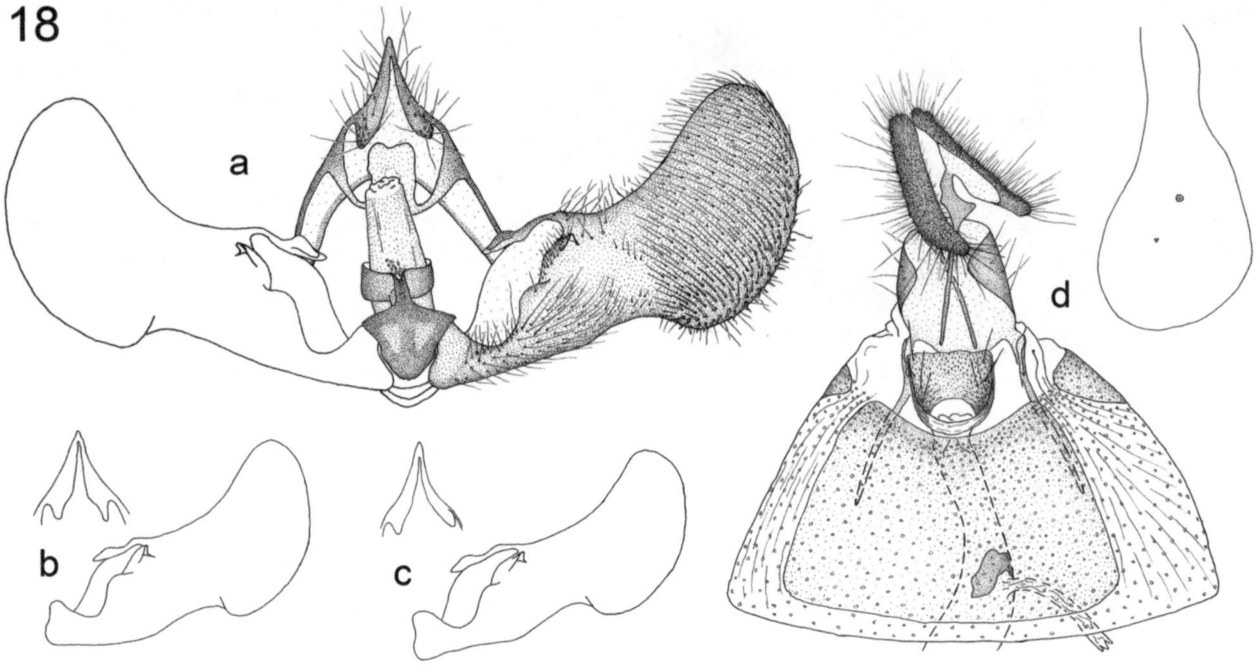


19

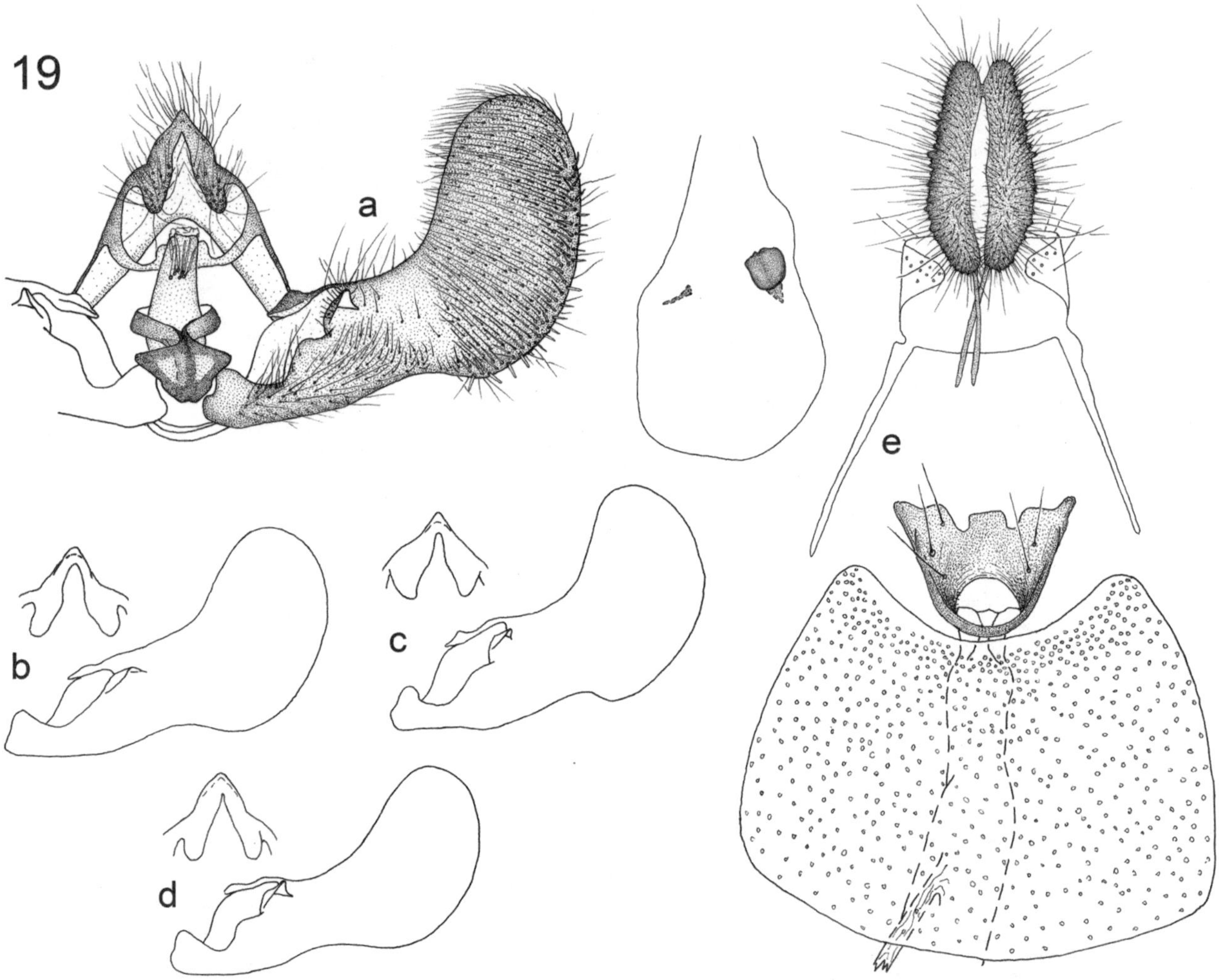


PLATE 8. **18.** ***Epiblema desertana*:** a, male; b–c, valva/uncus/socii; d, female. **19.** ***Epiblema rudei*:** a, male; b–d, valva/uncus/socii; e, female.

20

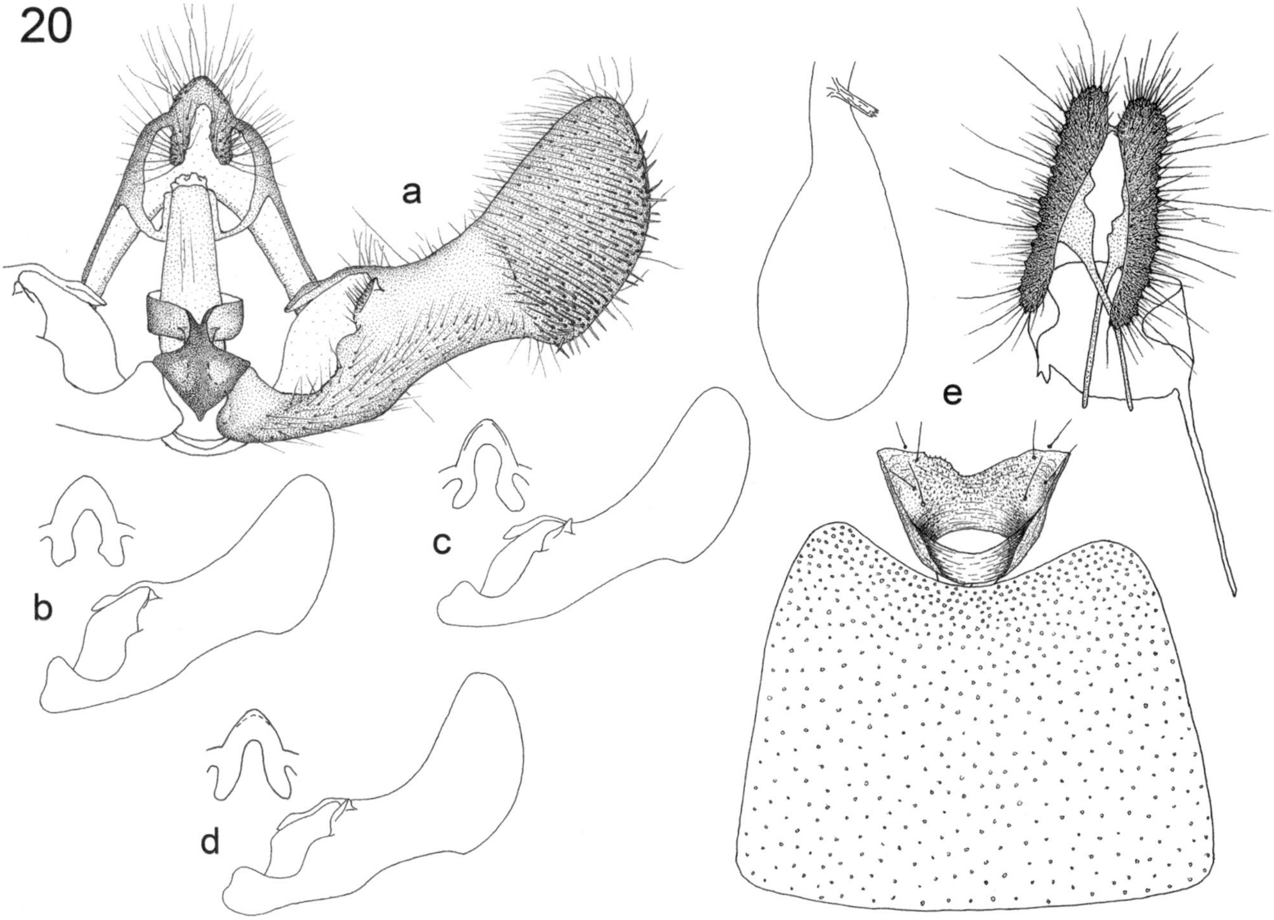


21

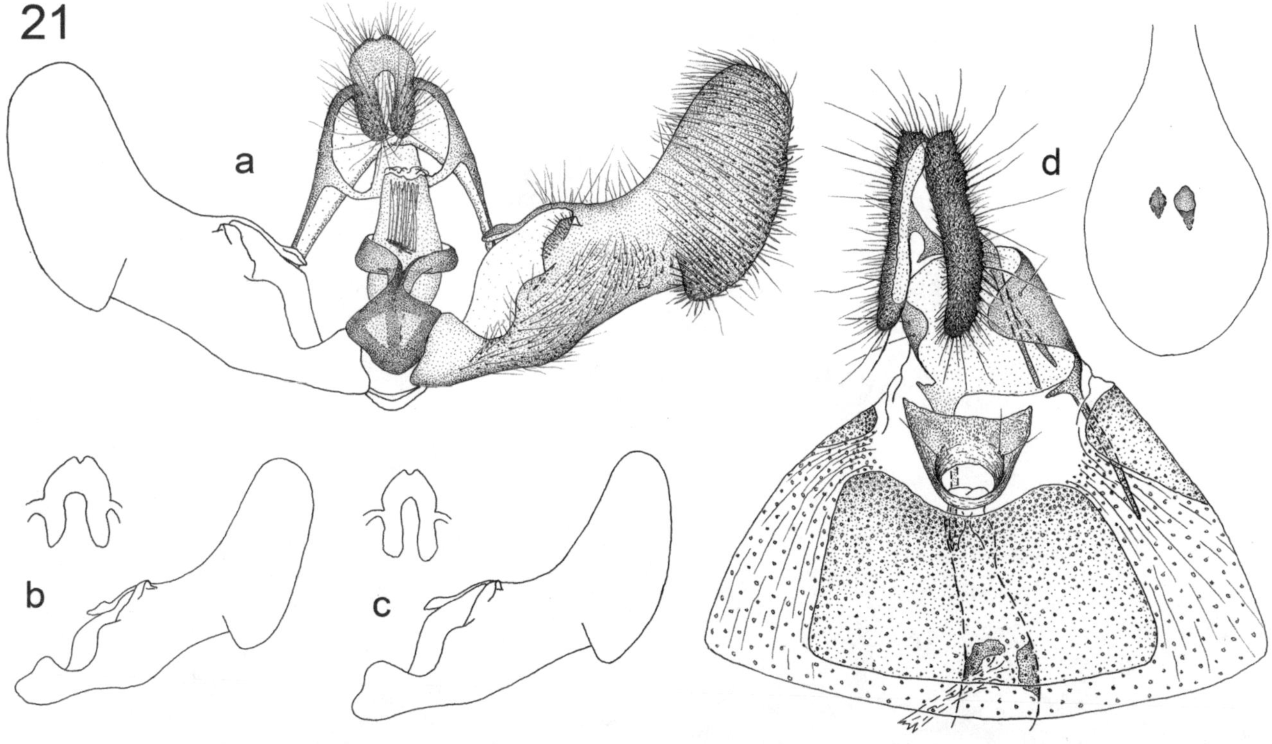


PLATE 9. **20. *Epiblema discretivana*:** a, male; b–d, valva/uncus/socii; e, female. **21. *Epiblema obfuscana*:** a, male; b–c, valva/uncus/socii; d, female.

PLATE 10. **22. *Epiblema radicana*:** a, male; b–c, valva/uncus/socii; d, female. **23. *Epiblema hirsutana*:** a, male; b, valva/uncus/socii; c, valva. **24. *Epiblema lyllana*:** a, male; b–c, valva/uncus/socii; d, female.

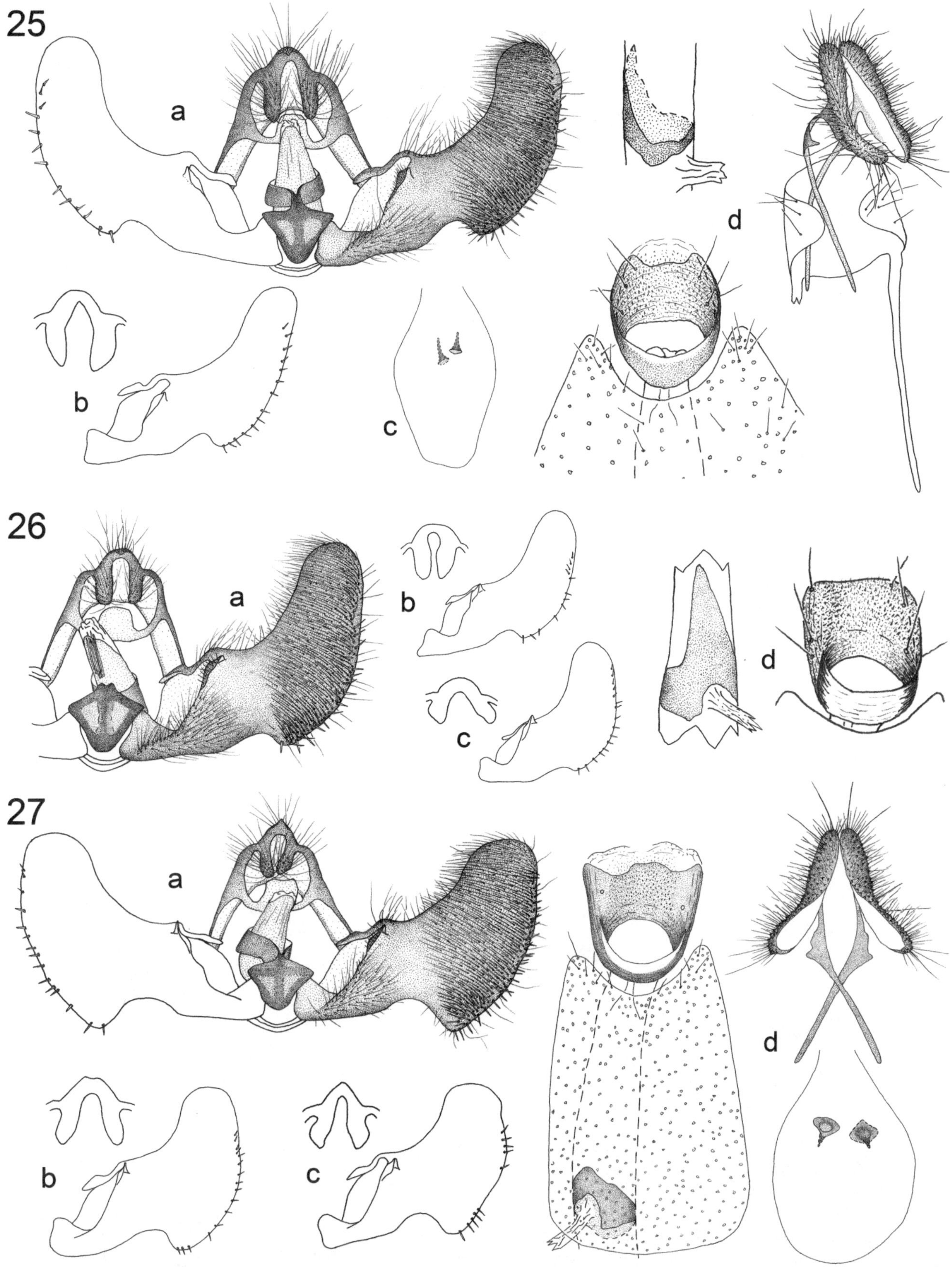


PLATE 11. **25. *Epiblema deverrae*:** a, male; b, valva/uncus/socii; c–d, female. **26. *Epiblema periculosana*:** a, male; b–c, valva/uncus/socii; d, sterigma/ductus bursae. **27. *Epiblema arctica*:** a, male; b–c, valva/uncus/socii; d, female.

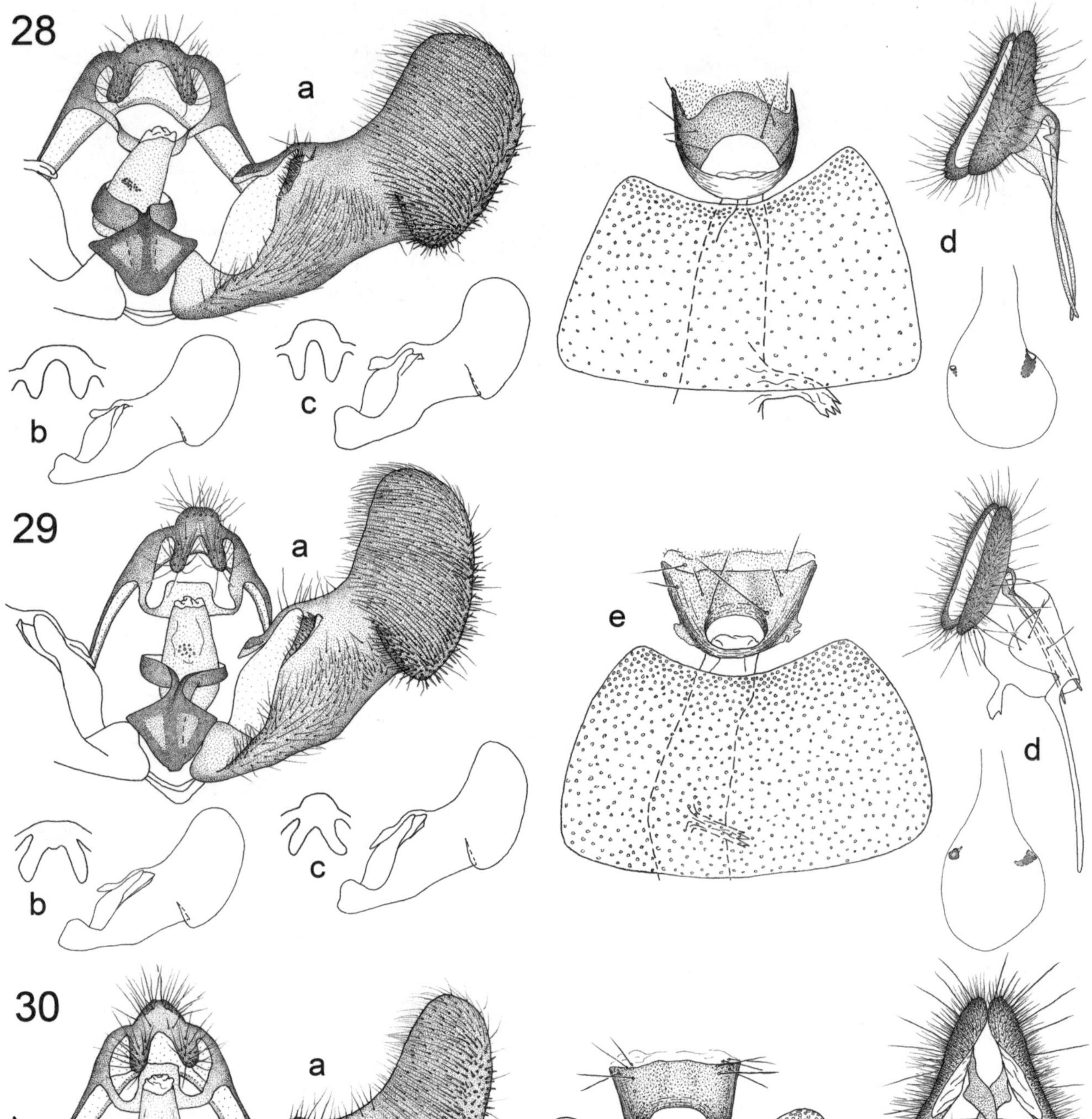


PLATE 12. **28.** ***Epiblema walsinghami*:** a, male; b–c, valva/uncus/socii; d, female. **29.** ***Epiblema iowana*:** a, male; b–c, valva/uncus/socii; d–e, female. **30.** ***Epiblema coracinana*:** a, male; b, valva/uncus/socii; c, female.

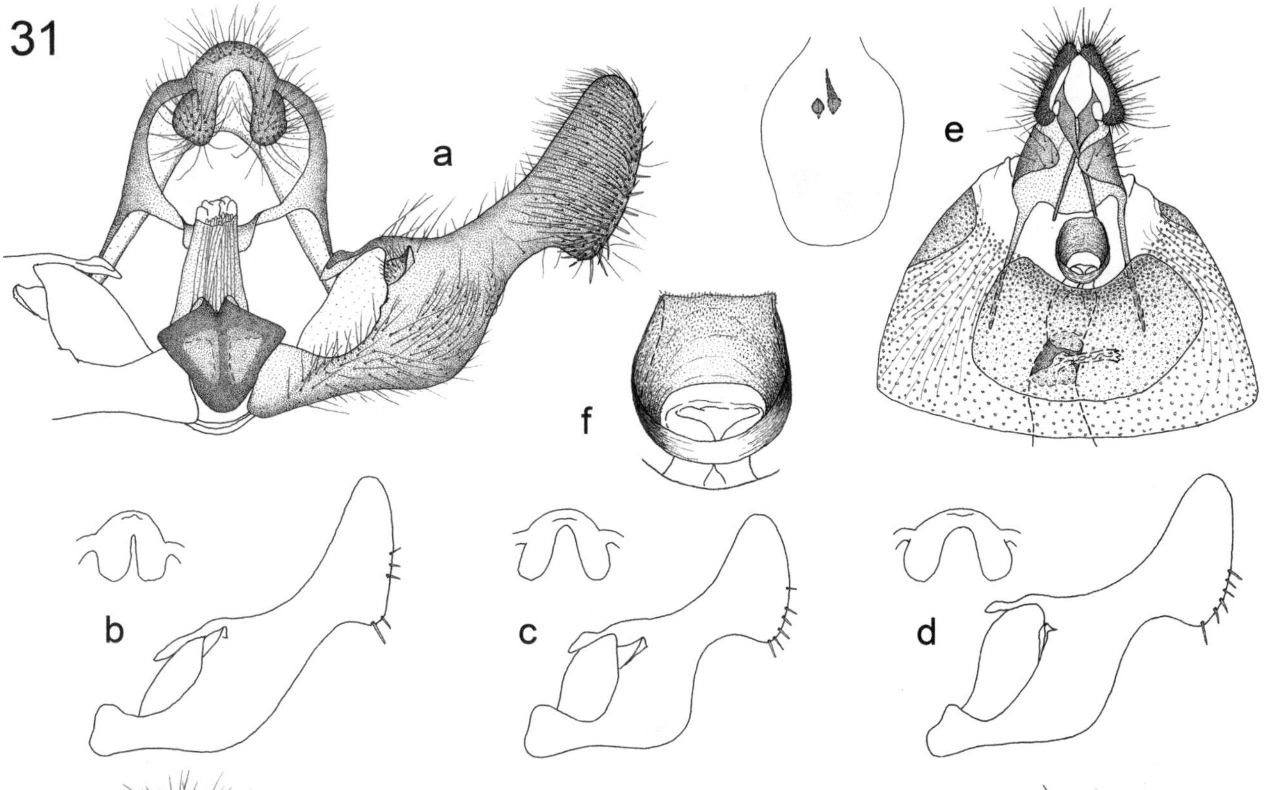


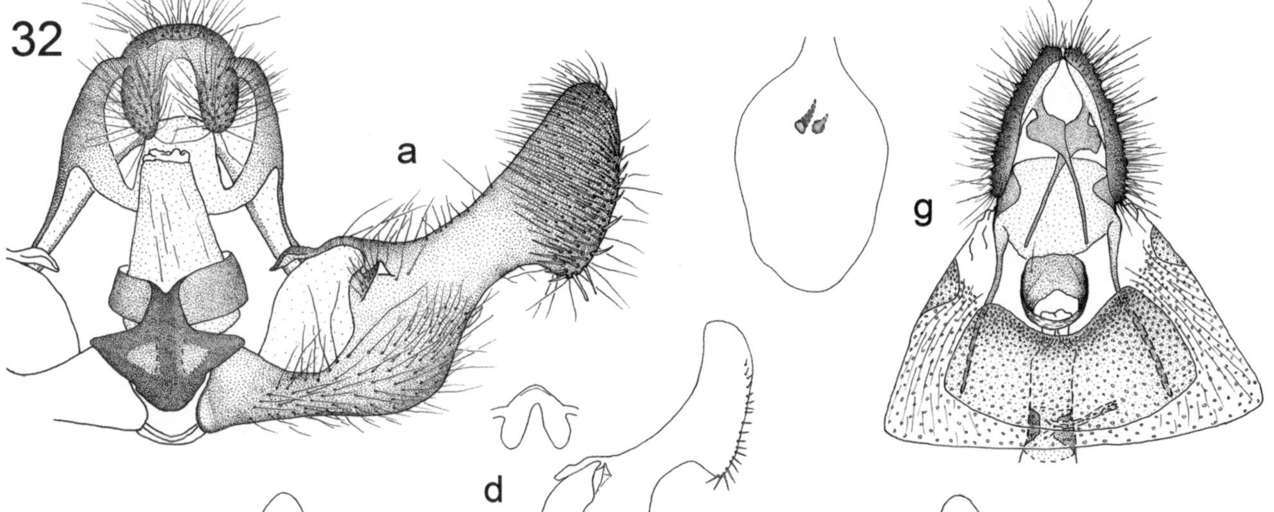


PLATE 13. **31.** ***Epiblema otiosana*****:** a, male; b–d, valva/uncus/socii; e, female; f, sterigma.
**32.** ***Epiblema dorsisuffusana*****:** a, male; b–f, valva/uncus/socii; g, female.

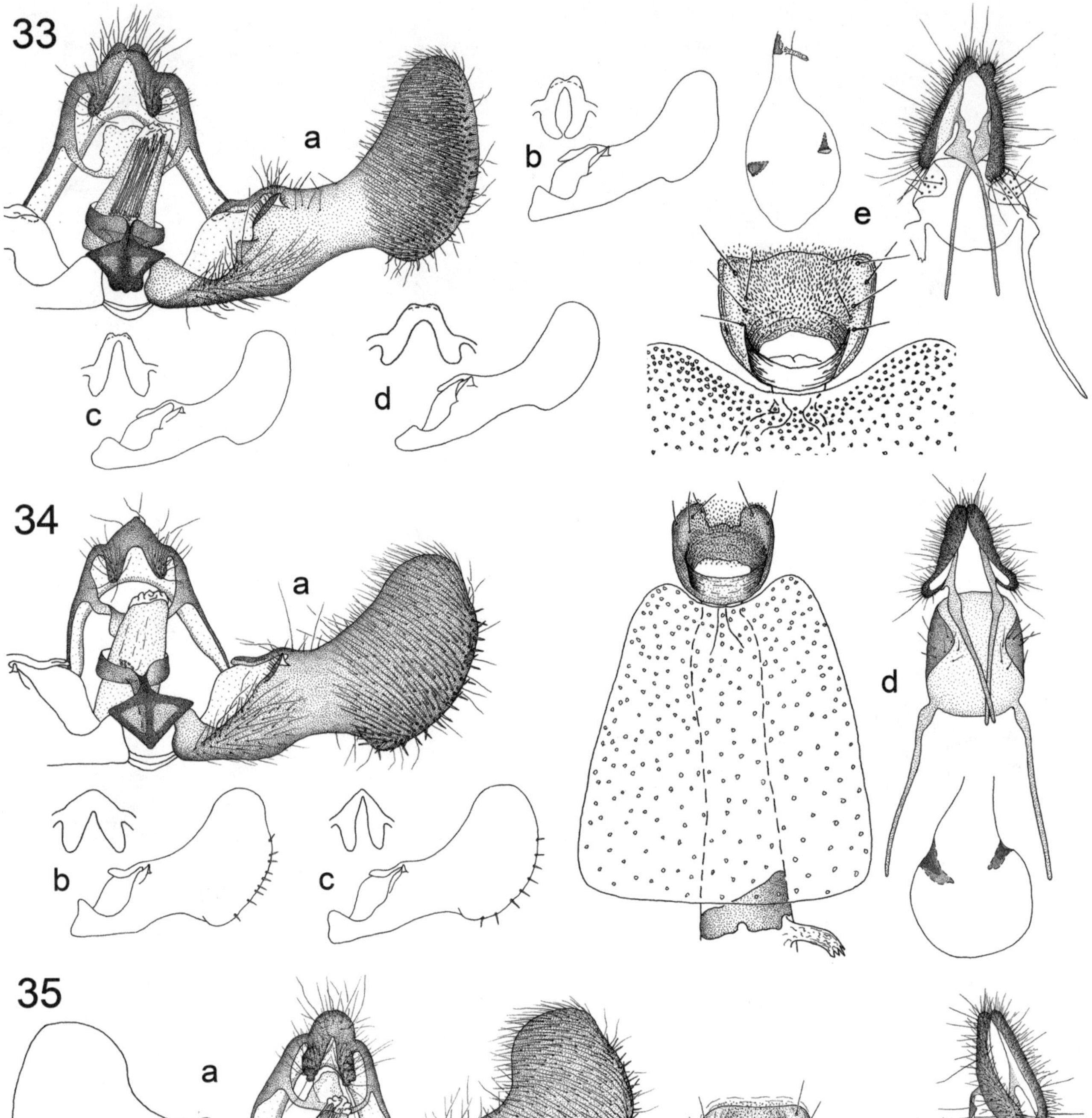


PLATE 14. **33. *Epiblema boxcana*:** a, male; b–d, valva/uncus/socii; e, female. **34. *Epiblema resumptana*:** a, male; b–c, valva/uncus/socii; d, female. **35. *Epiblema macneilli*:** a, male; b–c, valva/uncus/socii; d, female.

36

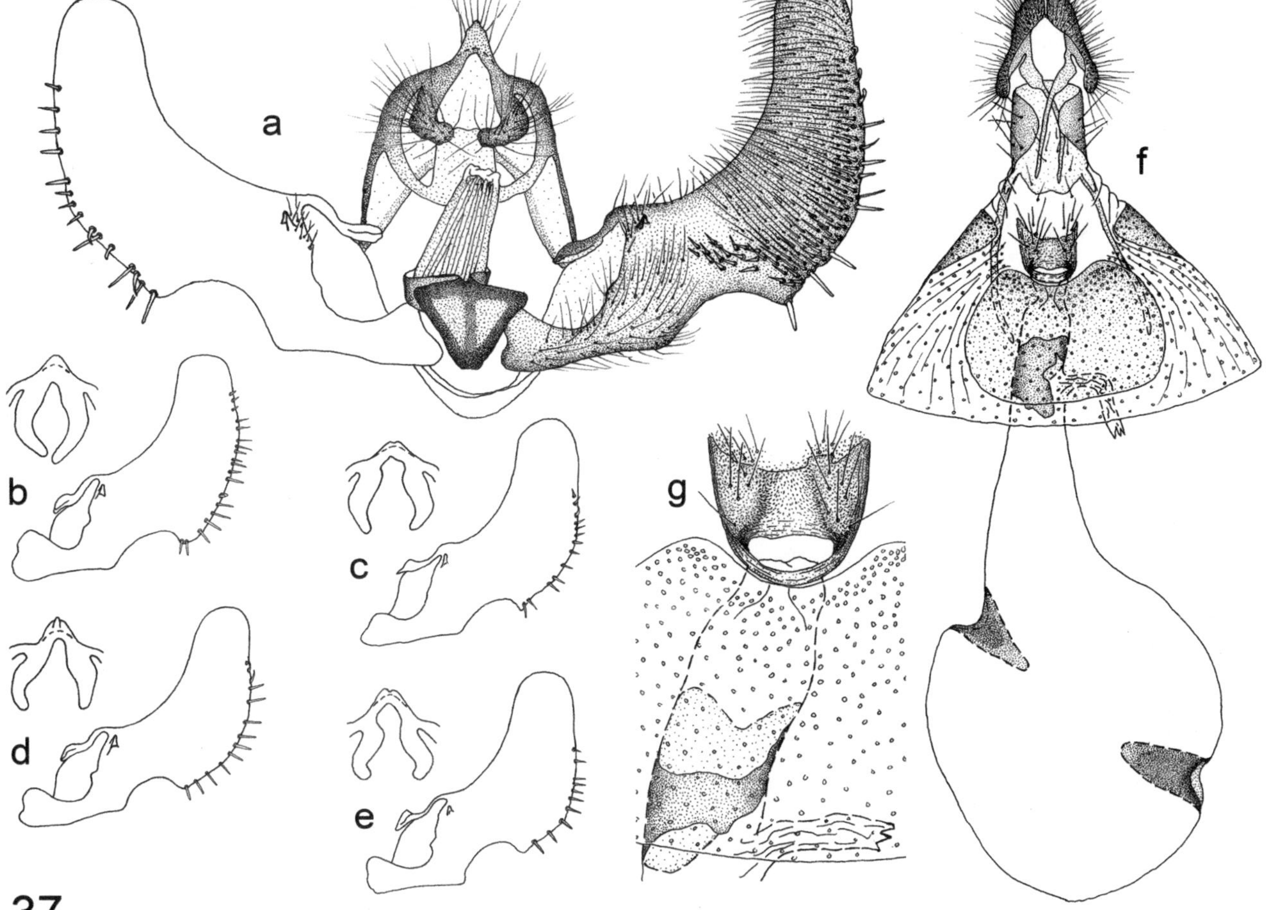


37

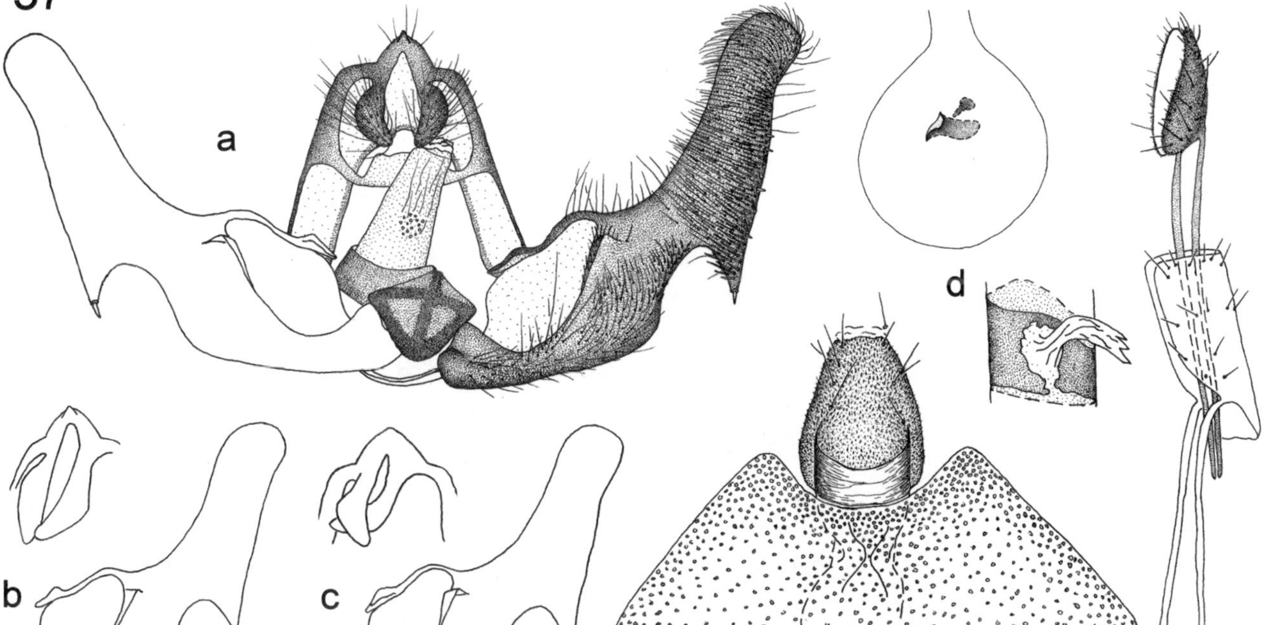


PLATE 15. **36. *Epiblema brightonana*:** a, male; b–e, valva/uncus/socii; f, female; g, sterigma/ductus bursae. **37. *Epiblema browni*:** a, male; b–c, valva/uncus/socii; d, female.

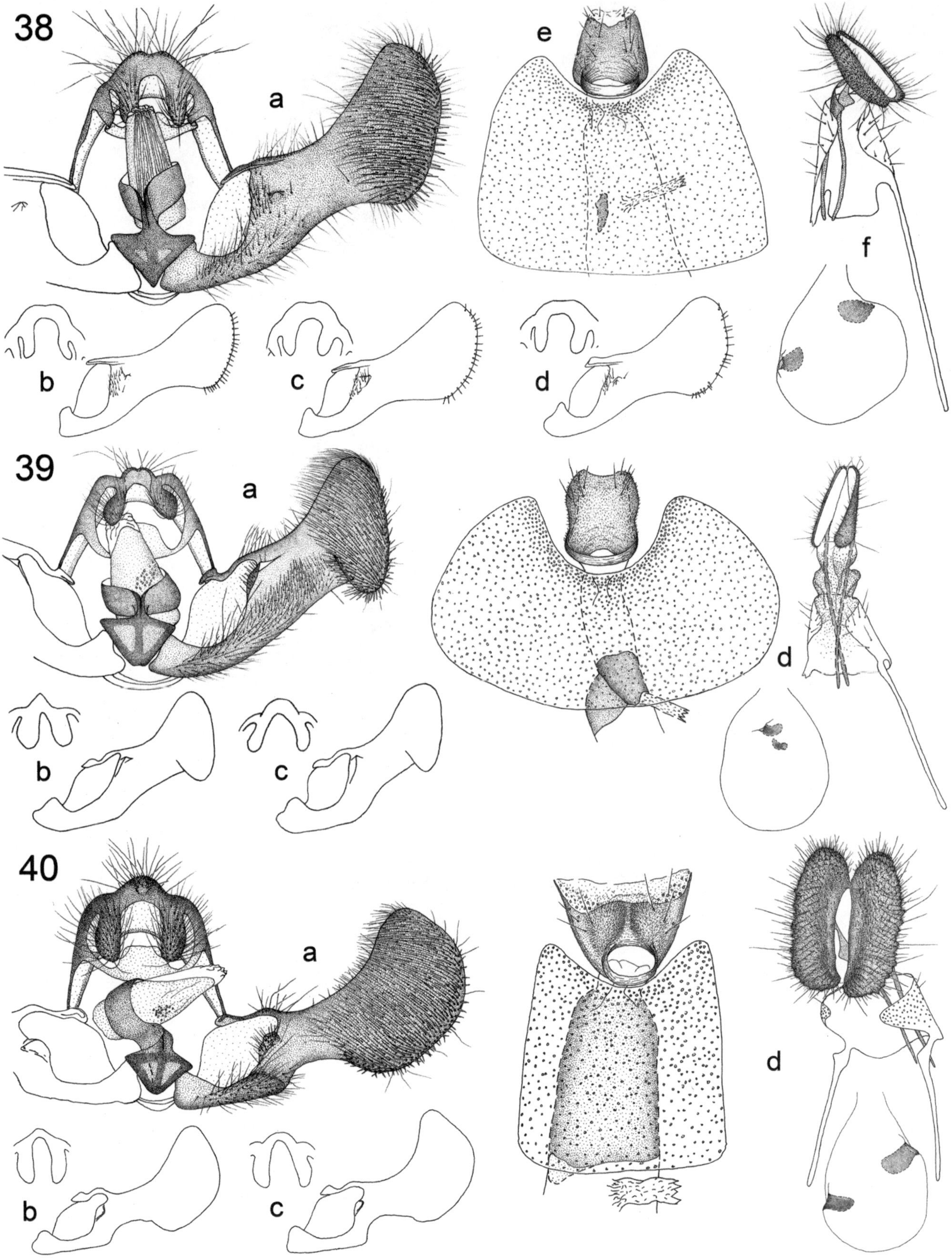


PLATE 16. **38.** ***Epiblema fulminana*****:** a, male; b–d, valva/uncus/socii; e–f, female. **39.** ***Epiblema alba*****:** a, male; b–c, valva/uncus/socii; d, female. **40.** ***Epiblema tandana*****:** a, male; b–c, valva/uncus/socii; d, female.

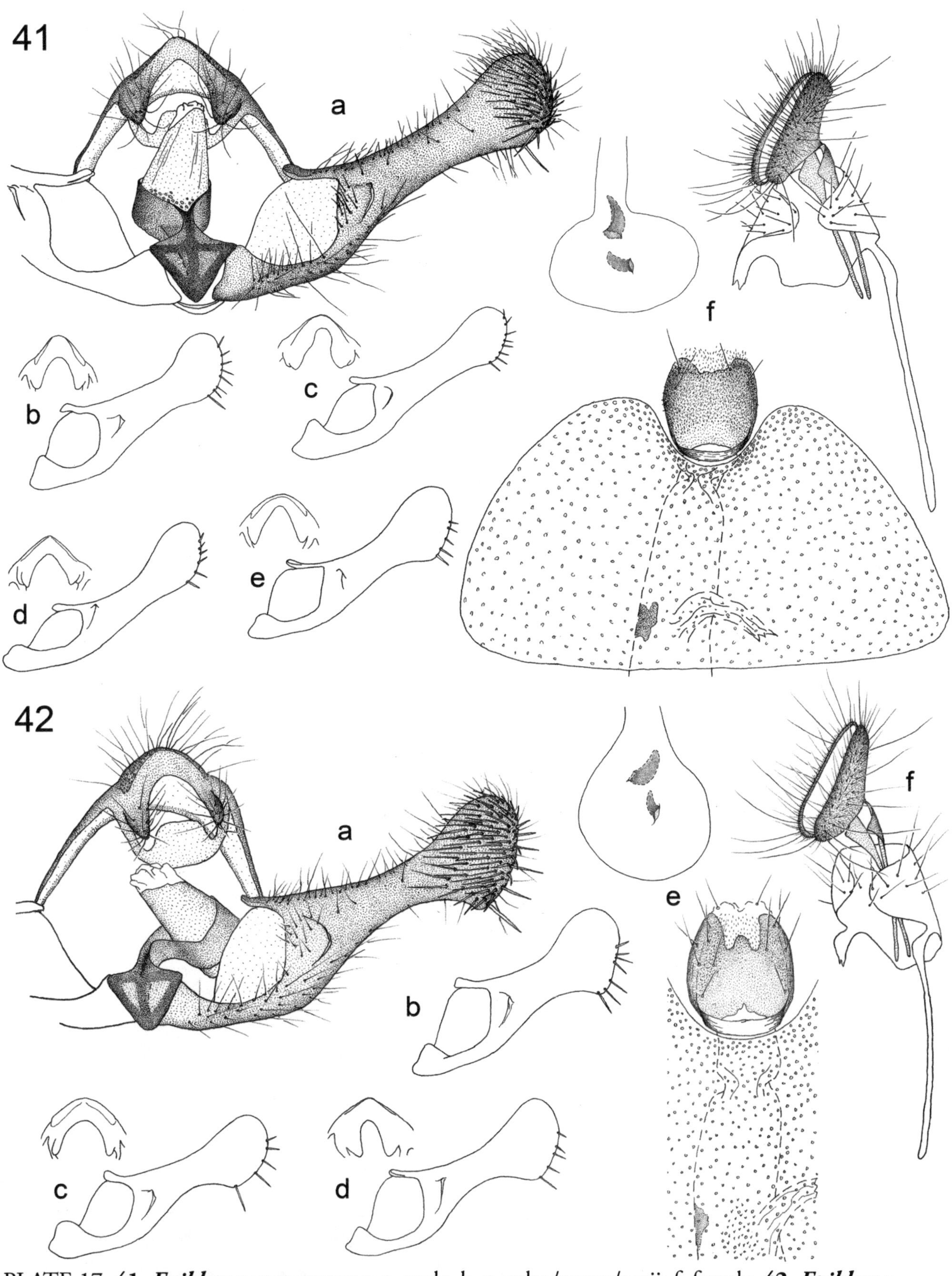


PLATE 17. **41.** ***Epiblema numerosana*:** a, male; b–e, valva/uncus/socii; f, female. **42.** ***Epiblema deflexana*:** a, male; b–d, valva/uncus/socii; e–f, female.

43

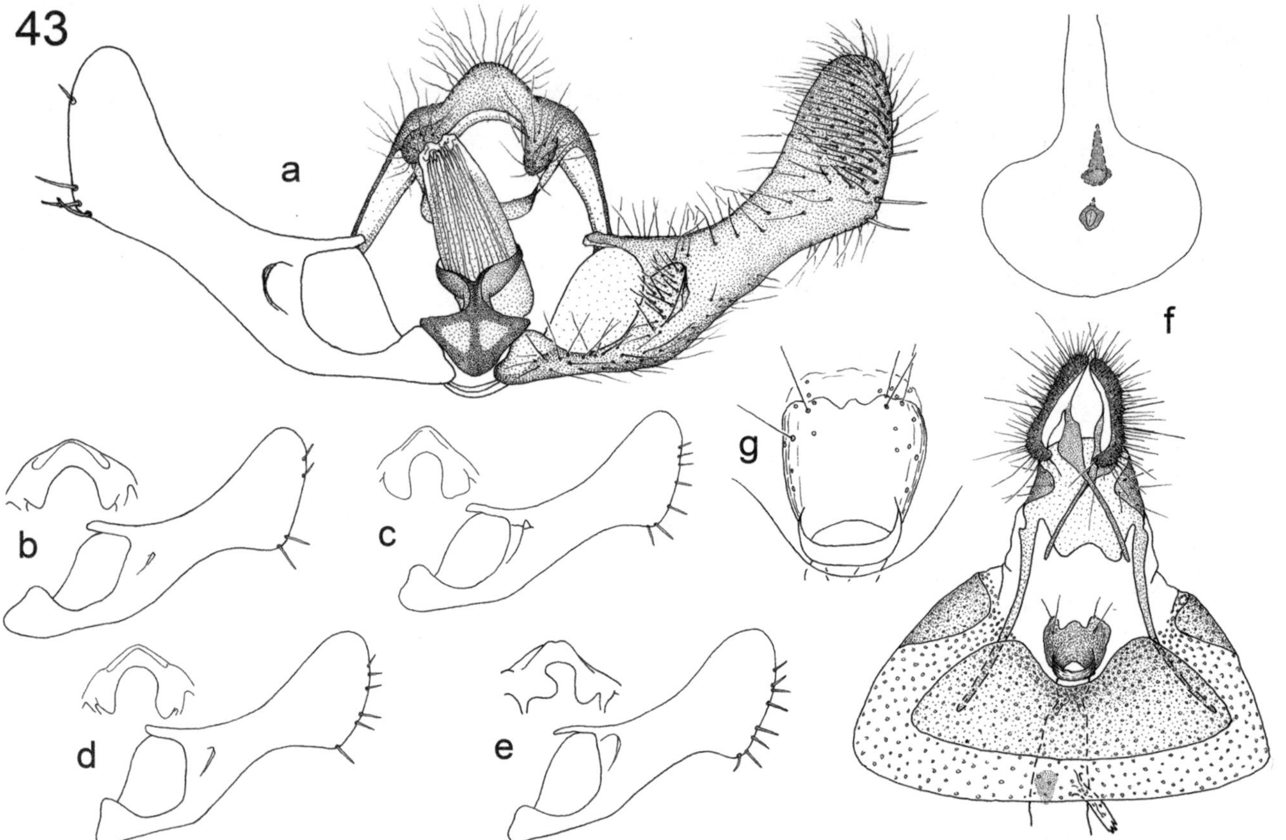


44

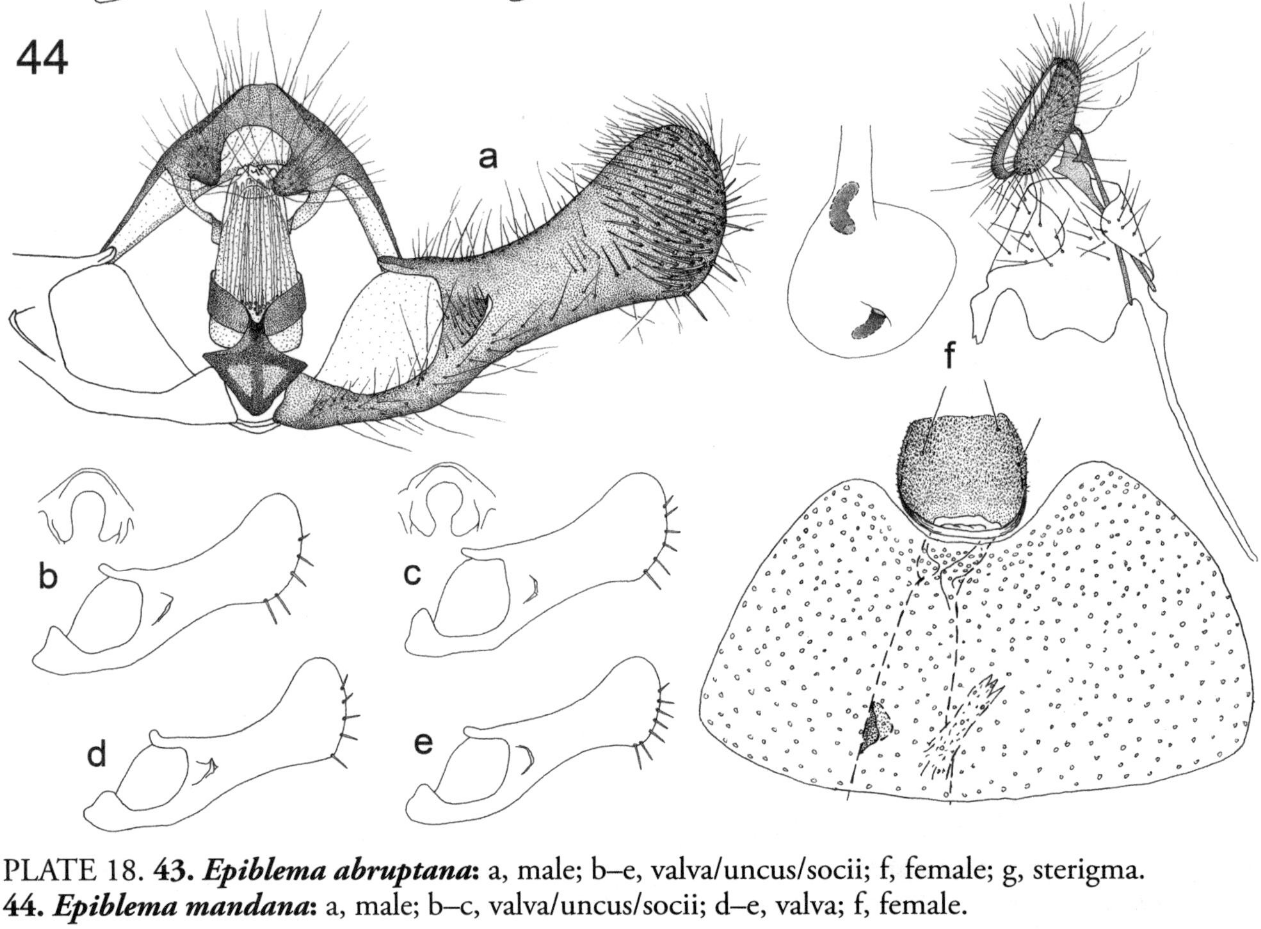


PLATE 18. **43.** ***Epiblema abruptana*:** a, male; b–e, valva/uncus/socii; f, female; g, sterigma. **44.** ***Epiblema mandana*:** a, male; b–c, valva/uncus/socii; d–e, valva; f, female.

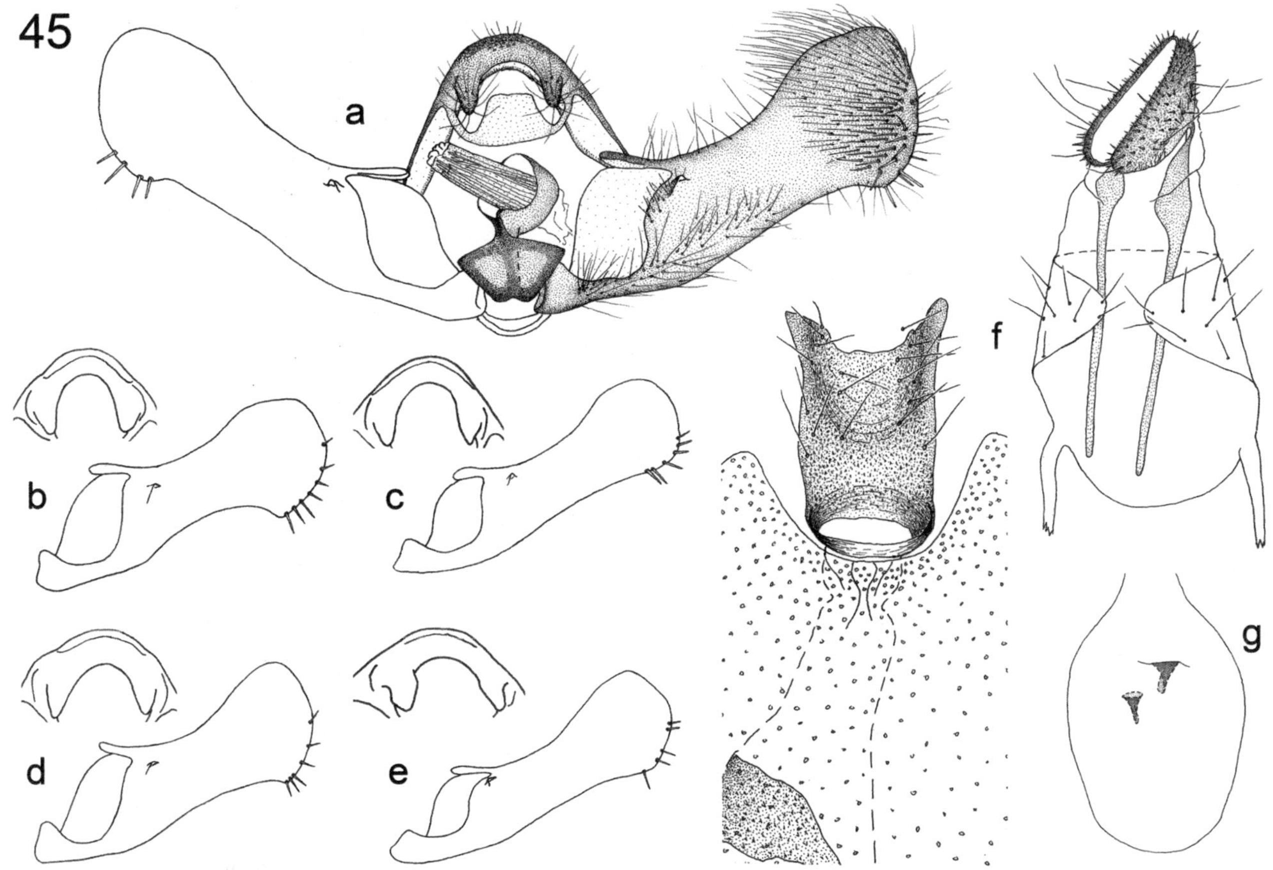


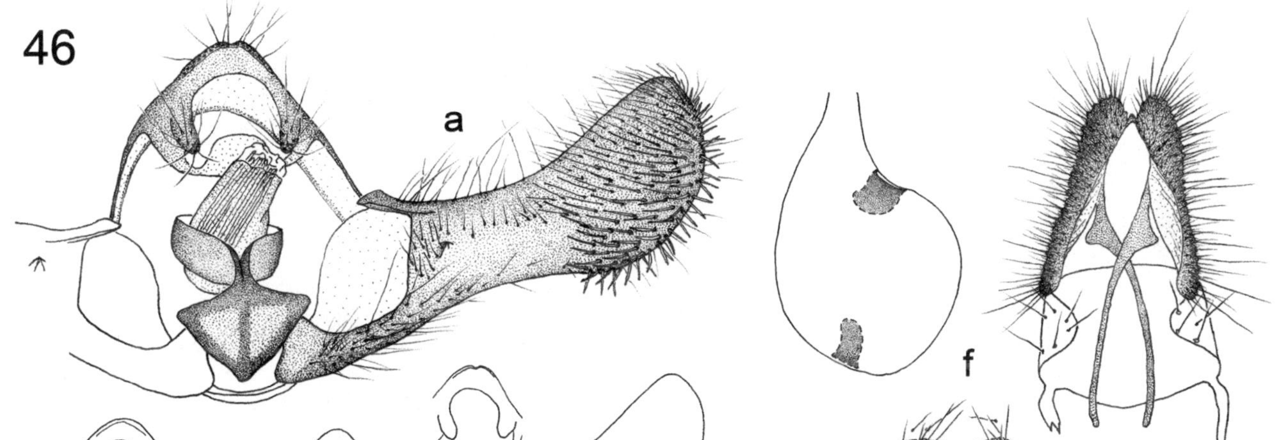


PLATE 19. **45.** ***Epiblema exacerbatricana*****:** a, male; b–e, valva/uncus/socii; f–g, female.
**46.** ***Epiblema glaseri*****:** a, male; b–e, valva/uncus/socii; f, female.

47

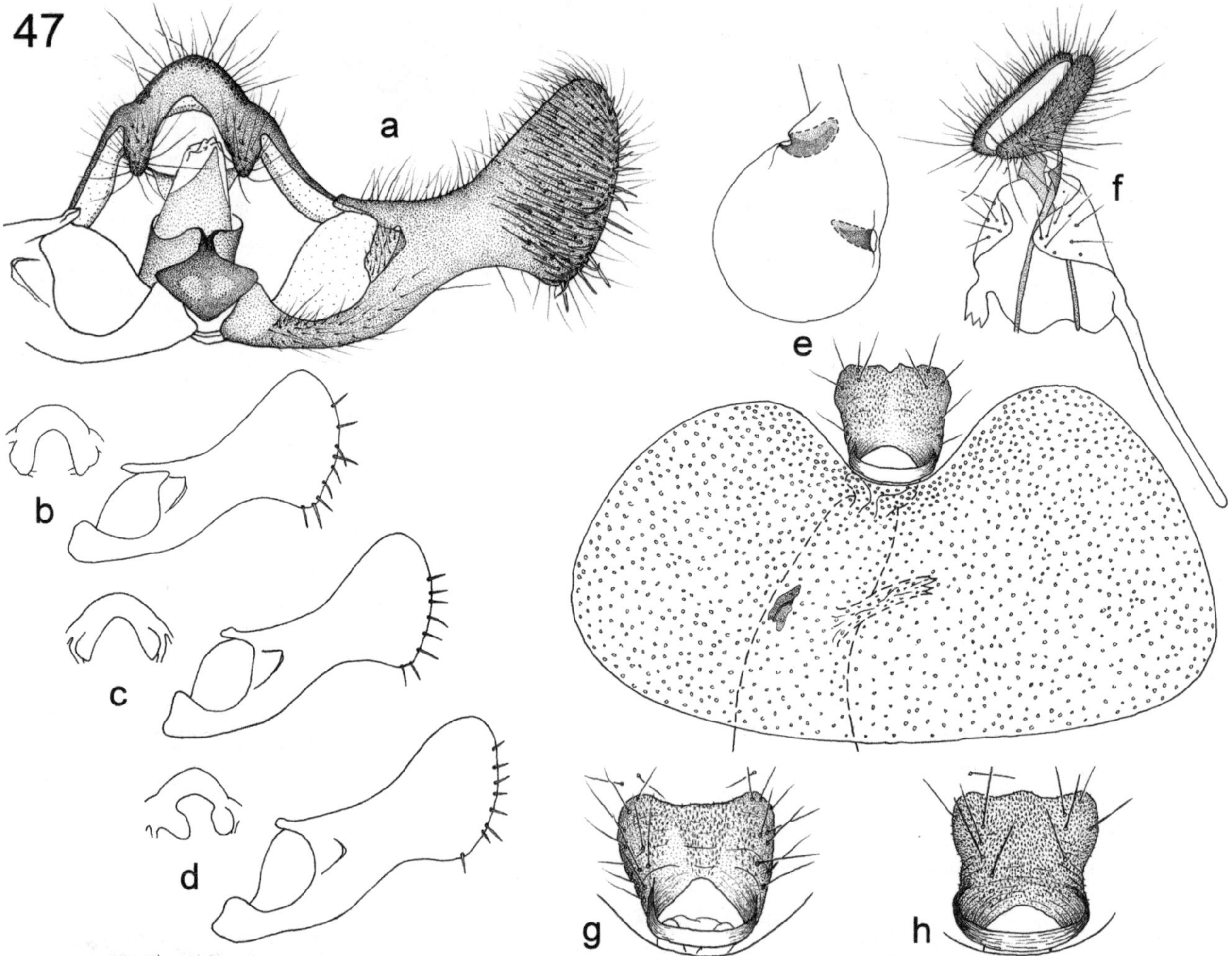


48

PLATE 20. **47. *Epiblema praesumptiosa*:** a, male; b–d, valva/uncus/socii; e–f, female; g–h, sterigma. **48. *Epiblema grossbecki*:** a, male; b–d, valva/uncus/socii; e, female.

49

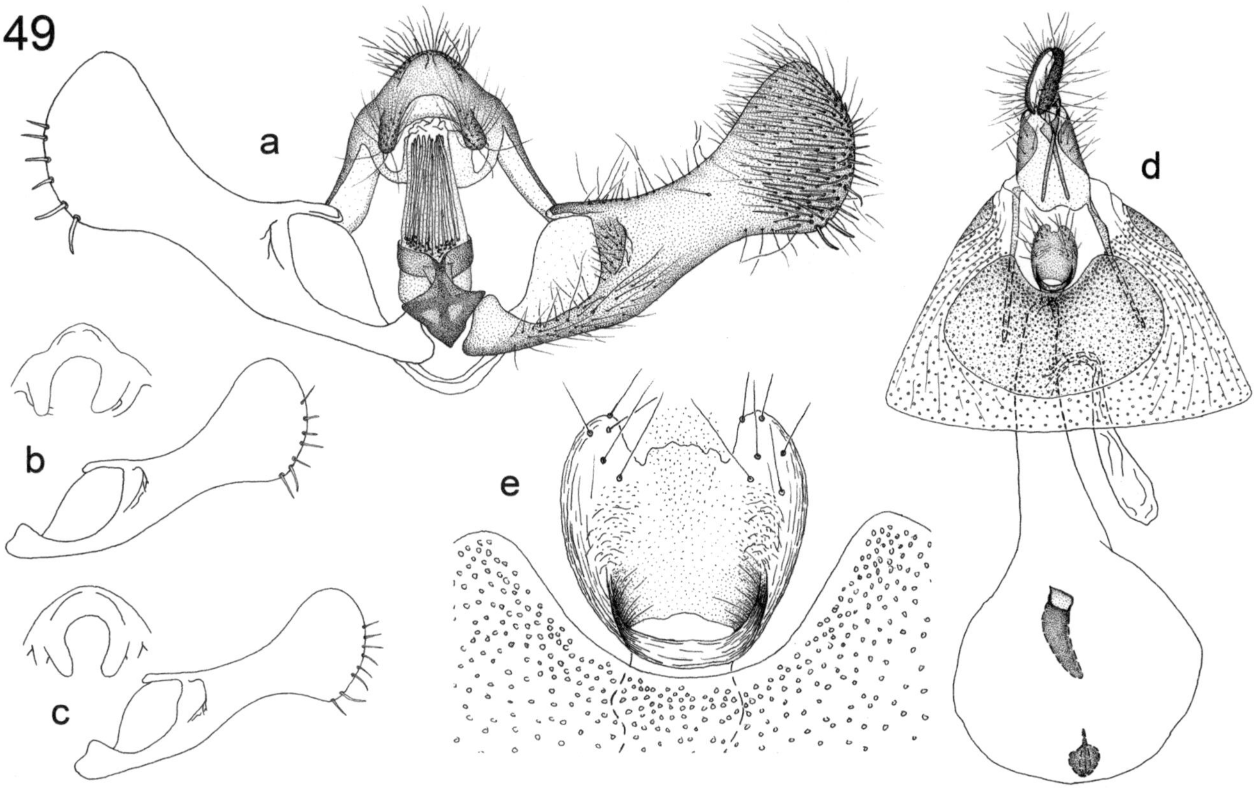


50

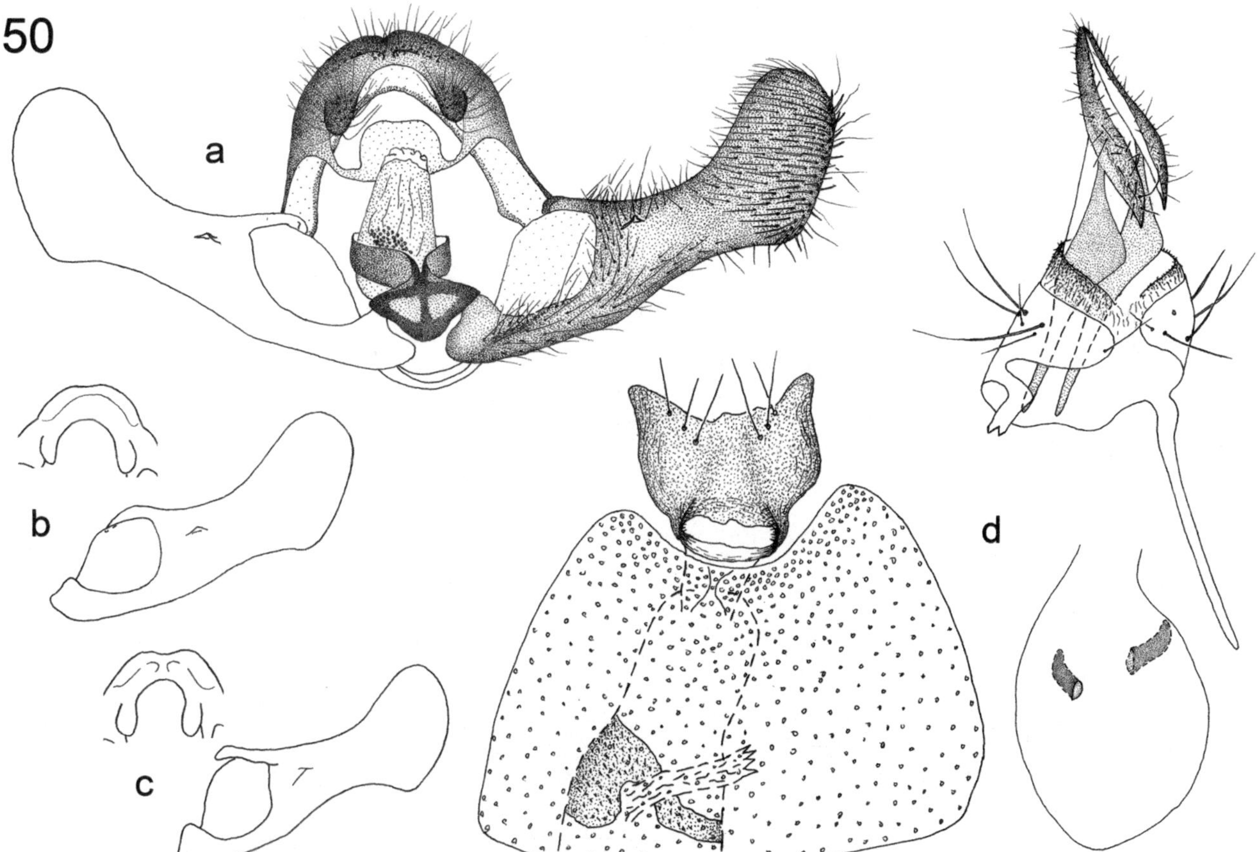


PLATE 21. **49.** ***Epiblema chromata*:** a, male; b–c, valva/uncus/socii; d, female; e, sterigma.
**50.** ***Epiblema insidiosana*:** a, male; b–c, valva/uncus/socii; d, female.

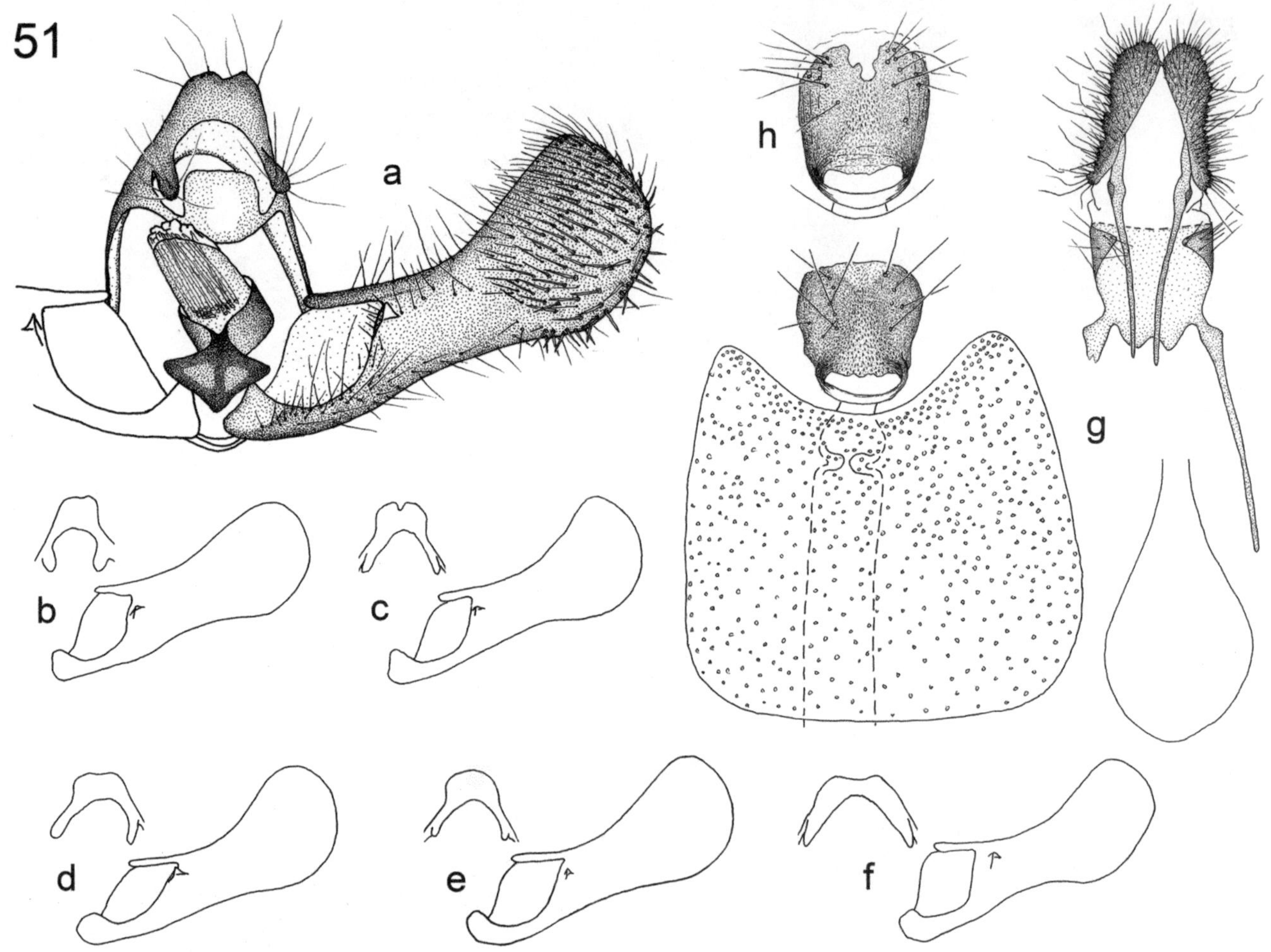


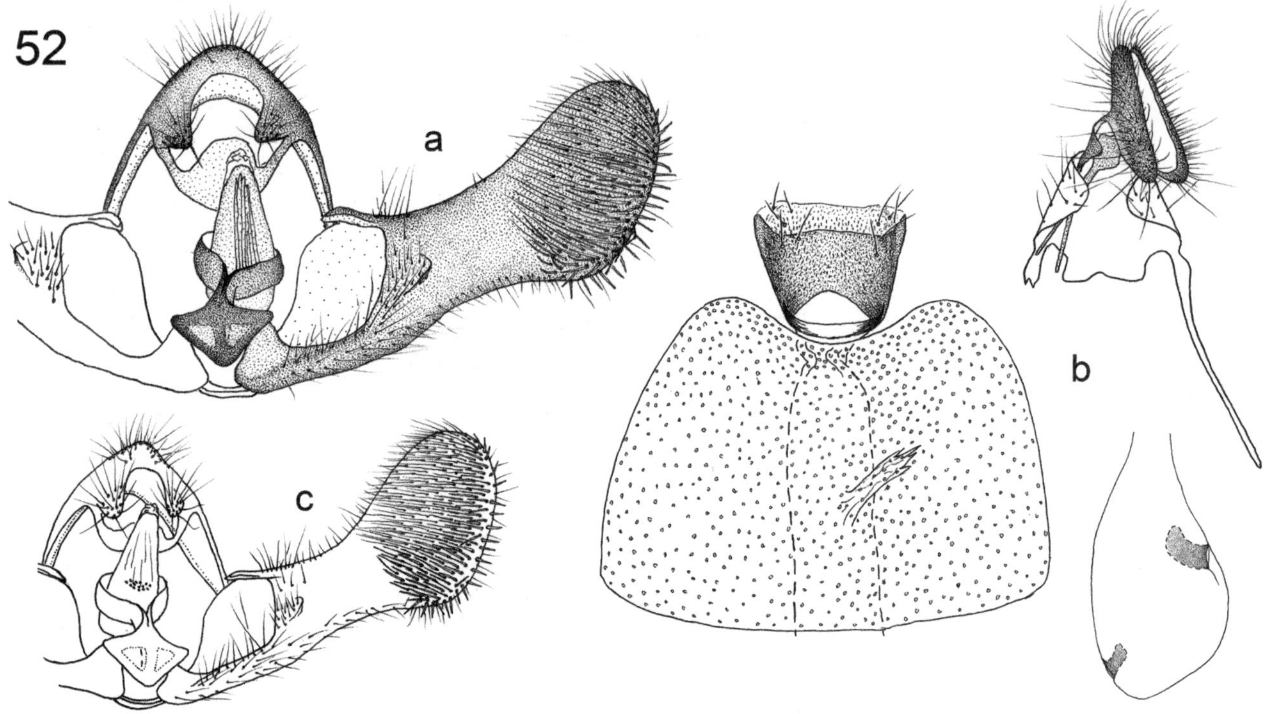


PLATE 22. **51.** ***Epiblema separationis*:** a, male; b–f, valva/uncus/socii; g, female; h, sterigma.
**52.** ***Epiblema colfaxana*:** a, male; b, female; c, male.

53

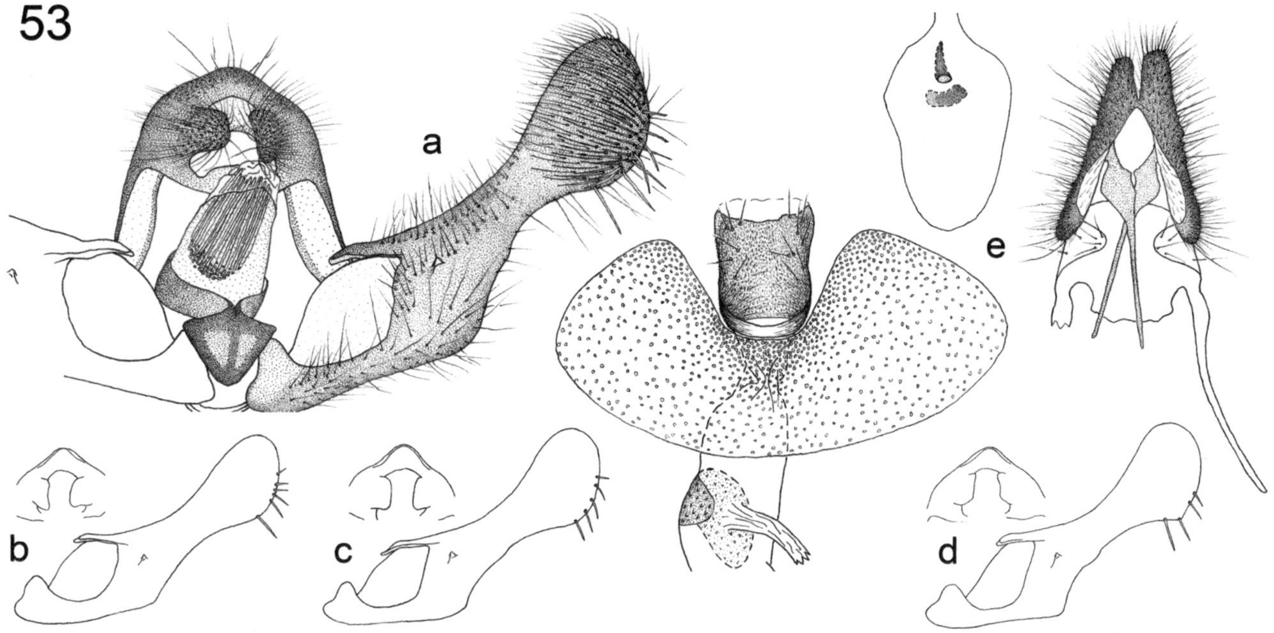


54

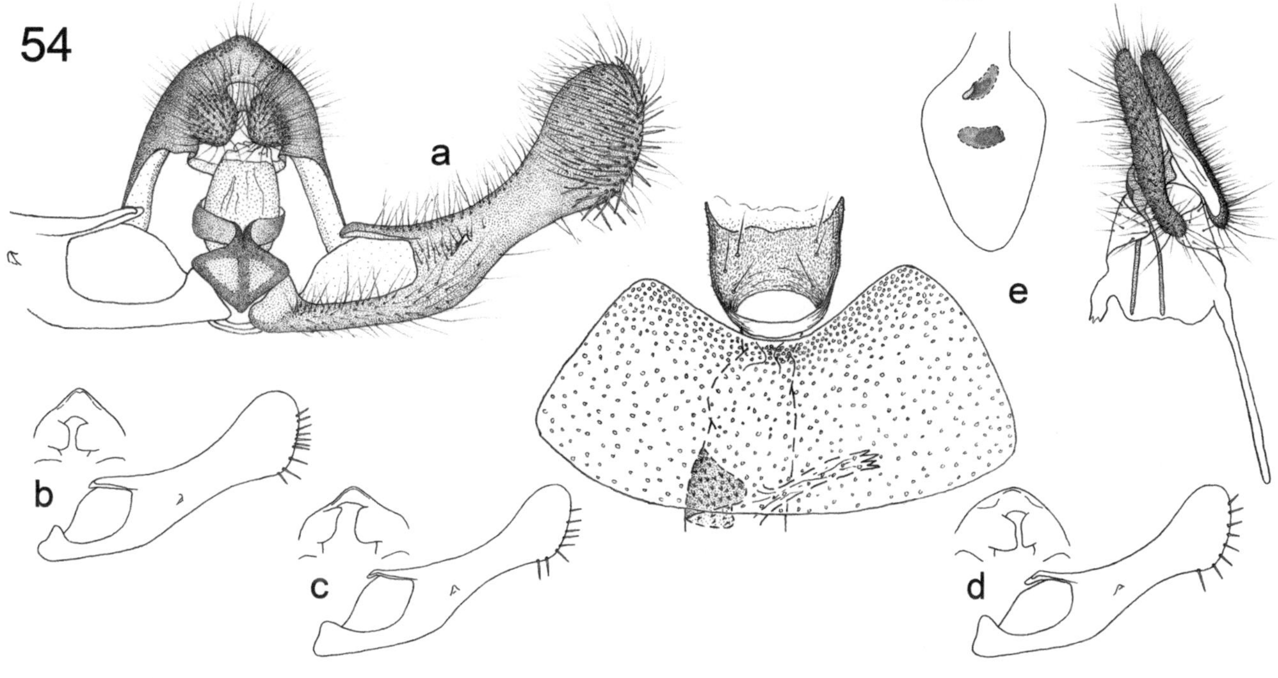


55

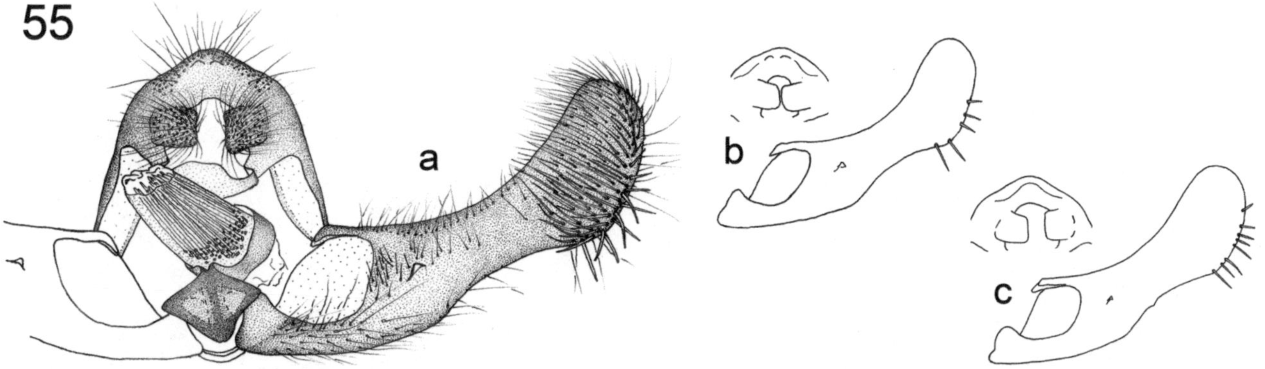


PLATE 23. **53.** ***Sonia constrictana*:** a, male; b–d, valva/uncus/socii; e, female. **54.** ***Sonia paraplesiana*:** a, male; b–d, valva/uncus/socii; e, female. **55.** ***Sonia roseomaculana*:** a, male; b–c, valva/uncus/socii.

56

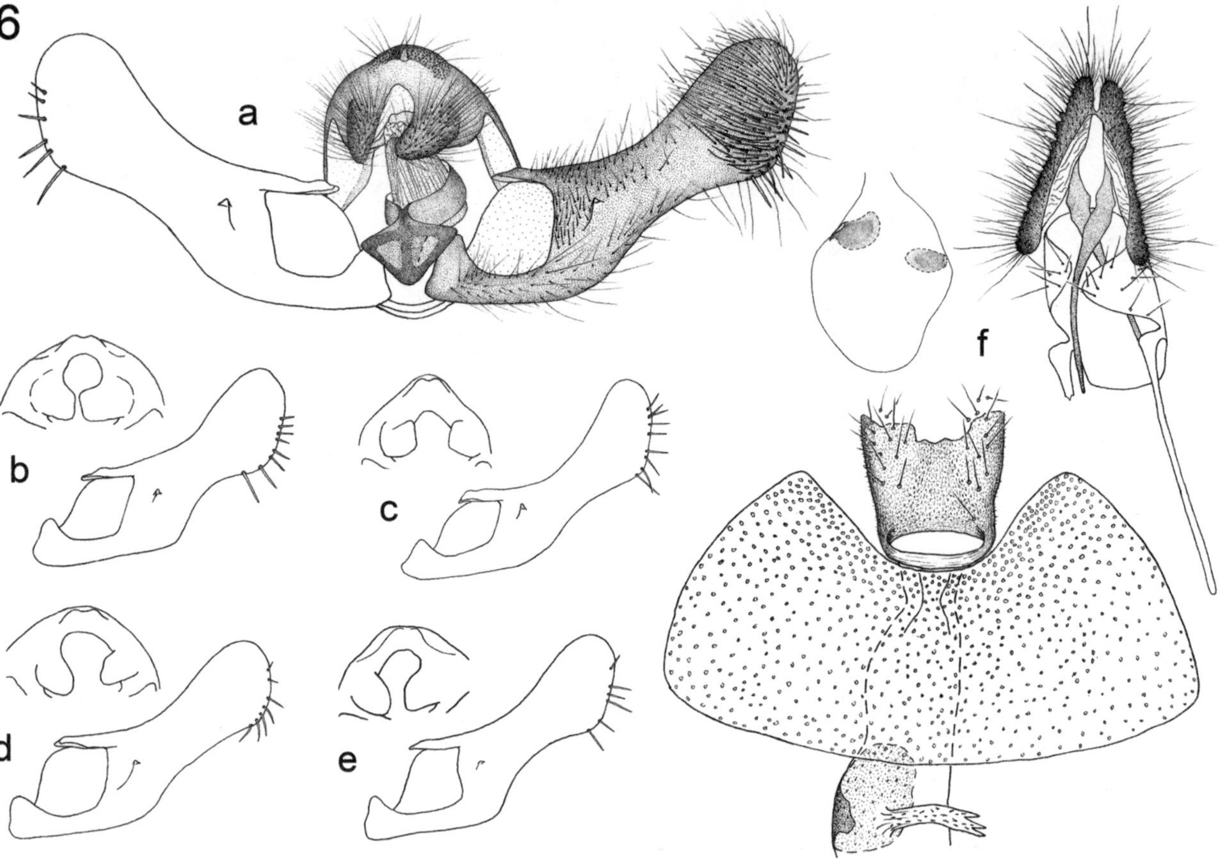


57

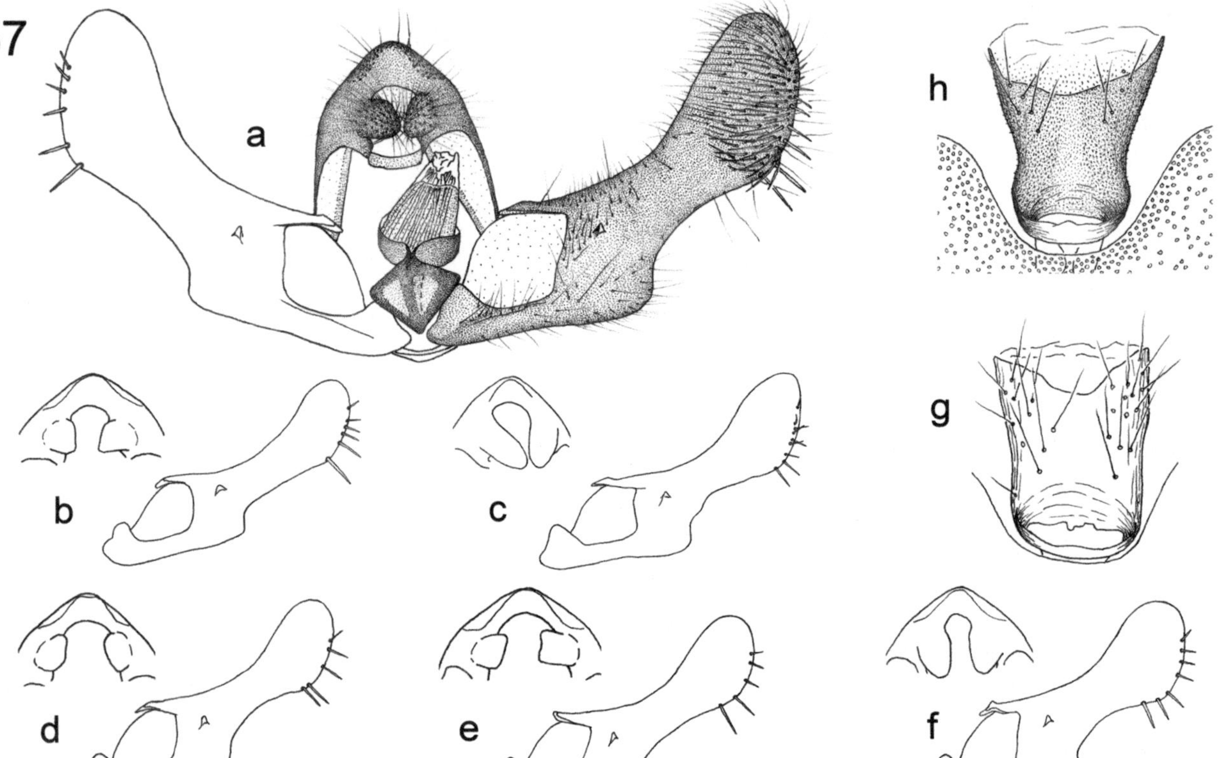


PLATE 24. **56. *Sonia ferruginana*:** a, male; b–e, valva/uncus/socii; f, female. **57. *Sonia griseolana*:** a, male; b–f, valva/uncus/socii; g–h, sterigma.

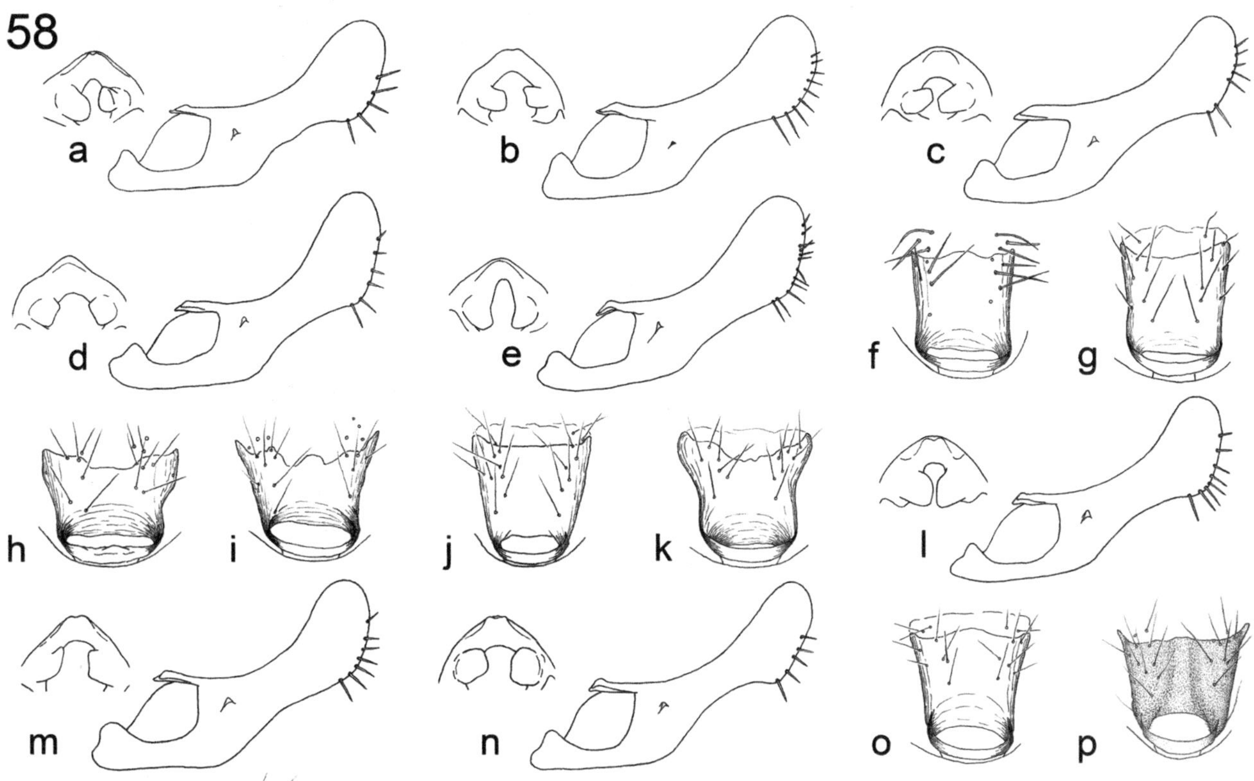

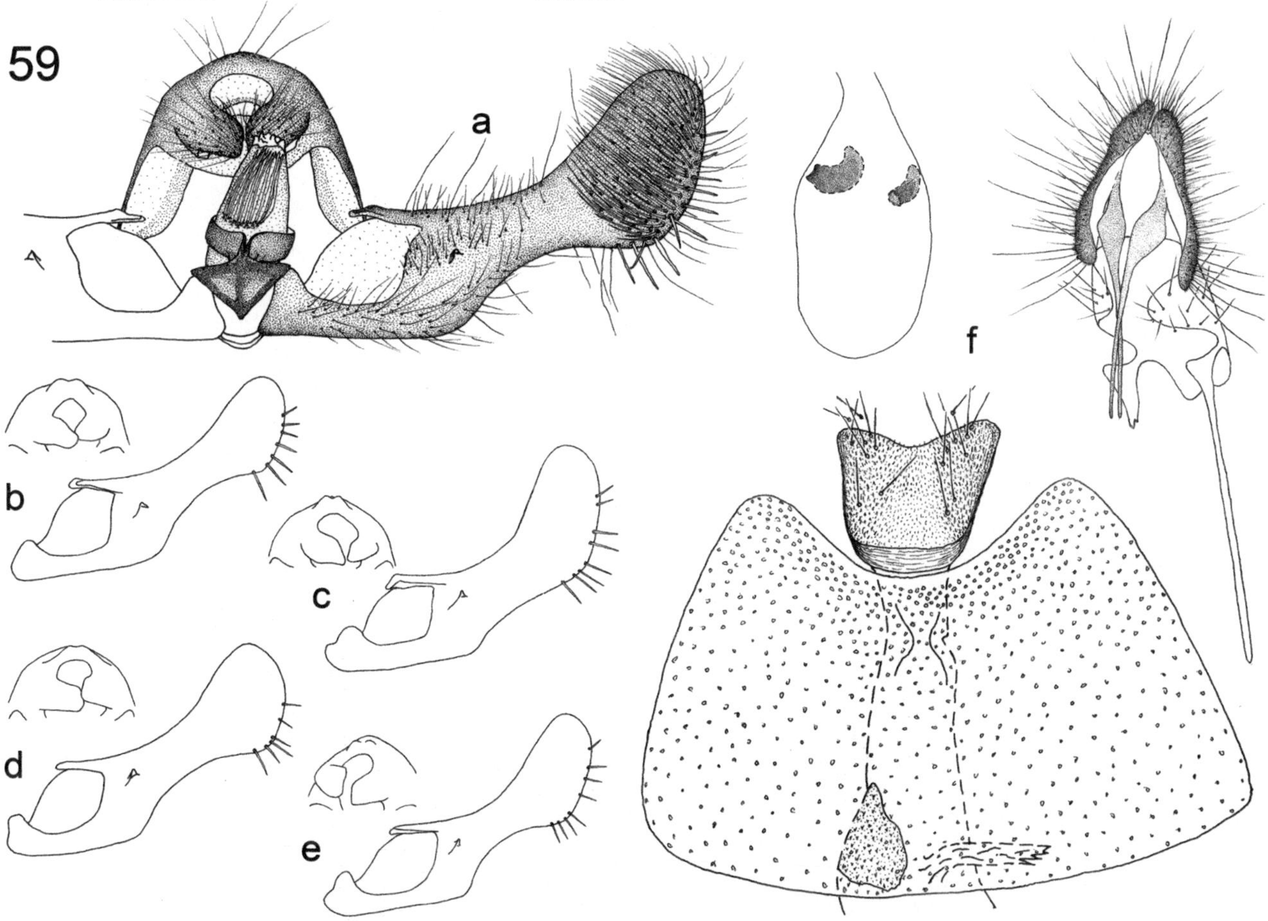

PLATE 25. **58. *ferruginana* complex gray:** a–e, valva/uncus/socii; f–k, sterigma. ***ferruginana* complex red:** l–n, valva/uncus/socii; o–p, sterigma. **59. *Sonia canadana*:** a, male; b–e, valva/uncus/socii; f, female.

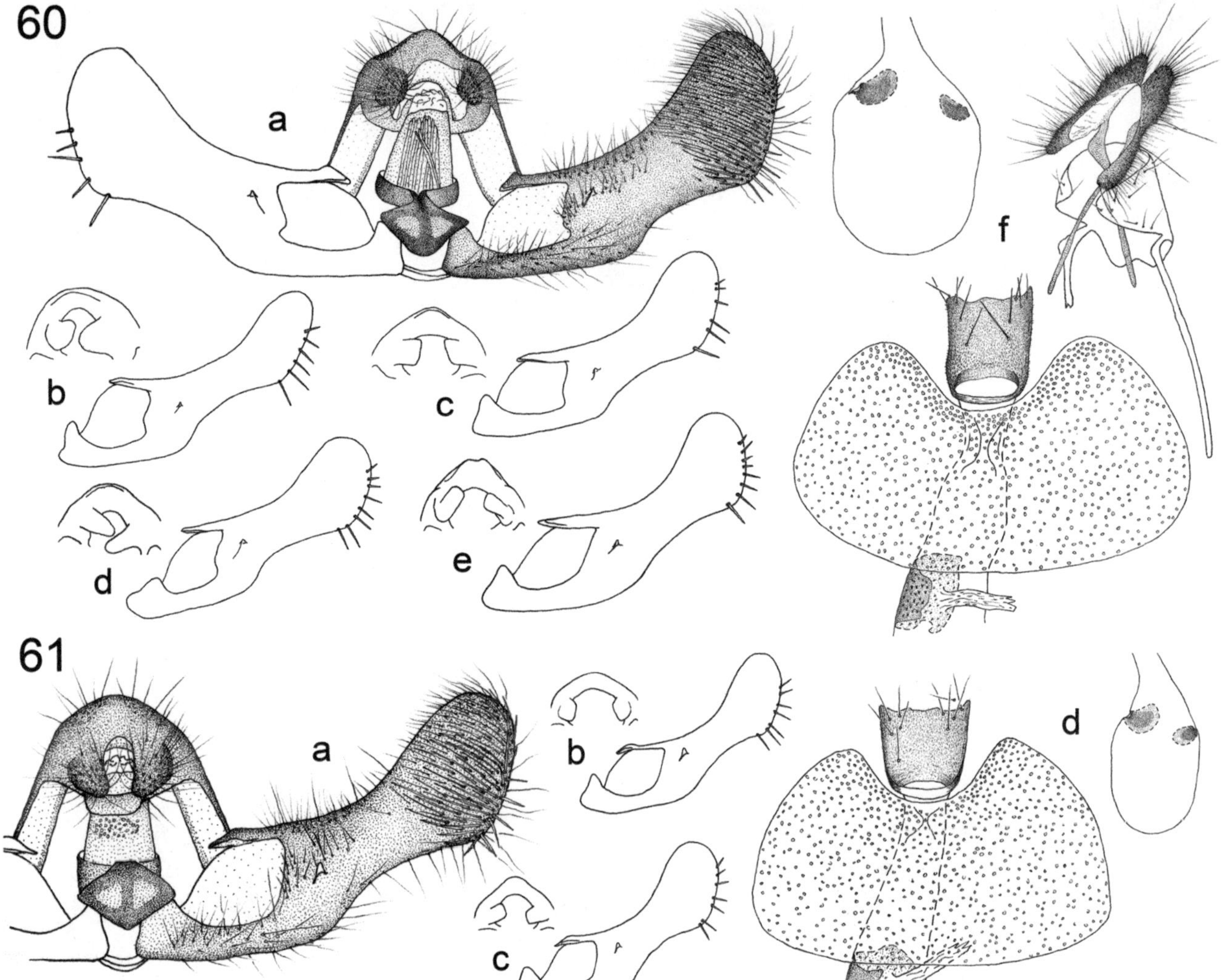


PLATE 26. **60.** ***Sonia nigrana*:** a, male; b–e, valva/uncus/socii; f, female. **61.** ***Sonia albimacula*:** a, male; b–c, valva/uncus/socii; d, female. **62.** ***Sonia parva*:** a, male; b–c, valva/uncus/socii; d, female. **63.** ***canadana*** **complex:** a–b, valva/uncus/socii.

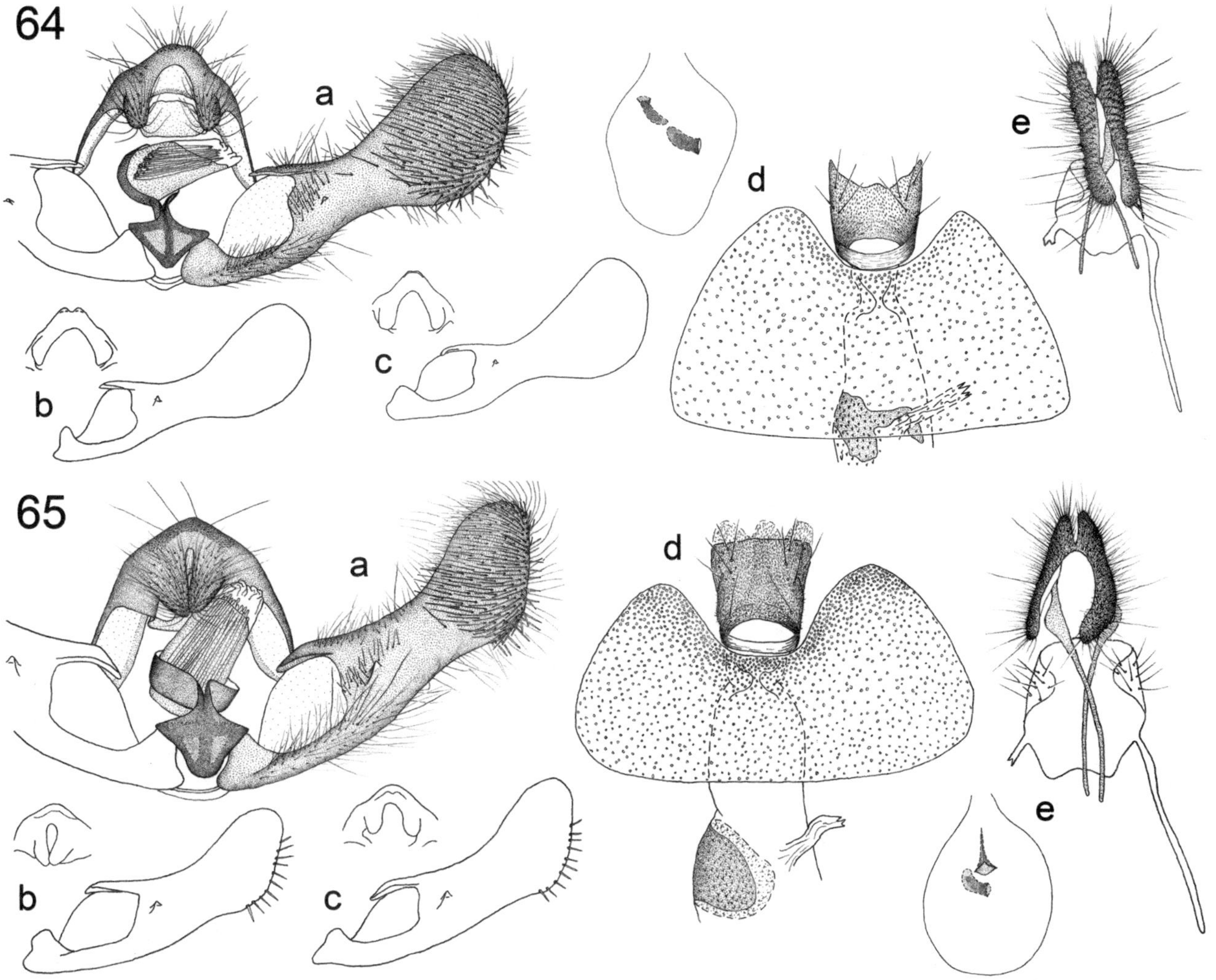


PLATE 27. **64. *Sonia divaricata*:** a, male; b–c, valva/uncus/socii; d–e, female. **65. *Sonia comstocki*:** a, male; b–c, valva/uncus/socii; d–e, female. **66. *Sonia filiana*:** a, male; b–c, valva/uncus/socii; d, sterigma/sternum7/ductus bursae; e, female.

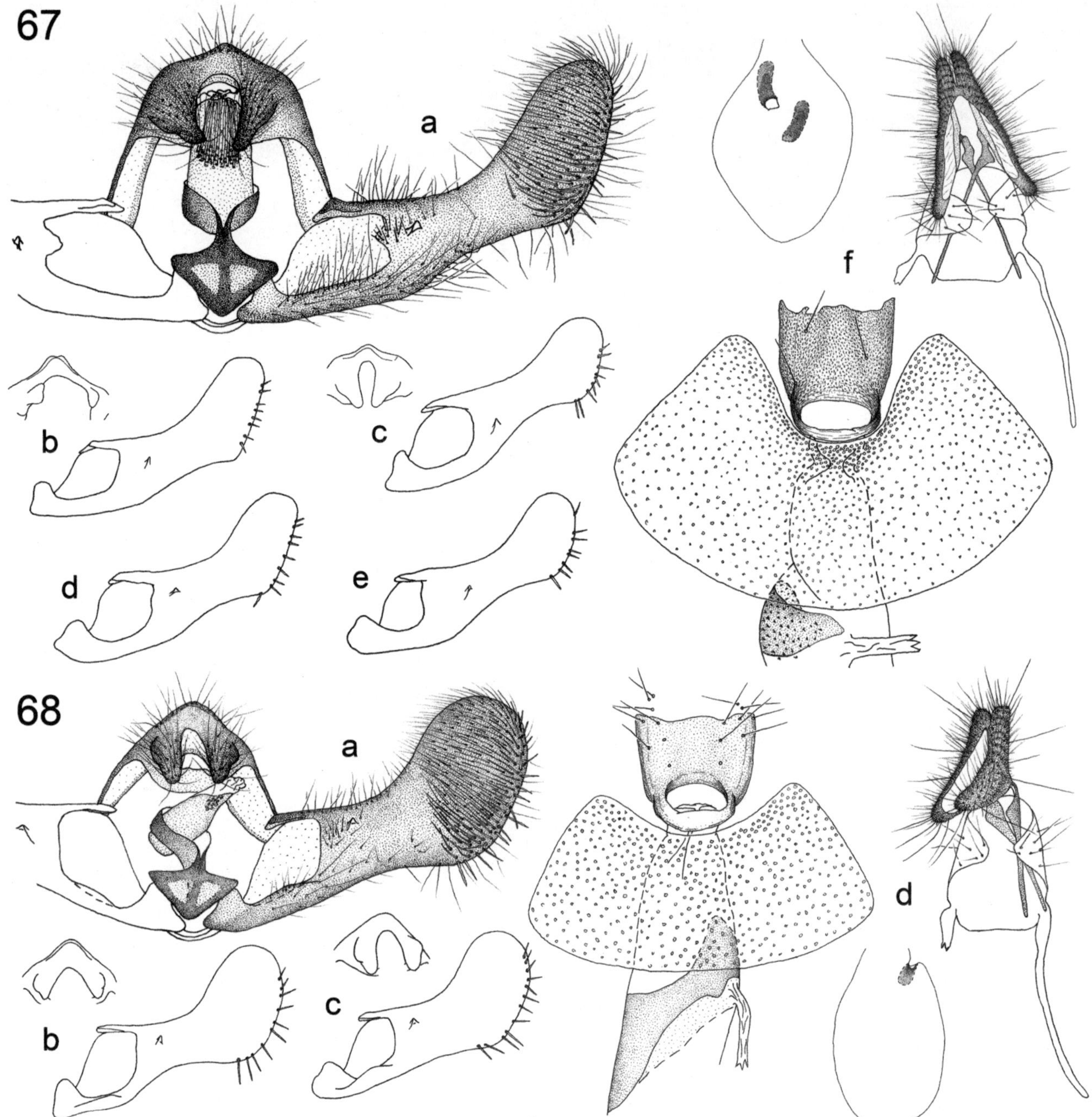


PLATE 28. **67.** ***Sonia vovana*****:** a, male; b–c, valva/uncus/socii; d–e, valva; f, female. **68.** ***Sonia hodgesi*****:** a, male; b–c, valva/uncus/socii; d, female. **69.** ***Sonia gibsoni*****:** male.

PLATE 29. **70.** ***Suleima helianthana*:** a–b, male; c–f, uncus/socii; g–m, valva/uncus/socii; n–p, female; q–s, sterigma.

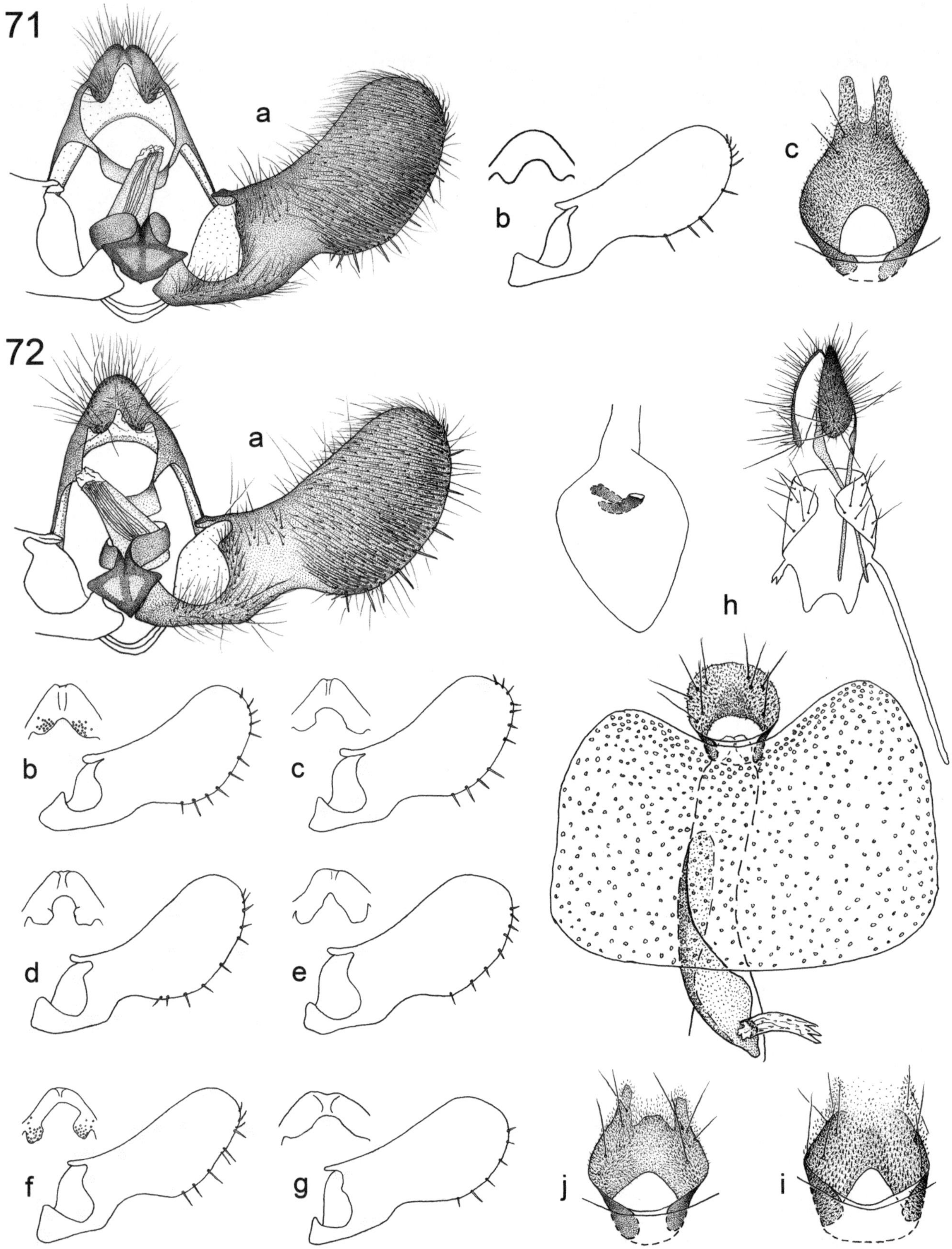


PLATE 30. **71.** ***Suleima lagopana*:** a, male; b, valva/uncus/socii; c, sterigma. **72.** ***Suleima powelli*:** a, male; b–g, valva/uncus/socii; h, female; i–j, sterigma.

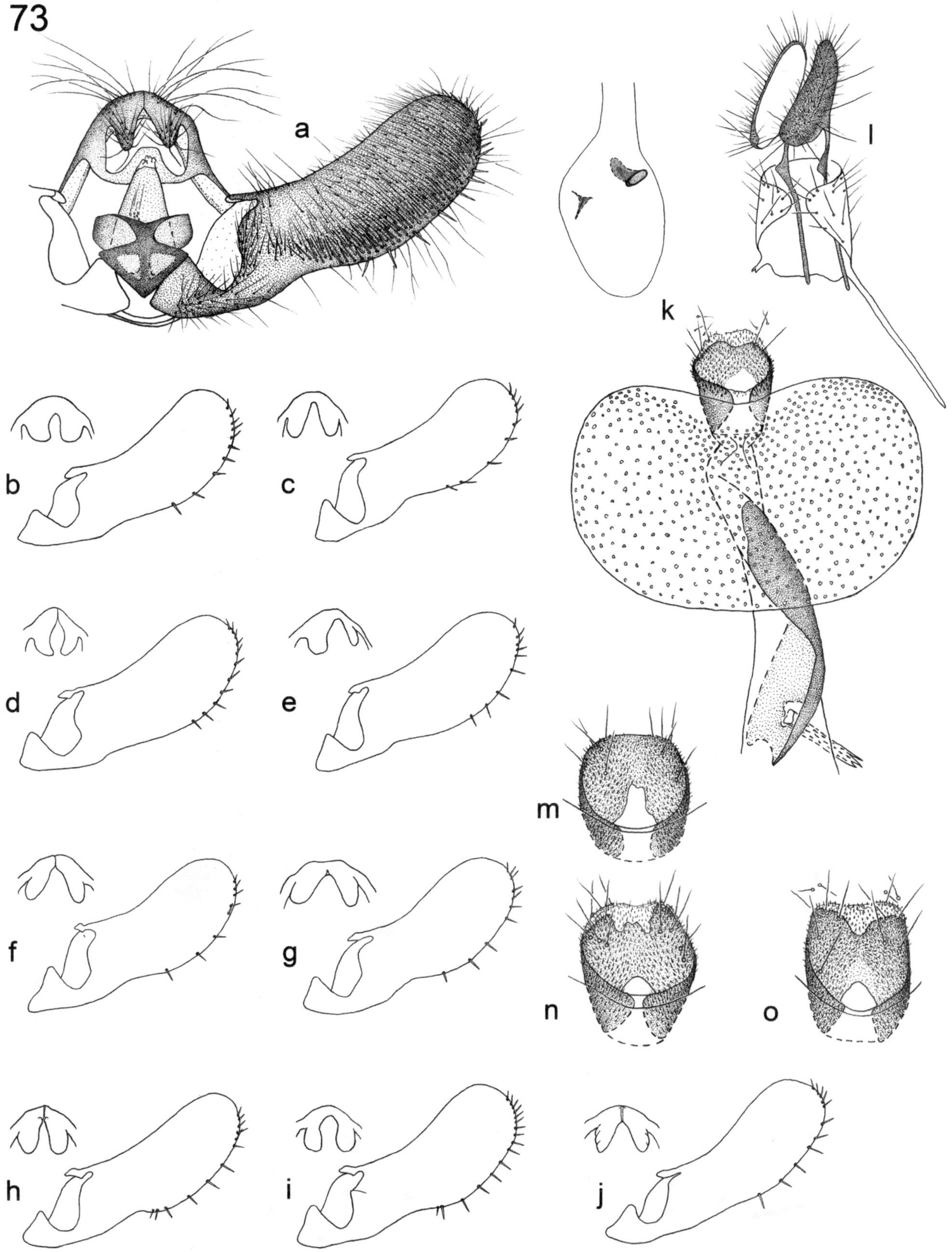


PLATE 31. **73. *Suleima baracana*:** a, male; b–j, valva/uncus/socii; k–l, female; m–o, sterigma.

74

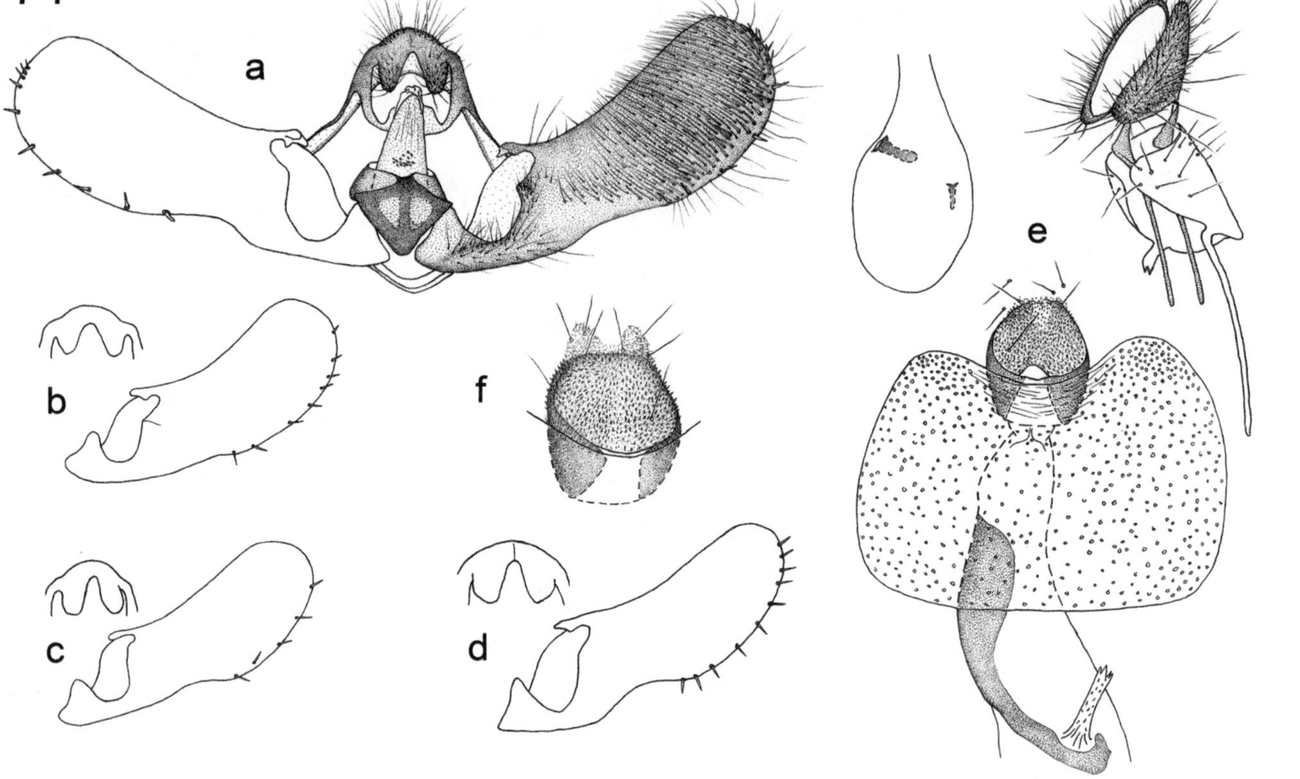


75

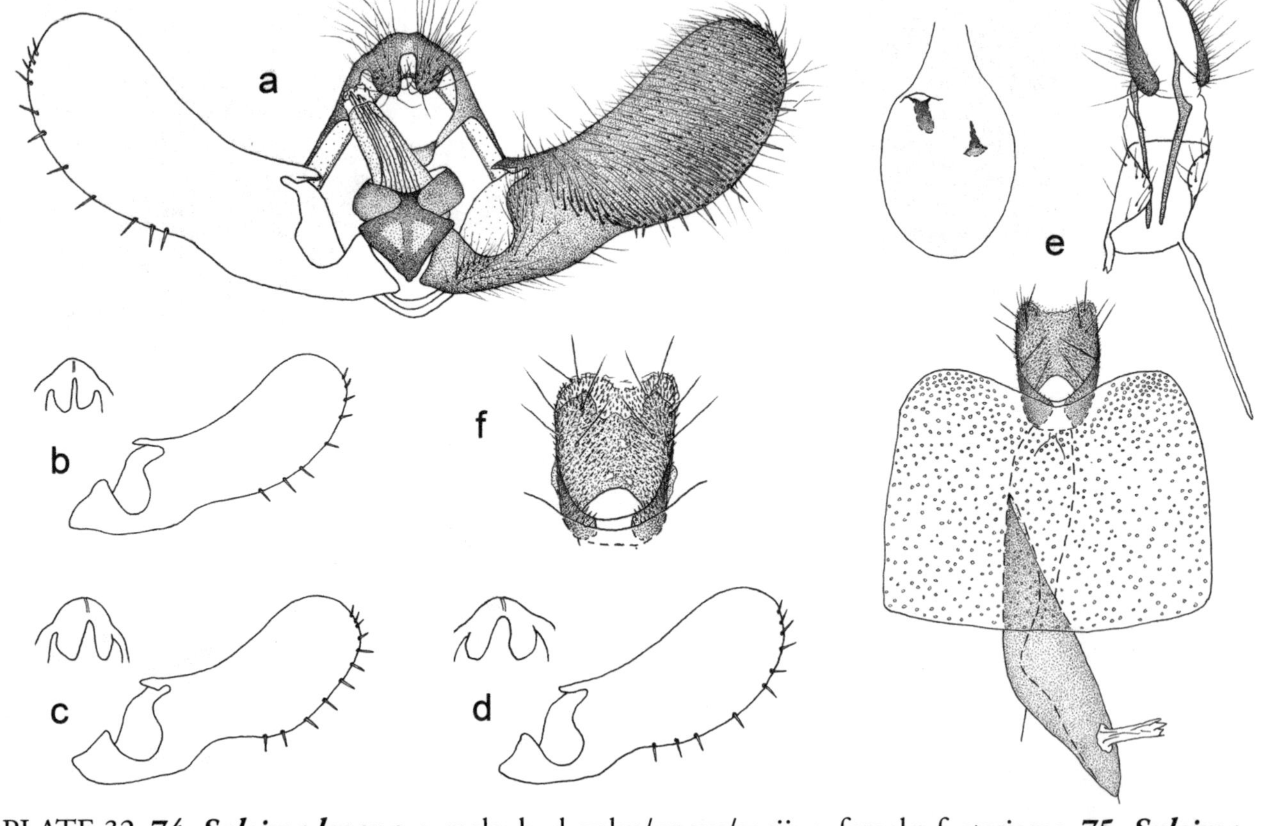


PLATE 32. **74.** ***Suleima leeana*:** a, male; b–d, valva/uncus/socii; e, female; f, sterigma. **75.** ***Suleima blanchardi*:** a, male; b–d, valva/uncus/socii; e, female; f, sterigma.

76

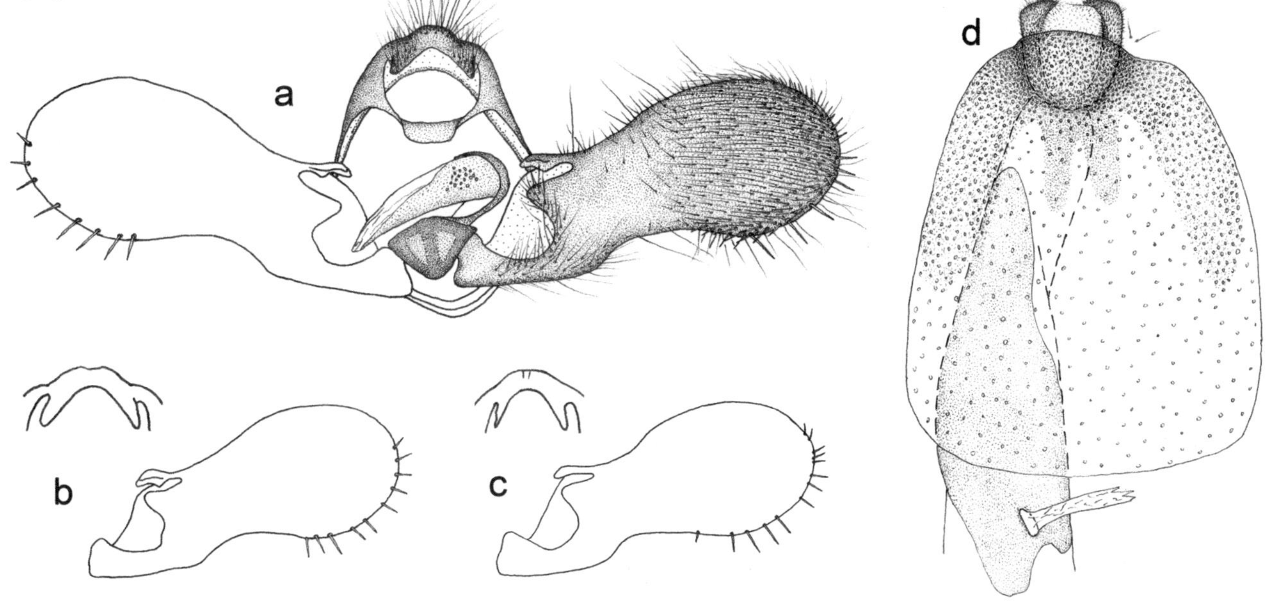


77

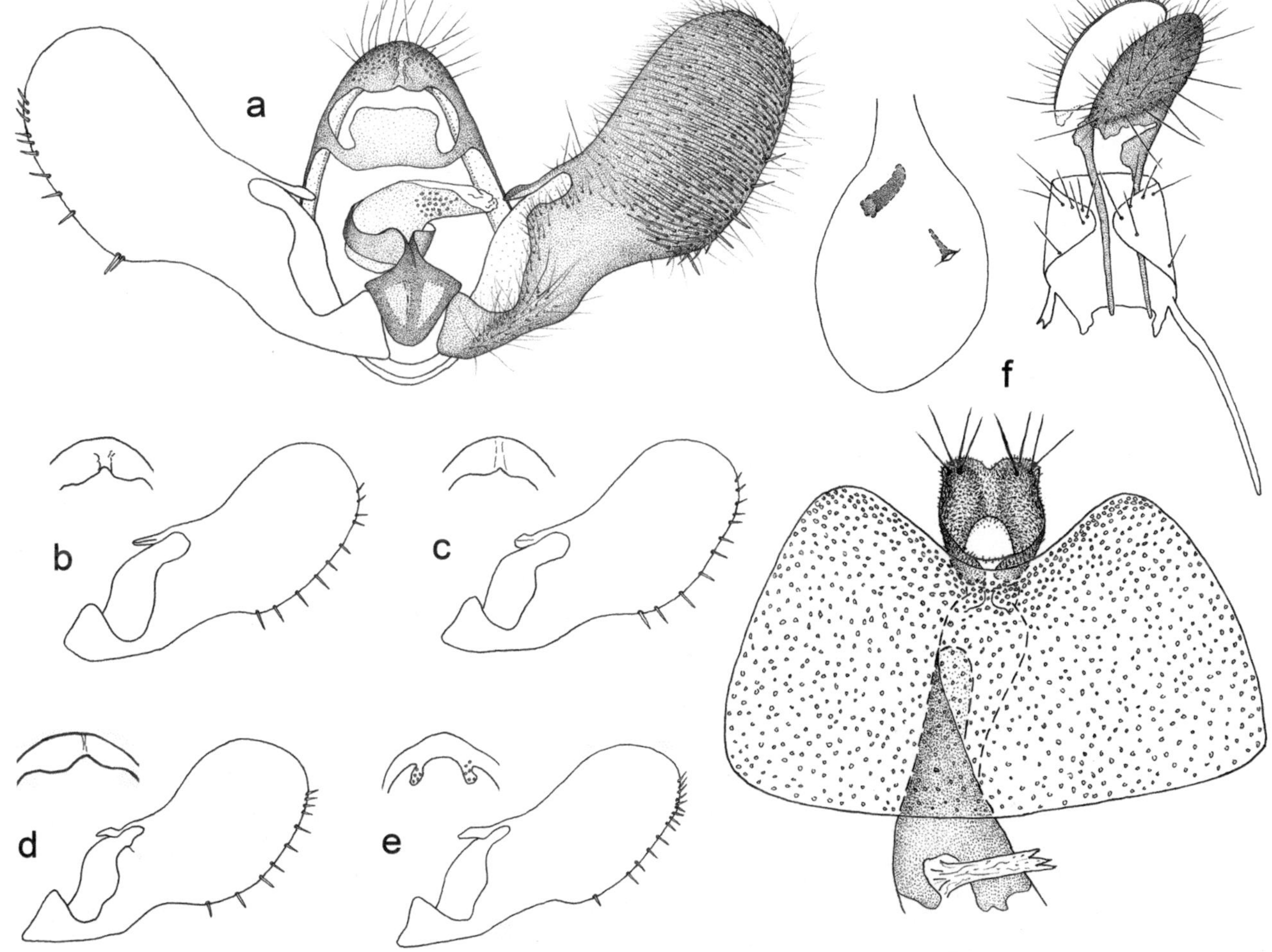


PLATE 33. **76. *Suleima skinnerana*:** a, male; b–c, valva/uncus/socii; d, sterigma/sternum 7/ductus bursae. **77. *Suleima flavidana*:** a, male; b–e, valva/uncus/socii; f, female.

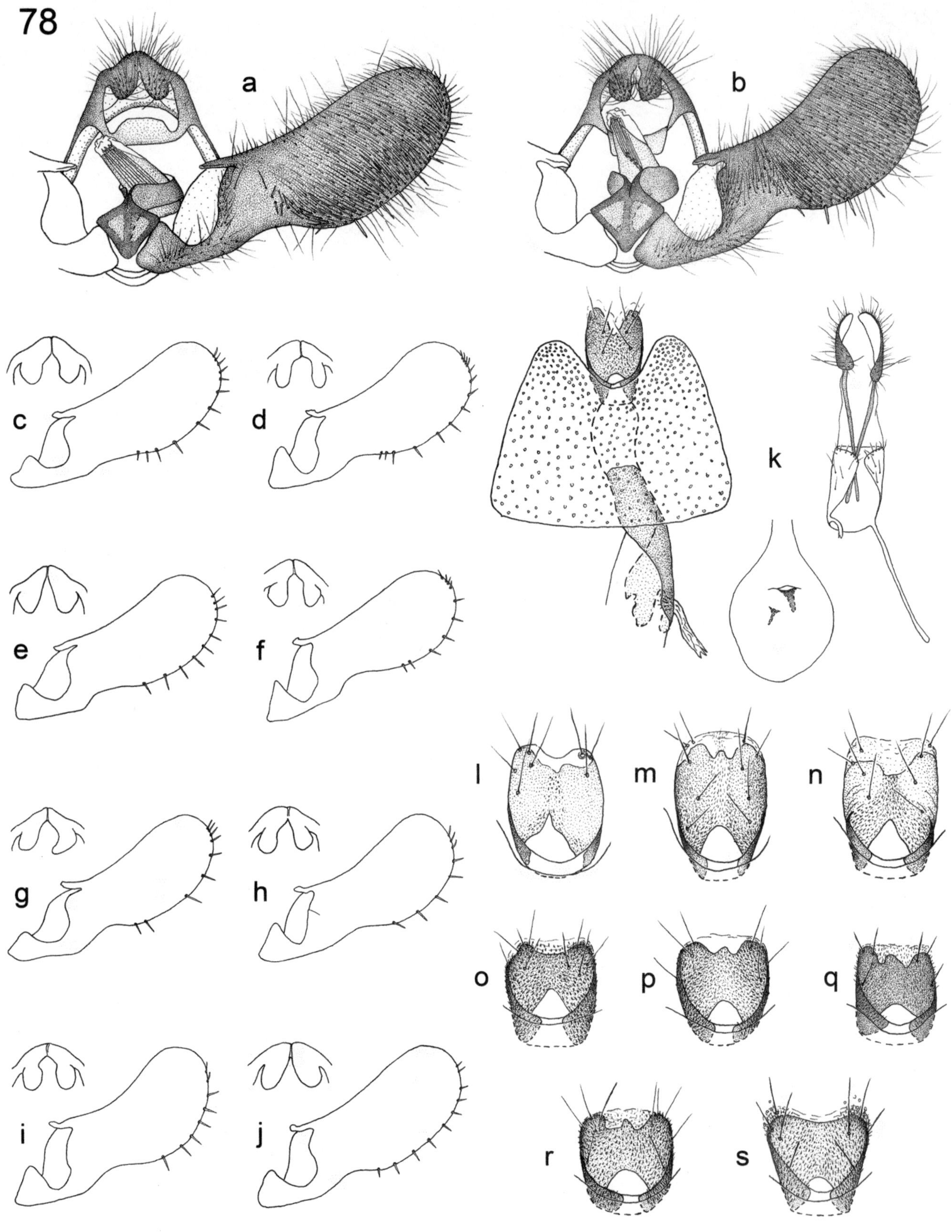


PLATE 34. **78.** ***Suleima daracana*:** a–b, male; c–j, valva/uncus/socii; k, female; l–s, sterigma.

79

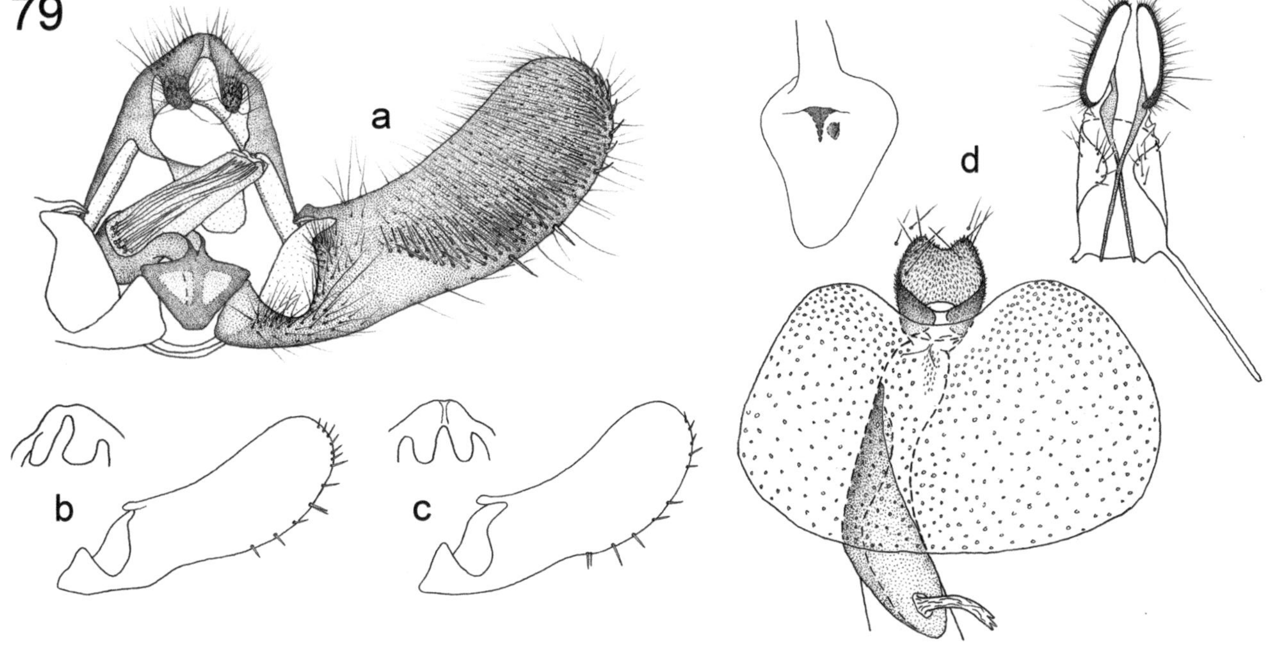


80

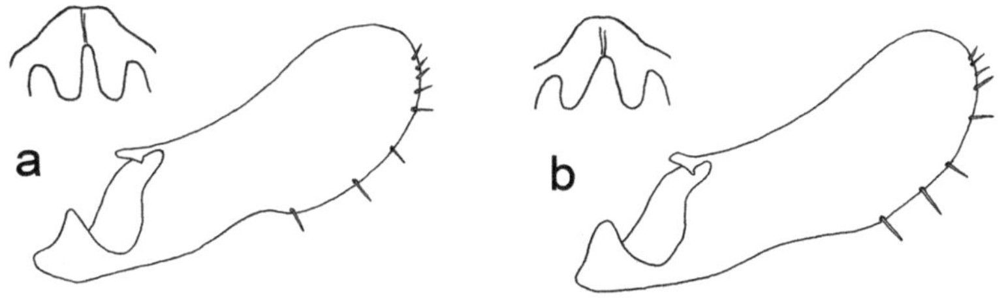


81

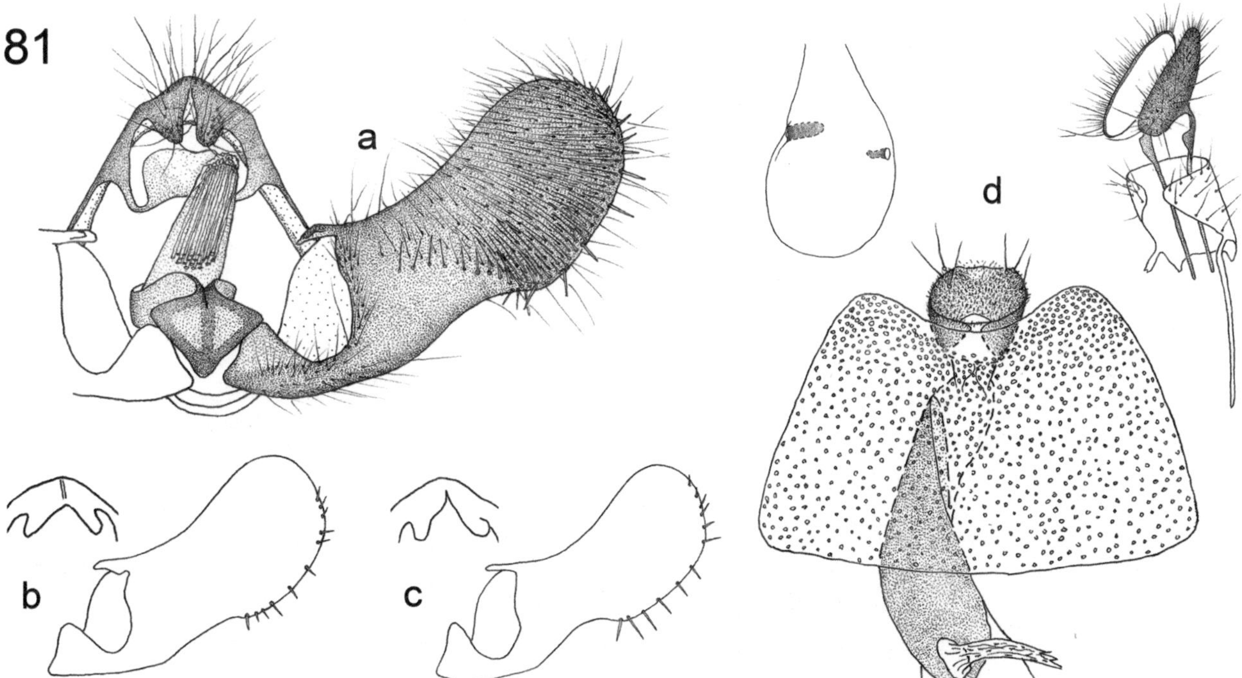


PLATE 35. **79.** ***Suleima cinerodorsana*****:** a, male; b–c, valva/uncus/socii; d, female. **80. unresolved specimens:** a–b, valva/uncus/socii. **81.** ***Suleima mendaciana*****:** a, male; b–c, valva/uncus/socii; d, female.

82

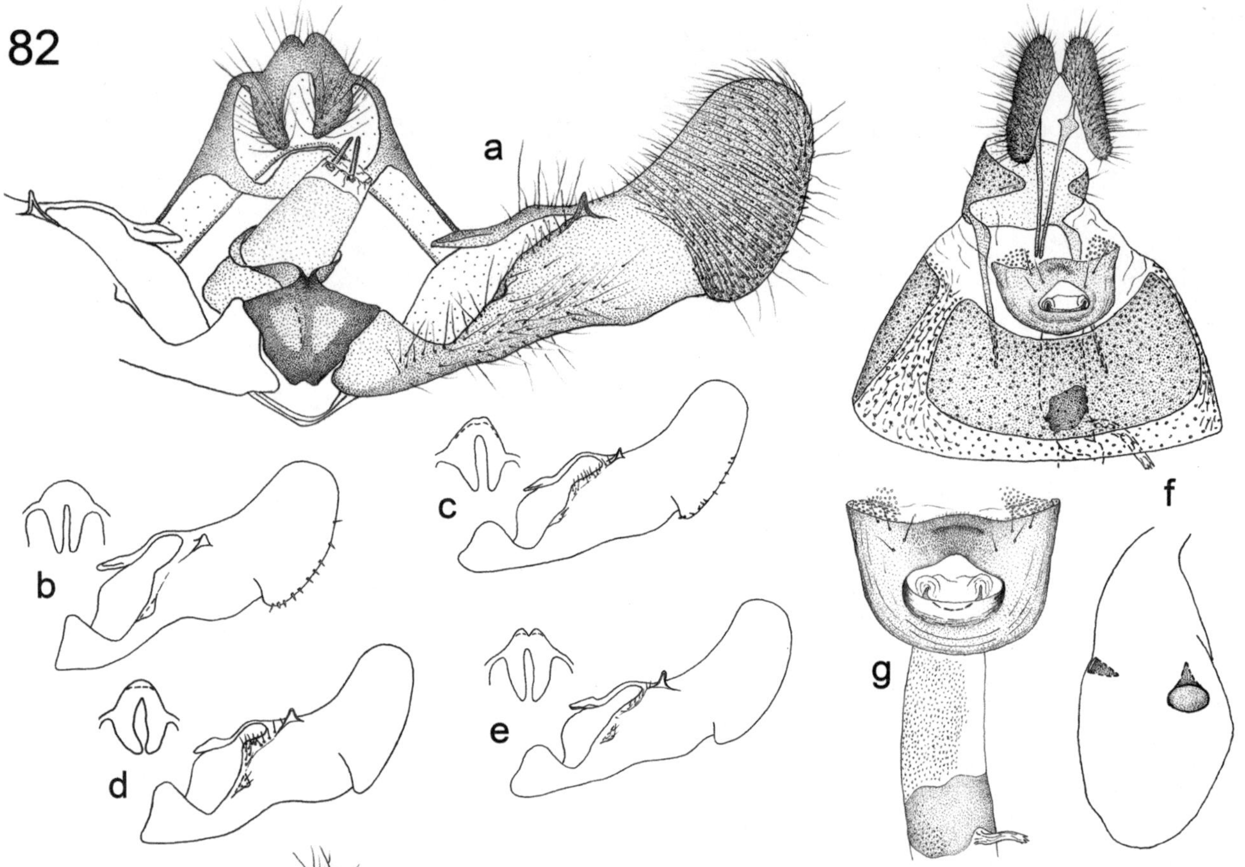


83

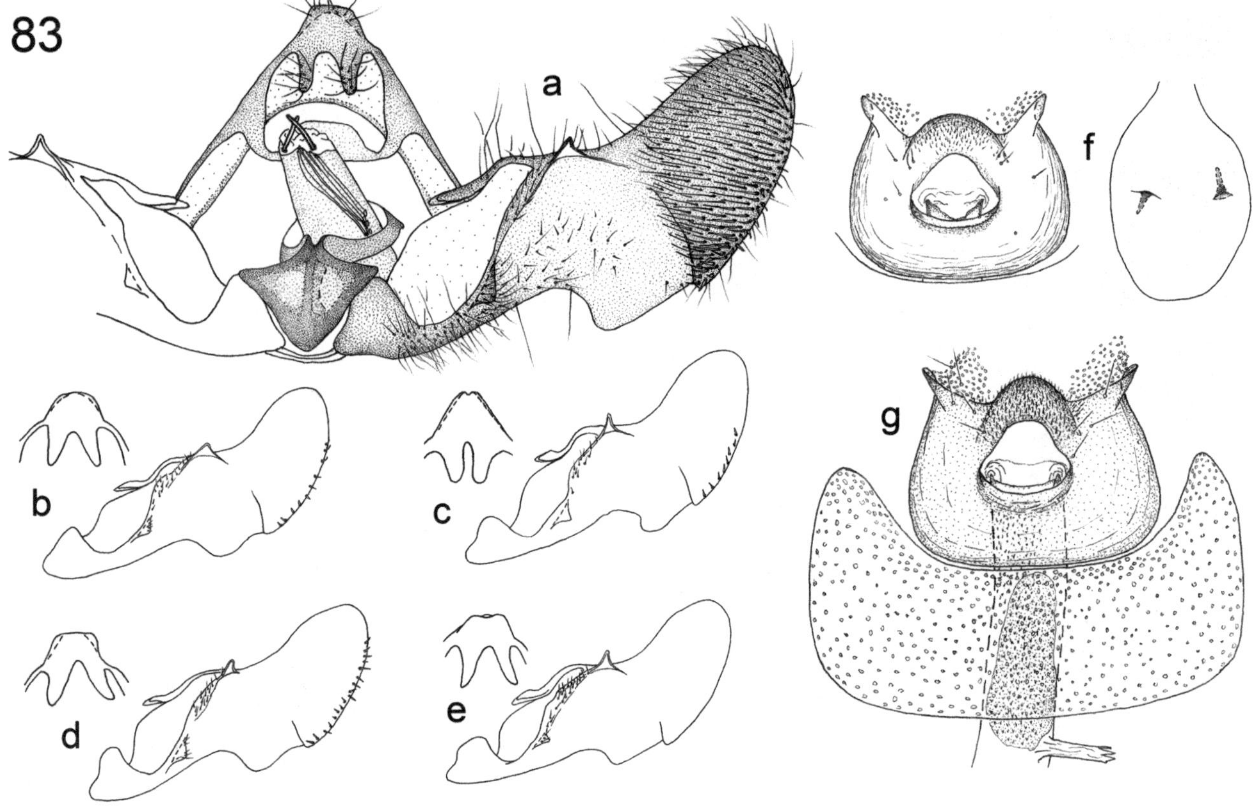


PLATE 36. **82.** ***Notocelia culminana*:** a, male; b–e, valva/uncus/socii; f, female; g, sterigma/ductus bursae. **83.** ***Notocelia lobana*:** a, male; b–e, valva/uncus/socii; f, sterigma/corpus bursae; g, sterigma/sternum 7/ductus bursae.

84

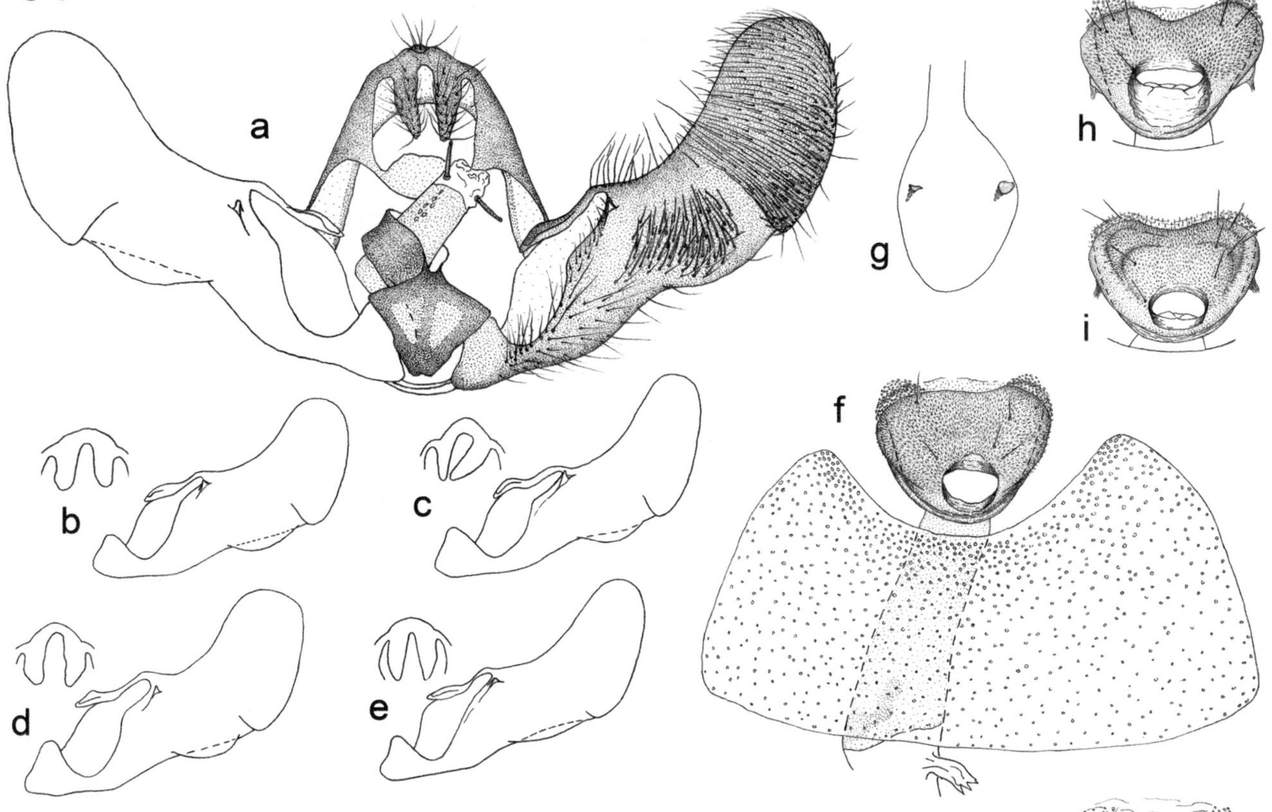


85

PLATE 37. **84.** ***Notocelia illotana*:** a, male; b–e, valva/uncus/socii; f–g, female; h–i, sterigma.
**85.** ***Notocelia cynosbatella*:** a, male; b–e, valva/uncus/socii; f, female; g–h, sterigma.

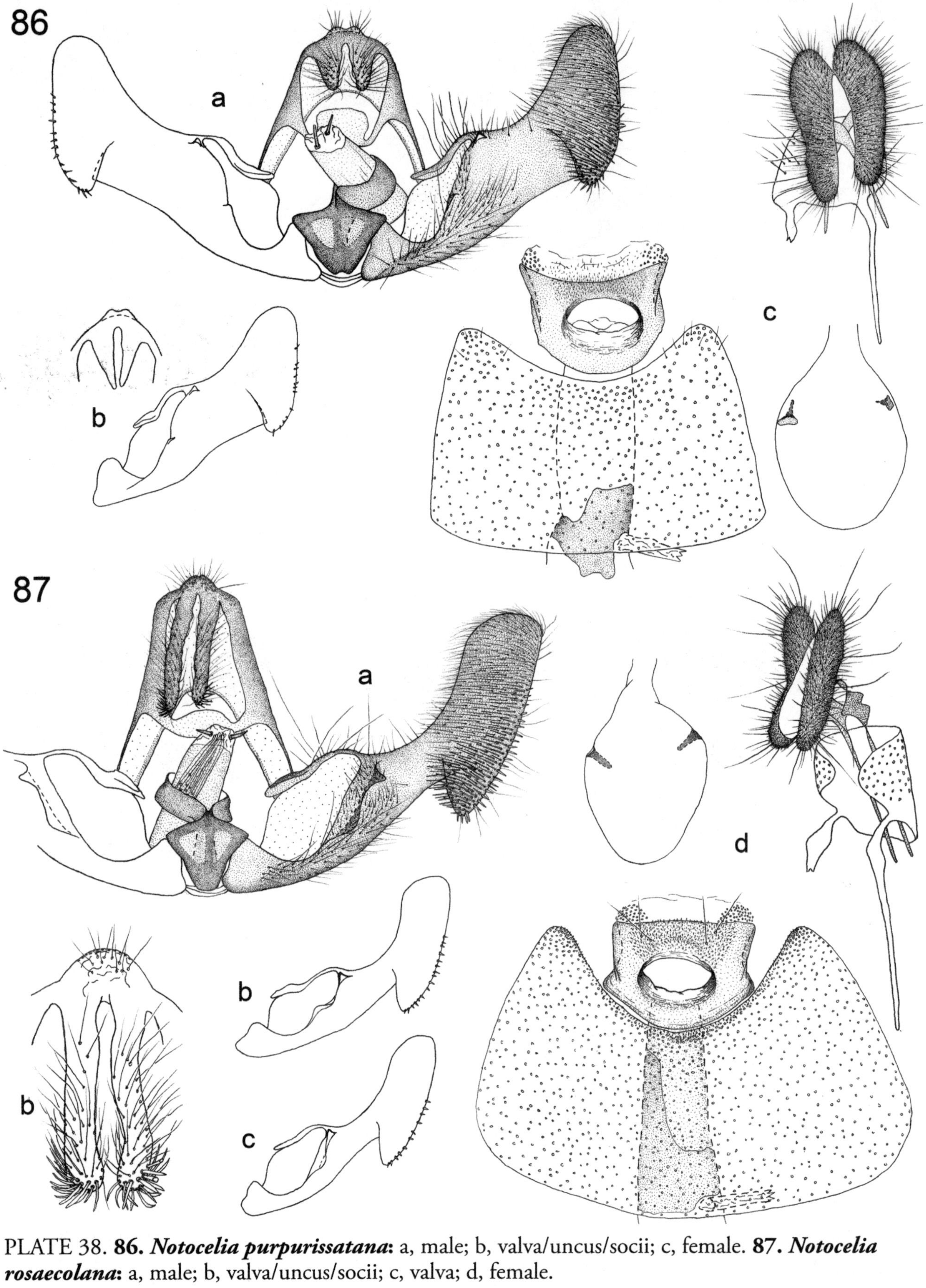


PLATE 38. **86. *Notocelia purpurissatana*:** a, male; b, valva/uncus/socii; c, female. **87. *Notocelia rosaecolana*:** a, male; b, valva/uncus/socii; c, valva; d, female.

# APPENDICES

## Appendix A

# Numerical Data

| | | Cornuti | | | FWL (mm) | | | MEAN | | | |
|---|---|---|---|---|---|---|---|---|---|---|---|
| **Sp #** | ***Epiblema* Type 1** | **Range** | **Avg** | **n** | **Range** | **Avg** | **n** | **AR** | **CFR** | **SA°** | **NR** |
| 1 | *strenuana* | 23 - 34 | 27 | 7 | 4.0 - 9.0 | 7.1 | 40 | 2.80 | 0.43 | 140 | 0.57 |
| 2 | *minutana* (typical) | 14 - 25 | 20 | 13 | 4.3 - 7.9 | 6.0 | 61 | 3.12 | 0.37 | 155 | 0.64 |
| 2 | *minutana* (pale) | 14 - 26 | 21 | 8 | 5.1 - 8.7 | 6.8 | 53 | 3.18 | 0.36 | 149 | 0.62 |
| 3 | *luctuosissima* | 9 - 16 | 11 | 7 | 4.7 - 6.8 | 6.1 | 40 | 2.82 | 0.38 | 158 | 0.73 |
| 4 | *adusta* | 12 - 24 | 17 | 6 | 4.4 - 6.3 | 5.3 | 35 | 2.86 | 0.39 | 157 | 0.64 |
| 5 | morphospecies | 9 - 18 | 13 | 3 | 5.0 - 6.3 | 5.6 | 5 | 3.14 | 0.36 | 157 | 0.67 |
| 6 | *ochraceana* | 6 - 8 | 7 | 2 | 5.5 | 5.5 | 2 | 2.87 | 0.40 | 146 | 0.64 |
| 7 | *sosana* | 13 - 26 | 17 | 9 | 5.5 - 9.1 | 7.3 | 48 | 3.04 | 0.40 | 154 | 0.76 |
| 8 | *cimarronensis* | 14 - 24 | 18 | 9 | 5.2 - 8.4 | 7.0 | 24 | 3.01 | 0.36 | 159 | 0.76 |
| 9 | *wielgusi* | 21 - 31 | 25 | 3 | 7.5 - 9.2 | 8.5 | 9 | 3.08 | 0.38 | 154 | 0.63 |
| 10 | *carolinana* | 17 - 32 | 26 | 6 | 6.9 - 10.9 | 9.2 | 31 | 2.61 | 0.65 | 130 | 0.59 |
| 11 | *arizonana* | 17 - 21 | 19 | 3 | 8.9 - 11.9 | 11.0 | 13 | 2.57 | 0.60 | 136 | 0.64 |
| 12 | *tripartitana* | 16 - 42 | 23 | 8 | 3.7 - 9.8 | 6.3 | 39 | 2.69 | 0.56 | 138 | 0.66 |
| 13 | *glenni* | 18 - 25 | 21 | 6 | 5.0 - 8.4 | 6.1 | 59 | 2.61 | 0.61 | 149 | 0.65 |
| 14 | *scudderiana* | 22 - 41 | 33 | 8 | 5.2 - 11.3 | 8.2 | 77 | 2.62 | 0.45 | 149 | 0.56 |
| 15 | *benignata* | 14 - 31 | 21 | 8 | 5.7 - 9.0 | 7.6 | 32 | 2.84 | 0.40 | 155 | 0.66 |
| 16 | *infelix* | 22 - 43 | 31 | 8 | 5.3 - 10.2 | 7.9 | 43 | 2.71 | 0.51 | 148 | 0.57 |
| 17 | *gibsoni* | 35 - 55 | 42 | 8 | 5.9 - 9.8 | 7.8 | 100 | 2.53 | 0.51 | 154 | 0.58 |
| 18 | *desertana* | 17 - 24 | 20 | 3 | 6.7 - 9.6 | 8.0 | 34 | 2.80 | 0.43 | 158 | 0.80 |
| 19 | *rudei* | 3 - 10 | 7 | 4 | 7.6 - 11.0 | 9.4 | 15 | 3.04 | 0.42 | 159 | 0.82 |
| 20 | *discretivana* | 0 - 5 | 2 | 3 | 5.8 - 8.4 | 6.9 | 28 | 2.58 | 0.43 | 150 | 0.67 |
| 21 | *obfuscana* | 13 - 22 | 18 | 5 | 6.9 - 9.3 | 8.0 | 29 | 2.57 | 0.57 | 150 | 0.67 |
| 22 | *radicana* | 6 - 11 | 8 | 3 | 6.7 - 8.4 | 7.3 | 11 | 2.74 | 0.35 | 160 | 0.83 |
| 23 | *hirsutana* | 9 - 15 | 12 | 3 | 5.8 - 7.2 | 6.4 | 9 | 2.58 | 0.36 | 134 | 0.75 |
| 24 | *lyallana* | 16 | 16 | 1 | 6.8 - 7.3 | 7.0 | 4 | 2.54 | 0.38 | 143 | 0.74 |
| 25 | *deverrae* | 16 | 16 | 2 | 5.0 - 7.1 | 6.2 | 13 | 2.61 | 0.39 | 145 | 0.72 |
| 26 | *periculosana* | 8 - 19 | 13 | 12 | 6.6 - 9.5 | 8.0 | 27 | 2.74 | 0.33 | 152 | 0.80 |
| 27 | *arctica* | 15 - 25 | 19 | 5 | 7.7 - 9.5 | 8.7 | 16 | 2.75 | 0.36 | 155 | 0.84 |
| 28 | *walsinghami* | 15 - 21 | 17 | 3 | 6.0 - 7.9 | 7.1 | 13 | 2.32 | 0.47 | 154 | 0.73 |
| 29 | *iowana* | 7 - 10 | 9 | 4 | 6.0 - 9.0 | 7.3 | 21 | 2.59 | 0.45 | 158 | 0.76 |
| 30 | *coracinana* | 23 - 30 | 27 | 2 | 5.6 - 7.4 | 6.6 | 8 | 2.66 | 0.57 | 149 | 0.64 |
| 31 | *otiosana* | 21 - 35 | 30 | 3 | 5.4 - 9.2 | 7.6 | 53 | 2.61 | 0.40 | 110 | 0.35 |
| 32 | *dorsisuffusana* | 20 - 39 | 28 | 6 | 6.9 - 9.0 | 8.1 | 24 | 2.68 | 0.40 | 106 | 0.36 |
| 33 | *boxcana* | 20 - 28 | 25 | 6 | 5.7 - 8.0 | 7.0 | 28 | 2.77 | 0.40 | 157 | 0.64 |
| 34 | *resumptana* | 8 - 15 | 11 | 8 | 4.4 - 6.7 | 5.6 | 30 | 2.78 | 0.34 | 148 | 0.78 |
| 35 | *macneilli* | 6 - 10 | 8 | 2 | 10.8 - 13.2 | 11.9 | 9 | 2.92 | 0.28 | 162 | 0.80 |
| 36 | *brightonana* | 12 - 24 | 18 | 7 | 5.2 - 7.9 | 6.5 | 42 | 2.32 | 0.42 | 122 | 0.70 |
| 37 | *browni* | 18 - 24 | 22 | 4 | 5.0 - 7.0 | 5.9 | 20 | 2.54 | 0.53 | 120 | 0.44 |

| | | | | | | | | | | | |
|---|---|---|---|---|---|---|---|---|---|---|---|
| 38 | *fulminana* | 60 - 88 | 77 | 3 | 7.3 - 14.0 | 11.3 | 43 | 2.62 | 0.28 | 155 | 0.70 |
| 39 | *alba* | 31 - 39 | 35 | 5 | 4.2 - 5.7 | 5.1 | 30 | 2.57 | | 149 | 0.67 |
| 40 | *tandana* | 23 - 30 | 27 | 3 | 6.6 - 10.2 | 8.7 | 20 | 2.37 | 0.37 | 115 | 0.53 |
| | ***Epiblema* Type 2** | | | | | | | | | | |
| 41 | *numerosana* | 26 - 56 | 39 | 13 | 5.2 - 7.9 | 6.5 | 35 | 2.74 | 0.26 | 148 | 0.48 |
| 42 | *deflexana* | 31 - 50 | 41 | 6 | 5.8 - 7.2 | 6.3 | 9 | 2.75 | 0.25 | 149 | 0.50 |
| 43 | *abruptana* | 22 - 49 | 33 | 13 | 5.2 - 8.8 | 6.4 | 57 | 2.68 | 0.28 | 150 | 0.57 |
| 44 | *mandana* | 24 - 45 | 33 | 13 | 5.3 - 8.4 | 6.9 | 77 | 2.68 | 0.28 | 150 | 0.59 |
| 45 | *exacerbatricana* | 20 - 39 | 28 | 10 | 5.0 - 6.9 | 6.1 | 46 | 2.74 | 0.25 | 156 | 0.72 |
| 46 | *glaseri* | 28 - 60 | 39 | 17 | 5.2 - 7.9 | 6.1 | 45 | 2.66 | 0.25 | 166 | 0.75 |
| 47 | *praesumptiosa* | 26 -60 | 38 | 24 | 6.2 - 9.5 | 7.9 | 114 | 2.90 | 0.29 | 149 | 0.63 |
| 48 | *grossbecki* | 27 - 43 | 36 | 5 | 7.2 - 9.1 | 8.3 | 11 | 2.72 | 0.31 | 150 | 0.56 |
| 49 | *chromata* | 26 - 36 | 31 | 3 | 6.9 - 9.7 | 8.3 | 27 | 2.81 | 0.27 | 155 | 0.56 |
| 50 | *insidiosana* | 28 - 56 | 43 | 7 | 4.9 - 6.1 | 5.5 | 13 | 2.50 | 0.28 | 159 | 0.59 |
| 51 | *separationis* | 7 - 22 | 12 | 12 | 3.7 - 6.5 | 5.1 | 62 | 2.95 | 0.26 | 162 | 0.73 |
| 52 | *colfaxana* | 9 - 12 | 11 | 2 | 5.6 - 7.6 | 6.6 | 3 | 3.21 | 0.30 | 157 | 0.67 |
| | **Average** | | | | | **7.2** | | **2.74** | **0.39** | **148** | **0.65** |

| | | **Cornuti** | | | **FWL (mm)** | | | **MEAN** | | | |
|---|---|---|---|---|---|---|---|---|---|---|---|
| **Sp #** | ***Sonia*** | **Range** | **Avg** | **n** | **Range** | **Avg** | **n** | **AR** | **CFR** | **SA°** | **NR** |
| 53 | *constrictana* | 34 - 65 | 47 | 8 | 6.0 - 8.3 | 7.2 | 38 | 2.81 | 0.28 | 120 | 0.48 |
| 54 | *paraplesiana* | 34 - 63 | 50 | 19 | 4.2 - 8.5 | 6.3 | 132 | 2.67 | 0.28 | 143 | 0.51 |
| 55 | *roseomaculana* | 55 - 75 | 64 | 4 | 7.2 - 8.6 | 7.9 | 5 | 2.47 | 0.27 | 146 | 0.62 |
| 56 | *ferruginana* | 30 - 61 | 42 | 11 | 4.0 - 9.1 | 7.0 | 163 | 2.64 | 0.29 | 132 | 0.56 |
| 57 | *griseolana* | 30 - 60 | 42 | 12 | 5.5 - 8.2 | 6.8 | 23 | 2.64 | 0.27 | 105 | 0.51 |
| 58 | *ferruginana* cplx gray | 30 - 68 | 40 | 17 | 5.8 - 9.2 | 7.4 | 33 | 2.58 | 0.30 | 130 | 0.53 |
| 58 | *ferruginana* cplx red | 36 - 61 | 46 | 15 | 5.8 - 8.1 | 6.9 | 27 | 2.56 | 0.30 | 134 | 0.50 |
| 59 | *canadana* | 22 - 56 | 36 | 11 | 6.1 - 8.6 | 7.2 | 55 | 2.56 | 0.31 | 127 | 0.52 |
| 60 | *nigrana* | 22 - 38 | 32 | 11 | 5.0 - 8.0 | 6.4 | 76 | 2.61 | 0.30 | 131 | 0.58 |
| 61 | *albimacula* | 26 - 42 | 32 | 11 | 5.4 - 7.1 | 6.4 | 29 | 2.58 | 0.30 | 131 | 0.58 |
| 62 | *parva* | 30 - 42 | 34 | 7 | 5.0 - 6.3 | 5.7 | 21 | 2.69 | 0.30 | 124 | 0.53 |
| 63 | *canadana* cplx | 31 - 39 | 35 | 3 | 6.1 - 7.2 | 6.7 | 7 | 2.75 | 0.29 | 136 | 0.53 |
| 64 | *divaricata* | 24 - 50 | 39 | 5 | 5.9 - 9.2 | 8.0 | 25 | 2.67 | 0.27 | 154 | 0.56 |
| 65 | *comstocki* | 70 - 88 | 80 | 4 | 9.2 - 11.3 | 10.1 | 12 | 2.73 | 0.31 | 161 | 0.69 |
| 66 | *filiana* | 28 - 32 | 30 | 3 | 10.1 - 12.2 | 11.1 | 9 | 2.77 | 0.29 | 151 | 0.66 |
| 67 | *vovana* | 35 - 76 | 54 | 7 | 6.9 - 11.3 | 9.2 | 43 | 2.86 | 0.29 | 150 | 0.62 |
| 68 | *hodgesi* | 7 - 12 | 9 | 3 | 4.7 - 7.0 | 5.8 | 55 | 2.91 | 0.27 | 166 | 0.66 |
| 69 | *gibsoni* | 32 | 32 | 1 | 8.1 | 8.1 | 1 | 2.84 | 0.25 | 149 | 0.54 |
| | **Average** | | **41** | | | **7.5** | | **2.69** | **0.29** | **138** | **0.57** |

| | | Cornuti | | | FWL (mm) | | | MEAN | | | |
|---|---|---|---|---|---|---|---|---|---|---|---|
| **Sp #** | ***Suleima*** | **Range** | **Avg** | **n** | **Range** | **Avg** | **n** | **AR** | **CFR** | **SA°** | **NR** |
| 70 | *helianthana* | 15 - 40 | 24 | 15 | 5.3 - 10.8 | 7.7 | 48 | 2.97 | | 146 | 0.82 |
| 71 | *lagopana* | 26 | 26 | 1 | 8.1 - 10.5 | 9.6 | 9 | 2.91 | | 145 | 0.82 |
| 72 | *powelli* | 14 - 24 | 18 | 6 | 7.3 - 10.5 | 9.0 | 18 | 3.18 | | 141 | 0.81 |
| 73 | *baracana* | 16 - 35 | 24 | 16 | 6.2 - 10.9 | 8.5 | 135 | 3.08 | | 155 | 0.84 |
| 74 | *leeana* | 20 - 25 | 22 | 5 | 6.2 - 8.6 | 7.6 | 20 | 3.23 | | 153 | 0.84 |
| 75 | *blanchardi* | 22 - 29 | 25 | 6 | 5.8 - 7.7 | 6.7 | 23 | 3.02 | | 150 | 0.83 |
| 76 | *skinnerana* | 19 - 24 | 22 | 4 | 8.0 - 9.6 | 9.0 | 16 | 3.05 | | 140 | 0.81 |
| 77 | *flavidana* | 17 - 25 | 22 | 4 | 5.7 - 8.6 | 7.5 | 16 | 2.86 | | 152 | 0.82 |
| 78 | *daracana* | 12 - 24 | 18 | 12 | 6.3 - 9.2 | 8.0 | 56 | 3.28 | | 151 | 0.85 |
| 79 | *cinerodorsana* | 15 - 32 | 24 | 3 | 5.0 - 7.4 | 6.5 | 19 | 3.02 | | 154 | 0.86 |
| 80 | *morphospecies* | 19 - 26 | 23 | 2 | 7.7 - 7.8 | 7.8 | 2 | 3.07 | | 160 | 0.90 |
| 81 | *mendaciana* | 20 - 26 | 23 | 4 | 5.5 - 6.9 | 6.2 | 7 | 3.08 | | 146 | 0.79 |
| | **Average** | | **23** | | | **7.8** | | **3.06** | | **149** | **0.83** |

| | | Cornuti | | | FWL (mm) | | | MEAN | |
|---|---|---|---|---|---|---|---|---|---|
| **Sp #** | ***Notocelia*** | **Range** | **Avg** | **n** | **Range** | **Avg** | **n** | **AR** | **CFR** |
| 82 | *culminana* | 10 - 18 | 14 | 16 | 7.0 - 9.3 | 8.1 | 68 | 2.69 | 0.41 |
| 83 | *lobana* | 10 - 15 | 12 | 12 | 7.2 - 8.7 | 7.8 | 19 | 2.55 | 0.40 |
| 84 | *illotana* | 4 - 15 | 9 | 15 | 6.5 - 8.8 | 8.1 | 43 | 2.54 | 0.49 |
| 85 | *cynosbatella* | 8 - 12 | 9 | 6 | 6.1 - 9.8 | 8.6 | 22 | 2.54 | 0.48 |
| 86 | *purpurissatana* | 17 - 18 | 18 | 2 | 5.6 - 7.9 | 6.7 | 13 | 2.54 | 0.51 |
| 87 | *rosaecolana* | 12 - 20 | 17 | 4 | 7.4 - 9.6 | 8.5 | 44 | 2.51 | 0.41 |
| | **Average** | | | | | **8.0** | | **2.56** | **0.45** |

## Appendix B

# Genitalia Data

**1.** ***Epiblema strenuana*** (Walker)

a. Kentucky, Gallatin County, Markland Dam, 27 May 1990, L. D. Gibson, slide DJW 3781, LDG.
b. Virginia, Fairfax County, Mt. Vernon, 21 August 1976, J. A. Powell, slide DJW 3954, EME.
c. Texas, Tarrant County, Benbrook, 20 May 1991, J. A. Powell, slide DJW 3959, EME.
d. California, La Puerta Valley, July 1911, G. H. Field, slide DJW 3934, EME.
e. Kentucky, Gallatin County, Markland Dam, 21 June 1990, L. D. Gibson, slide DJW 3782, LDG.
f. Alabama, Baldwin County, Weeks Bay NERS, 22 June 2008, D. J. Wright, slide TMG 779, USNM.
g. Florida, Volusia County, Blue Spring State Park, 10 June 1986, Powell and Heppner, slide DJW 3947, EME.
h. Kentucky, Henderson County, Sloughs WMA, Sauerheber Unit, 22 August 1992, L. D. Gibson, slide DJW 3945, LDG.

**2.** ***Epiblema minutana*** (Kearfott)

a. Ohio, Wyandot County, Killdeer Plains, 16 August 2006, D. J. Wright, slide TMG 782, USNM.
b. Florida, Volusia County, Blue Spring State Park, 12 June 1986, Powell and Heppner, slide DJW 3948, EME.
c. California, Contra Costa County, Antioch NWR, 2 April 1991, J. A. Powell, slide DJW 3960, EME.
d. California, Contra Costa County, Point Molate, 12 December 1988, J. A. Powell, slide DJW 3938, EME. [pale phenotype, reared from *Ambrosia chamissonis*]
e. California, Los Angeles County, San Clemente Island, 23 March 1972, J. A. Powell, slide DJW 3961, EME. [pale phenotype]
f. Kentucky, McCracken County, Paducah Zip Track, 12 August 2008, D. J. Wright, slide TMG 781, USNM.
g. Ohio, Erie County, Resthaven WLA, 24 May 1991, L. D. Gibson, slide DJW 3956, LDG.
h. California, Contra Costa County, Antioch NWR, 15 August 1991, J. A. Powell, slide DJW 3946, EME. [reared from *Ambrosia psilostachya*]
i. Arkansas, Pope County, Pinery Creek WA, 24 May 1991, J. A. Powell, slide DJW 3950, EME.

**3.** ***Epiblema luctuosissima*** Blanchard

a. Kansas, Morton County, 7.5 mi N of Elkhart, 2 August 1999, D. J. Wright, slide 152252, USNM.
b. Colorado, Yuma County, Bonny Reservoir SRA, 5 August 1996, D. J. Wright, slide 152253, USNM.
c. Texas, Padre Island National Seashore, 24 June 1976, A. and M. E. Blanchard, slide 90282, USNM.
d. Colorado, Baca County, Picture Canyon Picnic Area, 6 August 2005, D. J. Wright, slide 152251, USNM.
e. Colorado, Yuma County, Bonny Reservoir SRA, 5 August 1996, D. J. Wright, slide 152249, USNM.
f. Kansas, Morton County, 7.5 mi N of Elkhart, 2 August 1999, D. J. Wright, slide 152250, USNM.

**4.** ***Epiblema adusta*** Wright and Gilligan

a. PTP. Mississippi, Jackson County, Grand Bay Savanna, 30 April 2004, Brown, Lee and Martinez, slide DJW 4013, MEM.
b. PTP. North Carolina, Carteret County, Fort Macon State Park, 19 July 1998, J. B. Sullivan, slide DJW 3957, USNM.
c. PTP. Texas, Nueces County, North Padre Island, 24 August 1979, A. and M. E. Blanchard, slide DJW 3970, USNM.
d. PTP. North Carolina, Carteret County, Sunset Point, N Beaufort, 17 July 1999, J. B. Sullivan, slide DJW 3958, USNM.
e. PTP. Alabama, Baldwin County, Plash Island, 16–19 June 1984, R. and B. Brown, slide DJW 4017, MEM.

**5.** **Morphospecies**

a. Mexico, Baja California, Norte, Arroyo Catavina, 1 April 1976, P. Rude, slide DJW 4201, EME.
b. Mexico, Baja California, 1 mi E Santo Domingo, 18 March 1972, J. A. Powell, slide DJW 4185, EME.

**6.** ***Epiblema ochraceana*** Fernald

a. HTP. Florida, Palm Beach County, Palm Beach, slide DJW 3925, USNM.
b. Florida, Palm Beach County, Palm Beach, slide 70744, USNM.

**7.** ***Epiblema sosana*** (Kearfott)

a. Utah, Tooele County, Stockton, 20 June 1904, T. Spalding, slide DJW 3857, USNM.
b. California, San Bernardino County, Kelso Dunes, 29 June 1979, J. A. Powell, slide DJW 3940, EME.
c. California, San Bernardino County, Cucamonga, 16 January 1969, R. Goeden and D. Ricker, slide DJW 3859, USNM.
d. Arizona, Yuma County, 1.27 mi SE of I-8, Frontage Road, 8 February 2014, S. Lee, slide DJW 4000, ASUHIC.
e. Utah, Tooele County, Stockton, 1 August 1904, T. Spalding, slide DJW 3858, USNM.
f. Utah, Tooele County, Stockton, 1 August 1904, T. Spalding, slide 70748, USNM.

**8.** ***Epiblema cimarronensis*** Wright and Gilligan

a. PTP. Texas, Coleman County, Santa Anna, 19 May 1985, E. C. Knudson, slide DJW 3863, USNM.
b. PTP. California, San Diego County, Banner, 10 August 1970, R. Goeden and D. Ricker, slide DJW 4635, USNM. [reared from *Ambrosia confertiflora*]
c. PTP. Texas, Cottle County, Matador WMA, 30 July 1983, E. C. Knudson, slide ECK 1062, MGCL.
d. PTP. Kansas, Morton County, 7.5 mi N of Elkhart, 25 August 2000, D. J. Wright, slide DJW 3763, USNM.

**9.** ***Epiblema wielgusi*** Wright and Gilligan

a. PTP. Arizona, Maricopa County, Thunderbird Regional Mountain Park, 28 April 1975, R. and J. Wielgus, slide DJW 3996, ASUHIC.
b. PTP. Arizona, Maricopa County, Thunderbird Regional Mountain Park, 20 April 1975, R. and J. Wielgus, slide DJW 3995, ASUHIC.
c. PTP. Arizona, Maricopa County, Thunderbird Regional Mountain Park, 20 March 1976, R. S. Wielgus, slide DJW 3861, USNM.
d. HTP. Arizona, Maricopa County, Thunderbird Regional Mountain Park, 6 March 1977, R. S. Wielgus, slide DJW 3862, USNM.
e. PTP. Arizona, Maricopa County, Thunderbird Regional Mountain Park, 20 April 1975, R. and J. Wielgus, slide DJW 3997, USNM.

**10.** ***Epiblema carolinana*** (Walsingham)

a. Kentucky, Bullitt County, Route 480, 6.9 mi E Route 61, 8 July 1988, L. D. Gibson, slide DJW 3792, LDG.
b. Maryland, Plummer's Island, 22 July 1972, G. F. Hevel, slide 88659, USNM.
c. Illinois, Cook County, Palos Park, 5 August 1917, slide 72688, USNM.
d. Ohio, Adams County, 1 mi SE of Lynx, 12 July 1998, D. J. Wright, slide 152219, USNM.
e. Wyoming, Albany County, Fox Creek, 2 August 2003, J. S. Nordin, slide 152220, USNM.
f. Iowa, Pocahontas County, Kalsow Prairie, 18 July 2003, D. J. Wright, slide DJW 3814, USNM.
g. Iowa, Pocahontas County, Kalsow Prairie, 18 July 2003, D. J. Wright, slide 152221, USNM.
h. Ohio, Adams County, 1 mi SE of Lynx, 5 July 2002, D. J. Wright, slide DJW 3815, USNM.

**11.** ***Epiblema arizonana*** Powell

a. PTP. Arizona, Coconino County, Oak Creek Canyon, 30 June 1936, G. P. Englehardt, slide 70793, USNM. [reared from *Rudbeckia laciniata*]
b. HTP. Arizona, Coconino County, Oak Creek Canyon, 28 June 1936, G. P. Englehardt, slide 72844, USNM. [reared from *Rudbeckia laciniata*]
c. PTP. Arizona, Coconino County, Oak Creek Canyon, 26 June 1936, G. P. Englehardt, slide DJW 3991, USNM. [reared from *Rudbeckia laciniata*]
d. PTP. Arizona, Coconino County, Oak Creek Canyon, 24 June 1936, G. P. Englehardt, slide DJW 3798, USNM.
e. ALTP. Arizona, Coconino County, Oak Creek Canyon, 26 June 1936, G. P. Englehardt, slide 70792, USNM. [reared from *Rudbeckia laciniata*]

**12.** ***Epiblema tripartitana*** (Zeller)

a. Texas, Shelby County, 3 mi E of Liberty, 14 April 1924, I. J. Bottimer, slide 94179, USNM. [reared from heads and stems of *Rudbeckia maxima*]
b. New Mexico, Guadalupe County, 5 mi S of Santa Rosa, 27 May 1995, R. A. Watkins, slide 152295, USNM.
c. Colorado, Morgan County, County Road I, 28 July 1995, D. J. Wright, slide 152293, USNM.
d. Illinois, Putnam County, 19 June 1967, M. O. Glenn, slide 94178, USNM.
e. Texas, Shelby County, 3 mi E of Liberty, 8 April 1924, I. J. Bottimer, slide 94180, USNM. [reared from heads and stems of *Rudbeckia maxima*]
f. Kansas, Morton County, 7.5 mi N of Elkhart, 26 August 2000, D. J. Wright, slide 152296, USNM.

**13.** ***Epiblema glenni*** Wright

a. Ohio, Adams County, 1 mi SE of Lynx, 8 June 1989, D. J. Wright, slide 152240, USNM.
b. PTP. North Carolina, Macon County, Highlands, 3 July 1958, R. W. Hodges, slide 94175, USNM.
c. PTP. Ohio, Adams County, 1 mi SE of Lynx, 10 September 1968, D. J. Wright, slide 152239, USNM.
d. PTP. Illinois, Putnam County, 10 July 1969, M. O. Glenn, slide 94174, USNM.
e. PTP. Illinois, Putnam County, 21 July 1965, M. O. Glenn, slide 94173, USNM.
f. PTP. Ohio, Adams County, 1 mi SE of Lynx, 12 August 1998, D. J. Wright, slide 152235, USNM.

**14.** ***Epiblema scudderiana*** (Clemens)

a. Ohio, Erie County, Resthaven WLA, 24 May 1991, D. J. Wright, slide 152280, USNM.
b. Kansas, Morton County, 7.5 mi N of Elkhart, 2 August 1999, D. J. Wright, slide 152282, USNM.
c. Iowa, Pocahontas County, Kalsow Prairie, 25 June 1997, D. J. Wright, slide 152279, USNM.
d. Florida, Charlotte County, Punta Gorda, April 1952, H. Ramstadt, slide 95185, USNM. [dark phenotype]
e. Iowa, Pocahontas County, Kalsow Prairie, 22 June 2000, D. J. Wright, slide 152281, USNM
f. Georgia, Chatham County, Savannah, August 1943, C. H. Hoffman, slide 95193, USNM. [dark phenotype, reared from Goldenrod stem]

**15. *Epiblema benignata* McDunnough**

a. Iowa, Pocahontas County, Kalsow Prairie, 25 June 1997, D. J. Wright, slide 152216, USNM.
b. PTP. Canada, Manitoba, Aweme, 14 June 1923, N. Criddle, slide 70764, USNM.
c. Washington, Whitman County, Wilma, 28 March 1934, J. F. G. Clarke, slide 70763, USNM. [reared from stalks of *Artemisia vulgaris*]
d. California, San Bernardino County, San Bernardino, 1–7 June, slide 70758, USNM.
e. Washington, Whitman County, Wilma, 9 April 1934, J. F. G. Clarke, slide 124845, USNM. [reared from stalks of *Artemisia vulgaris*]

**16. *Epiblema infelix* Heinrich**

a. Kentucky, Powell County, Tunnel Ridge Road, 28 April 1989, D. J. Wright, slide 152247, USNM.
b. South Carolina, Berkeley County, Wedge Plantation, 22 April 1974, D. C. Ferguson, slide 94176, USNM.
c. Kentucky, Laurel County, Bolton Branch, 18 May 1996, D. J. Wright, slide 152248, USNM.
d. Kentucky, Laurel County, Bolton Branch, 18 May 1996, D. J. Wright, slide 152245, USNM.
e. Kentucky, Powell County, Tunnel Ridge Road, 28 April 1989, L. D. Gibson, slide LDG 188, LDG.
f. Kentucky, Owsley County, 3 mi NE of Booneville, 23 May 1981, L. D. Gibson, slide LDG 193. LDG.
g. PTP. North Carolina, Polk County, Tryon, 4 July 1904, W. F. Fiske, slide 70802, USNM.

**17. *Epiblema gibsoni* Wright and Covell**

a. PTP. Ohio, Adams County, 1 mi SE of Lynx, 5 July 1996, D. J. Wright, slide 152227, USNM.
b. Virginia, Smyth County, 5 mi E of Marion, 24 July 1970, J. W. Brown, slide 119171, USNM.
c. PTP. Indiana, Lake County, Ivanhoe D&S, 17 June 2000, R. Panzer, slide 152229, USNM. [interfascial spot reduced]
d. PTP. Kentucky, Rowan County, Route 1274, 2 mi E of Route 519, 16 July 1994, L. D. Gibson, LDG 195, LDG.
e. PTP. Ohio, Adams County, 1 mi SE of Lynx, 19 August 1998, D. J. Wright, slide 152233, USNM.
f. PTP. Ohio, Adams County, 1 mi SE of Lynx, 3 August 2000, D. J. Wright, slide 152232, USNM.

**18. *Epiblema desertana* (Zeller)**

a. Ohio, Wyandot County, Route T-108, 2.5 mi S of Route 294, 30 June 1990, D. J. Wright, slide 152222, USNM.
b. Ohio, Summit County, Springfield Bog Park, 17 June 2011, J. D. Hooper, slide DJW 3827, USNM.
c. Kentucky, Bullitt County, Route 480, 28 May 1989, L. D. Gibson, slide DJW 3790, LDG.
d. Kentucky, Boone County, Big Bone Lick State Park, 25 June 1986, L. D. Gibson, slide DJW 3791, LDG.

**19. *Epiblema rudei* Powell**

a. PTP. California, San Benito/Fresno County line, 21 April 1967, P. A. Rude, slide DJW 3836, USNM. [reared from *Gutierrezia*]
b. PTP. California, San Bernardino County, Colton, 13 March 1910, F. X. Williams, slide JAP 233, EME.
c. Texas, Hemphill County, Canadian, 13 March 1969, A. and M. E. Blanchard, slide 69846, USNM.
d. Texas, Cameron County, Laguna Atascosa, 3 March 1973, A. and M. E. Blanchard, slide DJW 3837, USNM.
e. California, Kings County, 2 mi S of Kettleman, 17 March 1975, J. A. Powell, slide DJW 3924, EME. [reared from *Gutierrezia*]

**20. *Epiblema discretivana* Heinrich**

a. Louisiana, Bossier Parish, Barksdale AFB, 9 May 1996, D. M. Pollock, slide DJW 3834, USNM.
b. Florida, Palm State Park, 3 April 1930, F. M. Jones, slide 70769, USNM.
c. Texas, Harris County, Sheldon, 10 April 1919, Johnson, slide 70770, USNM.
d. Florida, Lee County, Sanibel Island, 10 July 1982, E. C. Knudson, slide DJW 3835, USNM.
e. Florida, Manatee County, Oneco, 28 March 1957, J. G. Franclemont, slide DJW 3832, USNM.

**21. *Epiblema obfuscana* (Dyar)**

a. Kentucky, Boone County, Big Bone State Park, 18 May 1980, L. D. Gibson, slide DJW 3789, LDG.
b. Washington, Whitman County, Alamota, 20 March 1934, J. F. G. Clarke, slide 70782, USNM.
c. Pennsylvania, Allegheny County, Oak Station, 2 June 1911, F. Marloff, slide 70775, USNM.
d. Kentucky, Gallatin County, Markland Dam, 11 June 1989, L. D. Gibson, slide DJW 3788, LDG.

**22. *Epiblema radicana* (Walsingham)**

a. California, San Mateo County, San Bruno Mountains, 17 March 1968, P. A. Opler, slide DJW 3923, EME.
b. California, San Mateo County, San Bruno Mountains, 7 April 1962, J. A. Powell, slide DJW 3922, EME.
c. California, San Mateo County, San Bruno Mountains, 7 April 1962, J. A. Powell, slide JAP 1277, EME.
d. California, San Mateo County, San Bruno Mountains, 10 April 1962, J. A. Powell, slide DJW 3921, EME.

**23. *Epiblema hirsutana* (Walsingham)**

a. California, Alameda County, Berkeley Hills, 17 April 1962, J. A. Powell, slide DJW 3917, EME.
b. California, slide 70799, Walsingham, USNM. [Fernald Collection]
c. California, Alameda County, Berkeley Hills, 3 April 1959, Chemsak and Powell, slide JAP 550, EME.

**24. *Epiblema lyallana* McDunnough**

a. PTP. Quebec, Gaspe County, Mont Lyall, 2 August 1933, W. J. Brown, slide 89924, USNM.
b. HTP. Quebec, Gaspe County, Mont Lyall, 25 July 1933, W. J. Brown, slide TOR 4250, CNC.
c. Canada, Alberta, Lake Louise, 23 August 1988, G. S. Walley, slide TOR 2358, CNC. [determination uncertain]
d. ALTP. Quebec, Gaspe County, 27 July 1933, E. B. Watson, slide DJW 4034, USNM.

**25. *Epiblema deverrae* Brown**

a. California, Monterey County, Big Creek Reserve, 14–16 June 1991, Hsu and Powell, slide DJW 3920, EME.
b. PTP. California, Monterey County, Big Creek Reserve, 4–6 June 1982, J. A. Powell, slide DJW 3919, EME.
c. PTP. California, Sonoma County, Cazadero, 6 June 1979, J. A. Powell, slide DJW 3918, EME.
d. California, Monterey County, Big Creek Reserve, 7–9 June 1993, Hsu and Powell, slide DJW 3926, EME.

**26. *Epiblema periculosana* Heinrich**

a. Oregon, Crook County, Ochoco National Forest, 3 July 1967, J. H. Baker, slide DJW 3869, USNM.
b. Colorado, Chaffee County, Cottonwood Pass, 16 July 1982, R. W. Hodges, slide DJW 3872, USNM.
c. Canada, Alberta, Banff, 25 August, O. Bryant, slide 70800, USNM.
d. Canada, Alberta, Waterton Lakes, 26 June 1923, J. McDunnough, slide RLB 84, CNC.

**27. *Epiblema arctica* Miller**

a. PTP. Alaska, Eagle Summit, 2 July 1979, Opler and Powell, slide DJW 3927, EME.
b. PTP. Alaska, Schrader Lake, 2 July 1973, K. W. Phillip, slide DJW 3822, USNM.
c. PTP. Alaska, 4 mi N of Cantwell, 27 June 1979, Opler and Powell, slide JAP 4460, EME.
d. PTP. Alaska, Eagle Summit, 2 July 1979, Opler and Powell, slide RLB 1144, USNM.

**28. *Epiblema walsinghami* (Kearfott)**

a. Illinois, Putnam County, 5 May 1965, M. O. Glenn, slide DJW 3908, USNM.
b. Pennsylvania, Allegheny County, Oak Station, 13 May 1916, slide 70798, USNM.
c. New Jersey, Hemlock Falls, 29 April, slide 95203, USNM.
d. New Jersey, Passaic County, Great Notch, 10 May 1914, slide 95202, USNM.

**29. *Epiblema iowana* McDunnough**

a. Missouri, Benton County, 3 mi NW of Warsaw, 5 April 1978, J. R. Heitzman, slide DJW 3841, USNM.
b. Illinois Putnam County, 3 June 1950, M. O. Glenn, slide 26399, USNM. [PTP: *E. naoma*, reared from roots of *Ratibida pinnata*]
c. Ohio, Erie County, Resthaven WLA, 24 May 1991, D. J. Wright, slide DJW 3842, USNM.
d. Illinois Putnam County, 24 April 1948, M. O. Glenn, slide DJW 3843, USNM.
e. Illinois Putnam County, 28 April 1948, M. O. Glenn, slide 26397, USNM. [PTP: *E. naoma*]

**30. *Epiblema coracinana* Wright and Gilligan**

a. PTP. Pennsylvania, Allegheny County, Oak Station, F. Marloff, 21 May 1915, slide MS 13292, CMNH.
b. PTP. Pennsylvania, Allegheny County, Oak Station, F. Marloff, 5 July 1908, slide DJW 4306, USNM.
c. HTP. Pennsylvania, Allegheny County, Oak Station, F. Marloff, 16 June 1908, slide DJW 4304, CMNH.

**31. *Epiblema otiosana* (Clemens)**

a. Ohio, Wyandot County, Killdeer Plains, 15 June 1995, D. J. Wright, slide 152267, USNM.
b. Oklahoma, Marshall County, Willis, University of Oklahoma Biological Station, 26 June 1953, D. R. Davis, slide 70812, USNM.
c. Kentucky, Boone County, Middle Creek Road, 8 June 1984, slide DJW 3786, L. D. Gibson, LDG.
d. North Carolina, Polk County, Tryon, 30 June 1904, W. F. Fiske, slide 70811, USNM.
e. Kentucky, Henderson County, Sloughs WMA, Sauerhaber Unit, 22 August 1992, L. D. Gibson, slide DJW 3785, LDG.
f. Kentucky, Henderson County, Sloughs WMA, Sauerhaber Unit, 22 August 1992, D. J. Wright, slide 152268, USNM.

**32. *Epiblema dorsisuffusana* (Kearfott)**

a. Iowa, Pocahontas County, Kalsow Prairie, 18 June 1992, D. J. Wright, slide 152224, USNM.
b. Oklahoma, Murray County, Arbuckle Mountains, 19–30 July 1984, D. and M. Davis, slide DJW 4204, USNM.
c. North Carolina, Ashe County, Pond Mountain, Top Plateau, 1 July 2019, J. B. Sullivan, slide JBS 10977, JBS.
d. Ohio, Adams County, 1 mi SE of Lynx, 5 July 1996, D. J. Wright, slide DJW 4203, USNM.
e. Maryland, Washington County, Indian Springs WMA, 21 June 2004, J. Glaser, slide DJW 4002, USNM.
f. North Carolina, Durham County, New Hope Bottomlands, 5 June 2022, S. P. Hall, slide JBS 9588, JBS
g. Arkansas, Washington County, Devil's Den State Park, 4 July 1966, R. W. Hodges, slide 124829, USNM.

**33. *Epiblema boxcana*** (Kearfott)

a. Kentucky, Boone County, Big Bone Lick State Park, 4 May 1981, L. D. Gibson, slide DJW 3780, LDG.
b. Ohio, Erie County, Resthaven WLA, 24 May 1991, D. J. Wright, slide 152215, USNM.
c. Illinois, Macon County, Decatur, slide 70723, USNM.
d. Washington, Whitman County, Snake River opposite Clarkston, 12 April 1934, J.F.G. Clarke, slide 70824, USNM. [reared from *Xanthium spaciosum*]
e. Illinois, Putnam County, 24 May 1946, M. O. Glenn, slide 124838, USNM.

**34. *Epiblema resumptana*** (Walker)

a. Connecticut, Windham County, 1.6 mi S of Hampton, 16 May 1994, D. L. Wagner, slide 152273, USNM.
b. Kentucky, Laurel County, Forest Service Road 615a, 18 April 1991, L. D. Gibson, slide LDG 190, LDG.
c. Oregon, Baker County, Baker, Spring Creek, 18 May 1967, J. H. Baker, slide DJW 3906, USNM.
d. Ohio, Hamilton County, Cincinnati, 2 June 1906, A. Braun, slide DJW 3905, USNM.

**35. *Epiblema macneilli*** Powell

a. PTP. California, Inyo County, Mono Pass, 3 September 1965, J. A. Powell, slide DJW 3856, USNM.
b. PTP. California, Inyo County, near Mono Pass, 30 August 1965, C. D. MacNeill, slide JAP 3867, EME.
c. PTP. California, Inyo County, near Mono Pass, 31 August 1965, P. A. Opler, slide DJW 4003, EME.
d. PTP. California, Inyo County, Mono Pass, 3 September 1965, P. Rude, slide DJW 3855, USNM.

**36. *Epiblema brightonana*** (Kearfott)

a. Slide 70803, USNM. [Type Collection of W. D. Kearfott]
b. Pennsylvania, Beaver County, New Brighton, 5 September 1907, F. A. Merrick, slide 70804, USNM.
c. Kentucky, Gallatin County, Markland Dam, 4 August 1989, L. D. Gibson, slide DJW 3796, LDG.
d. Virginia, Grayson County, Big Spring Bog Natural Area Preserve, 15 July 2002, Roble and Wilson, slide 118790, USNM.
e. Illinois, Putnam County, 13 August 1960, M. O. Glenn, slide 70871, USNM.
f. Ohio, Adams County, 1 mi SE of Lynx, 1 August 1997, D. J. Wright, slide 152217, USNM.
g. Kentucky, Gallatin County, Markland Dam, 22 June 1989, L. D. Gibson, slide DJW 3794, LDG.

**37. *Epiblema browni*** Wright and Gilligan

a. HTP. Mississippi, Oktibbeha County, Osborn Prairie, 30 August 1997, Brown and Koehn, slide DJW 4019, USNM.
b. PTP. Mississippi, Oktibbeha County, Osborn Prairie, 30 August 1997, Brown and Koehn, slide DJW 4021, MEM.
c. PTP. Louisiana, Bossier Parish, Bodcau WMA, 13 September 1996, R. L. Brown, slide DJW 4459, MEM.
d. PTP. Louisiana, Bossier Parish, Bodcau WMA, 13 September 1996, R. L. Brown, slide DJW 4020, MEM.

**38. *Epiblema fulminana*** (Walsingham)

a. Ohio, Adams County, 1 mi SE of Lynx, 16 July 1990, D. J. Wright, slide DJW 3884, USNM.
b. Kansas, Riley County, Konza Prairie, 23 July 1995, D. J. Wright, slide DJW 4136, USNM.
c. Kentucky, Bullitt County, Route 480, 7 mi E of Route 61, 22 July 1989, D. J. Wright, slide DJW 4137, USNM.
d. Iowa, Pocahontas County, Kalsow Prairie, 18 July 2003, D. J. Wright, slide DJW 4138, USNM.
e. Ohio, Wyandot County, Killdeer Plains, 21 July 1995, E. H. Metzler, slide DJW 3885, USNM.
f. Kentucky, Rowan County, Route 1274, 2 mi SW of Route 519, 8 July 2005, L. D. Gibson, slide DJW 4133, LDG.

**39. *Epiblema alba*** Wright and Gilligan

a. PTP. North Carolina, Pender County, Holly Shelter Game Land, 6 August 1997, J. B. Sullivan, slide DJW 4051, MEM.
b. PTP. Florida, Santa Rosa County, Florida A&M University Biological Station, 25–28 August 1986, J. B. Heppner, slide DJW 4049, MGCL.
c. HTP. Florida, Putnam County, Palatka, 23 August 1992, H. D. Baggett, slide DJW 4052, USNM.
d. PTP. Florida, Hernando County, Withlacoochee State Forest, 18 September 1990, W. L Adair Jr., slide DJW 4050, MGCL.

**40. *Epiblema tandana*** (Kearfott)

a. Maryland, Allegheny County, Warrior Mountain WMA, 28 June 2001, J. Glaser, slide DJW 3993, USNM.
b. Kentucky, Gallatin County, Markland Dam, 24 August 2006, L. D. Gibson, slide DJW 3992, USNM.
c. Ohio, Greene County, Wright-Patterson AFB, 12 June 1992, E. H. Metzler, slide 152290, USNM.
d. Pennsylvania, York County, near Dillsburg, 19 June 1969, G. Werschkow, slide DJW 3994, USNM.

**41. *Epiblema numerosana*** (Zeller)

a. Texas, Anderson County, Tennessee Colony, 23 August 1978, A. and M. E. Blanchard, slide DJW 3972, USNM.
b. Illinois, Mason County, Sand Ridge State Forest, 21 July 1985, R. Leuschner, slide DJW 4042, EME.
c. Texas, Anderson County, Tennessee Colony, 23 August 1978, A. and M. E. Blanchard, slide 90272, USNM.
d. Texas, Washington County, Brenham, 9 September 1979, E. C. Knudson, slide ECK 194, MGCL.
e. North Dakota, Ransom County, 1 mi SE of McLeod, 19 July 1963, J. R. Powers, slide DJW 4167, EME.
f. Texas, Anderson County, Tennessee Colony, 23 August 1978, A. and M. E. Blanchard, slide DJW 3968, USNM.

**42. *Epiblema deflexana*** Heinrich

a. HTP. Texas, Cameron County, Brownsville, 27 May 1917, A. Busck, slide 72861, USNM.
b. Texas, Bexar County, 6 May 1993, D. M. Pollock, slide DJW 4046, MEM.
c. Texas, Bexar County, 6 May 1993, D. M. Pollock, slide DJW 4008, MEM.
d. Mexico, Nuevo Leon, 4 mi W of Iturbide, 13–14 September 1976, Powell and Chemsak, slide DJW 3939, EME.
e. Mexico, Nuevo Leon, 4 mi W of Iturbide, 13–14 September 1976, Powell and Chemsak, slide DJW 4158, EME.
f. Texas, Bexar County, 6 May 1993, D. M. Pollock, slide DJW 4045, MEM.

**43. *Epiblema abruptana*** (Walsingham)

a. Ohio, Adams County, 1 mi SE of Lynx, 25 July 1998, D. J. Wright, slide 152214, USNM.
b. Indiana, Newton County, Beaver Lake Prairie Preserve, 18 July 1997, E. H. Metzler, slide DJW 4178, USNM.
c. Texas, Sabine County, Six Mile, 5–6 June 1987, E. C. Knudson, slide DJW 3847, MGCL.
d. Florida, Putnam County, Palatka, 25 September 1990, H. D. Baggett, slide DJW 4181, MEM.
e. Texas, Hays/Comal County, El Rancho Cima, 29 August 1975, A. and M. E. Blanchard, slide 90248, USNM.
f. Texas, Sabine County, Six Mile, 5–6 June 1987, E. C. Knudson, slide DJW 3845, MGCL.
g. Texas, Burnet County, Shovel Mountain, 16–23 June, slide DJW 3980, USNM.

**44. *Epiblema mandana*** (Kearfott)

a. Texas, Tyler County, Lake Steinhagen, 24 August 1996, E. C. Knudson, slide DJW 3975, MGCL.
b. Maryland, Dorchester County, Taylors Island, 14 June 1998, J. Glaser, slide DJW 3973, USNM.
c. Ohio, Adams County, 1 mi SE of Lynx, 27 June 1998, D. J. Wright, slide DJW 3779, USNM.
d. Maryland, Worcester County, Vaughn WMA, 20 June 2000, J. Glaser, slide DJW 3819, USNM.
e. Texas, Hardin County, Big Thicket National Preserve, 22 April 1995, E. C. Knudson, slide DJW 3849, MGCL.
f. Maryland, Dorchester County, Taylors Island WMA, 26 August 1999, J. Glaser, slide DJW 3974, USNM.

**45. *Epiblema exacerbatricana*** Heinrich

a. South Carolina, Chesterfield County, Patrick, 15 September 2012, J. Glaser, slide DJW 3812, USNM.
b. Florida, Levy County, Goethe State Forest, 12 October 2012, Hayden and Ponce, slide DJW 4163, MGCL.
c. Alabama, Baldwin County, 1 mi E of Oyster Bay, 13 October 1990, R. L. Brown, slide DJW 4194, MEM.
d. Massachusetts, Plymouth County, Myles Standish State Forest, 14 August 1990, D. L. Wagner, slide DJW 4193, MEM.
e. HTP. North Carolina, Moore County, 16–23 August, Southern Pines, slide 72862, USNM.
f. Florida, Marion County, Ocala National Forest, 16 October 1990, H. D. Baggett, slide DJW 3966, USNM.
e. Florida, Pinellas County, Saint Petersburg, October, slide DJW 3813, USNM.

**46. *Epiblema glaseri*** Wright and Gilligan

a. PTP. Florida, Highlands, Archbold Biological Station, 1 May 1975, J. B. Heppner, slide DJW 4078, MGCL.
b. PTP. Georgia, Emanuel County, Ohoopee Dunes, 16 September 2012, J. Glaser, slide DJW 3807, USNM.
c. PTP. Florida, Levy County, Goethe State Forest, 3 September 2011, Hayden and Romero, slide DJW 4080, MGCL.
d. PTP. Louisiana, Calcasieu Parish, 4 mi ESE of Buhler, 15–16 September 1993, Brown and Pollock, slide DJW 4120, MEM.
e. Florida, Highlands County, Archbold Biological Station, 26–29 March 2011, S. Johnson, slide DJW 4543, SJ.
f. PTP. Alabama, Baldwin County, 1 mi E of Oyster Bay, 4 September 1988, Brown and MacGown, slide DJW 4079, MEM.

**47. *Epiblema praesumptiosa*** Heinrich

a. Texas, Cameron County, Laguna Atascosa, 1 April 1978, A. and M. E. Blanchard, slide DJW 4433, USNM.
b. Texas, Cameron County, Brownsville, 8–15 March, slide 70739, USNM.
c. Mississippi, Jackson County, Grand Bay Savanna, 30 April 2004, Brown and Martinez, slide DJW 4033, EME.
d. North Carolina, Carteret County, Fort Macon State Park, 16 July 1996, J. B. Sullivan, slide DJW 4029, MEM.
e. PTP. Texas, Cameron County, Brownsville, 27 May 1917, A. Busck, slide 17708, USNM.
f. Alabama, Baldwin County, Oyster Bay, 4–8 September 1988, R. Brown and J. MacGown, slide DJW 4032, MEM.
g. North Carolina, Carteret County, Fort Macon State Park, 16 July 1996, J. B. Sullivan, slide DJW 4030, MEM.
h. Alabama, Baldwin County, Bon Secour NWR, 15 October 1996, T. L. Schiefer, slide DJW 3902, USNM.

**48.** ***Epiblema grossbecki*** Heinrich

a. PTP. Florida, Lee County, Fort Myers, 1 April 1912, slide DJW 3839, USNM.
b. Florida, Everglade, 8–15 April, slide DJW 3830, USNM.
c. PTP. Florida, Lee County, Fort Myers, 1 April 1912, slide 25558, USNM.
d. Florida, Sarasota County, Siesta Key, 19 November 1952, C. P. Kimball, slide 89889, USNM.
e. Florida, Everglade, 8–15 April, slide DJW 3829, USNM.

**49.** ***Epiblema chromata*** Miller

a. Texas, San Patricio County, Welder WLR, 30 June 1975, A. and M. E. Blanchard, slide DJW 3816, USNM.
b. PTP. Texas, Cameron County, San Benito, 16–23 March, slide 89797, USNM.
c. Texas, Cameron County, Brownsville, 11 March 1977, A. and M. E. Blanchard, slide DJW 3817, USNM.
d. Texas, Cameron County, Laguna Atascosa, 9 March 1975, A. and M. E. Blanchard, slide DJW 3818, USNM.
e. PTP. Texas, Kerr County, Kerrville, 11 June 1911, H. Lacey, slide 89798, USNM.

**50.** ***Epiblema insidiosana*** Heinrich

a. PTP. North Carolina, Moore County, Southern Pines, 8–15 July, slide 70750, USNM.
b. Florida, Hernando County, Croom WLA, 9 July 1991, W. L Adair Jr., slide 2122, MGCL.
c. Florida, Marion County, Marion Oaks, 17 May 1990, J. S. Kutis, slide DJW 4014, MEM.
d. PTP. North Carolina, Moore County, Southern Pines, 8–15 July, slide DJW 3965, USNM.

**51.** ***Epiblema separationis*** Heinrich

a. Texas, Jefferson County, Sabine Pass, 11 June 1994, E. C. Knudson, slide DJW 3911, MGCL.
b. Texas, San Patricio County, Sinton, 5 July 1975, A. and M. E. Blanchard, slide 69855, USNM.
c. Texas, Cameron County, Palmito Hill, 22 October 1984, E. C. Knudson, slide ECK 1002, MGCL.
d. PTP. Texas, Cameron County, Brownsville, 27 May 1917, A. Busck, slide DJW 3912, USNM.
e. Mississippi, Jackson County, Grand Bay Savanna, 30 April 2004, Lee, Brown, and Martinez, slide DJW 4055, MEM.
f. North Carolina, Carteret, Fort Macon State Park, 3 August 1997, J. B. Sullivan, slide DJW 3915, USNM.
g. Texas, Brazoria County, Churchill Bridge, 9 May 1982, E. C. Knudson, slide DJW 3913, MGCL.
h. North Carolina, Carteret, Fort Macon State Park, 3 August 1997, J. B. Sullivan, slide DJW 3916, USNM.

**52.** ***Epiblema colfaxana*** Wright and Gilligan

a. HTP. New Mexico, Colfax County, Exit 419 on I-25, 5 August 2008, T. M. Gilligan, slide DJW 3371, USNM.
b. PTP. Colorado, Cheyenne County, Wild Horse Post Office, 15 September 2007, Gilligan and Harp, slide DJW 2036, USNM.
c. California, Lassen County, Ravendale, 20 June 1959, G. I. Stage, slide DJW 4162, EME.

**53.** ***Sonia constrictana*** (Zeller)

a. Texas, Cameron County, Sabal Palm Sanctuary, 4–5 August 1994, E. C. Knudson, slide DJW 4104, MGCL.
b. Texas, Hidalgo County, Santa Ana NWR, 24 June 1988, A. and M. E. Blanchard, slide DJW 4105, MGCL.
c. Texas, San Patricio County, Welder WLA, 3 July 1975, A. and M. E. Blanchard, slide 97204, USNM.
d. Texas, Zampata County, Zampata, 18 September 1973, A. and M. E. Blanchard, slide 97203, USNM.
e. Texas, Hidalgo County, Bentsen-Rio Grande Valley State Park, 11 October 1980, E. C. Knudson, slide DJW 4106, MGCL.

**54.** ***Sonia paraplesiana*** Blanchard

a. Alabama, Baldwin County, Weeks Bay NERS, 22 June 2008, D. J. Wright, slide DJW 4091, USNM.
b. Kentucky, August, A. Busck, slide 70894, USNM.
c. PTP. Texas, Harris County, Houston, 5 June 1967, A. and M. E. Blanchard, slide 97588, USNM.
d. Florida, Highlands County, Archbold Biological Station, 2 May 1975, J. B. Heppner, slide 4102, MGCL.
e. Florida, Highlands County, Lake Placid, 30 April 1964, R. W. Hodges, slide DJW 4092, USNM.

**55.** ***Sonia roseomaculana*** Wright and Gilligan

a. PTP. Florida, Franklin County, Apalachicola National Forest, 1 September 1991, H. D. Baggett, slide DJW 4279, USNM.
b. HTP. Florida, Franklin County, Apalachicola National Forest, 1 September 1991, H. D. Baggett, slide DJW 4082, MEM.
c. PTP. Florida, Franklin County, Apalachicola National Forest, 1 September 1991, H. D. Baggett, slide DJW 4273, MEM.

**56.** ***Sonia ferruginana*** Wright and Gilligan

a. HTP. Kentucky, Bracken County, Meldahl Dam, 21 June 2007, L. D. Gibson, slide DJW 4064, USNM.
b. PTP. Kentucky, McCracken County, Paducah Zip Track, 12 August 2008, D. J. Wright, slide DJW 3799, USNM.
c. Alabama, Lawrence County, Prairie Grove Glade, 25 May 2004, S. Lee and J. MacGown, slide DJW 4063, MEM.
d. PTP. New York, Tompkins County, Ithaca, Snyder Heights, 6 August 1976, J. G. Franclemont, slide DJW 4069, MEM.
e. Kentucky, Gallatin County, Markland Dam, 11 June 1989, L. D. Gibson, slide LDG 38, LDG.
f. PTP. Ohio, Adams County, 1 mi SE of Lynx, 27 June 1998, D. J. Wright, slide DJW 3802, USNM.

**57.** ***Sonia griseolana*** Wright and Gilligan

a. HTP. Mississippi, Oktibbeha County, 6 mi SW of Starkville, 23 July 1984, R. L. and B. B. Brown, slide DJW 4119, USNM.
b. PTP. Texas, Newton County, Deweyville, 18 August 1979, E. C. Knudson, slide DJW 4117, MGCL.
c. PTP. Kentucky, McCracken County, Paducah Zip Track, 12 August 2008, L. D. Gibson, slide DJW 4061, LDG.
d. PTP. Kentucky, McCracken County, Paducah Zip Track, 16 June 2007, L. D. Gibson, slide DJW 4125, LDG.
e. PTP. Mississippi, Oktibbeha County, 6 mi SW of Starkville, 3 August 1986, R. L. and B. B. Brown, slide DJW 4126, MEM.
f. PTP. Louisiana, Bossier Parish, Barksdale AFB, 21 August 1996, D. M. Pollock, slide DJW 4118, MEM.
g. PTP. Texas, Tyler County, Big Thicket National Preserve, 20 June 1996, R. Leuschner, slide DJW 4280, EME.
h. PTP. Mississippi, Oktibbeha County, 6 mi SW of Starkville, 21 August 1984, R. L. and B. B. Brown, slide DJW 4123, MEM.

**58.** ***ferruginana* complex grayish phenotype**

a. Alabama, Baldwin County, Weeks Bay NER Reserve, 5 August 2000, R. L. Brown, slide DJW 4296, MEM.
b. Florida, Collier County, Fakahatchee Strand, 30 June 1984, H. D. Baggett, slide DJW 4070, MEM.
c. New Jersey, Cumberland County, Hansey Creek Road, 29 June 2019, S. Johnson, slide DJW 4542, USNM.
d. Kentucky, Boone County, Boone Cliffs Nature Preserve, 8 June 1984, L. D. Gibson, slide WEM 199903, LDG.
e. Ohio, Erie County, Resthaven WLA, 12 July 1991, L. D. Gibson, slide DJW 4271, LDG.
f. Florida, Collier County, Fakahatchee Strand, 30 June 1984, H. D. Baggett, slide DJW 4297, USNM.
g. New Jersey, Cumberland County, Hansey Creek Road, 18 July 2020, S. Johnson, slide DJW 4541, USNM.
h. Pennsylvania, Lackawanna County, 3 mi E Archbald, 29 June 2013, S. Johnson, slide DJW 4548, SJ.
i. New York, Albany County, Guilderland, 18 June 2021, M. Mello, slide DJW 4545, USNM.
j. Alabama, Baldwin County, Weeks Bay NER Reserve, 5 August 2000, R. L. Brown, slide DJW 4067, MEM.
k. Illinois, Putnam County, 10 August 1974, M. O. Glenn, slide WEM 179905, INHS.

***ferruginana* complex reddish phenotype**

l. Maryland, Worcester County, Isle of Wight, 9 July 1998, J. Glaser, slide DJW 4113, USNM.
m. Alabama, Baldwin County, 6 mi NW of Gulf Shores, 26 July 1985, R. L. and B. B. Brown, slide DJW 4114, MEM.
n. Florida, Putnam County, Palatka, 1 September 1990, H. D. Baggett, slide DJW 4134, MEM.
o. Florida, Highlands County, 6 mi SE Lake Placid, 5 May 1975, J. B. Heppner, slide DJW 4086, MGCL.
p. North Carolina, Jones County, Croatan National Forest, 10 August 2016, J. B. Sullivan, slide DJW 4437, JBS.

**59.** ***Sonia canadana*** McDunnough

a. Ohio, Adams County, 1 mi SE of Lynx, 21 August 1993, D. J. Wright, slide DJW 3883, USNM.
b. Ohio, Adams County, 1 mi SE of Lynx, 12 August 1998, D. J. Wright, slide DJW 4072, USNM.
c. New York, Tompkins County, Ithaca, Snyder Heights, 30 July 1976, J. G. Franclemont, slide RLB 1670, MEM.
d. Ohio, Wyandot County, Killdeer Plains, 16 August 2006, D. J. Wright, slide DJW 3882, USNM.
e. Ohio, Adams County, 1 mi SE of Lynx, 31 August 1989, L. D. Gibson, slide LDG 39, LDG.
f. Ohio, Erie County, Resthaven WLA, 19 August 1996, E. H. Metzler, slide DJW 3810, USNM.

**60.** ***Sonia nigrana*** Wright and Gilligan

a. PTP. Ohio, Adams County, 1 mi SE of Lynx, 2 July 2003, D. J. Wright, slide DJW 3803, USNM.
b. PTP. Kansas, Riley County, Konza Prairie, 23 July 1995, D. J. Wright, slide DJW 3909, USNM.
c. HTP. Ohio, Adams County, 1 mi SE of Lynx, 5 July 2002, D. J. Wright, slide DJW 4077, USNM.
d. PTP. Maryland, Washington County, Sideling Hill WMA, 24 August 2000, J. Glaser, slide DJW 3910, USNM.
e. PTP. Ohio, Adams County, 1 mi SE of Lynx, 3 August 1998, D. J. Wright, slide DJW 4286, USNM.
f. PTP. Ohio, Adams County, 1 mi SE of Lynx, 25 July 1997, D. J. Wright, slide DJW 4074, USNM.

**61.** ***Sonia albimacula*** Wright and Gilligan

a. HTP. Ohio, Adams County, 1 mi SE of Lynx, 27 June 1998, D. J. Wright, slide DJW 4394, USNM.
b. PTP. Massachusetts, Barnstable County, Barnstable, 3 July 1960, C. P. Kimball, slide DJW 4412, USNM.
c. PTP. Vermont, Caledonia County, Peacham, 11 July 2006, M. Sabourin, slide DJW 4536, CMNH.
d. PTP. Maine, Penobscot County, Lincoln, 15 July, slide DJW 4593, USNM.

**62.** ***Sonia parva*** Wright and Gilligan

a. HTP. Iowa, Pocahontas County, Kalsow Prairie, 18 July 2003, D. J. Wright, slide DJW 4088, USNM.
b. PTP. Ohio, Erie County, Resthaven WLA, 20 July 1990, D. J. Wright, slide DJW 4089, USNM.
c. PTP. Kentucky, Menifee County, Indian Creek, 6 July 1991, L. D. Gibson, slide DJW 4387, LDG.
d. PTP. Kentucky, Menifee County, Indian Creek, 6 July 1991, L. D. Gibson, slide DJW 4551, LDG.

**63.** ***canadana* complex**

a. Wisconsin, Sauk County, Spring Green Nature Conservancy Preserve, 26 June 1995, J.-F. Landry and V. Nazari, slide DJW 4111, CNC.
b. Canada, Manitoba, Manitoba Tall Grass Prairie Preserve, 17 July 2007, J.-F. Landry and V. Nazari, slide DJW 4290, CNC.

**64.** ***Sonia divaricata*** Miller

a. Kentucky, Laurel County, 0.5 mi E of Marsh Branch, 21 May 2009, L. D. Gibson, slide DJW 3887, LDG.
b. Arkansas, Logan County, Cove Lake, 14–20 May 1989, R. L. and B. B. Brown, slide WEM 269902, MEM.
c. Alabama, Dekalb County, DeSoto State Park, 18 May 1990, R. L. Brown, slide WEM 269901, MEM.
d. Tennessee, Blount County, Foothills Parkway, GSMNP, 9 June 2002, D. J. Wright, slide DJW 3888, USNM.
e. Kentucky, Laurel County, Forest Service Road 4176, 11 June 1993, D. J. Wright, slide DJW 3889, USNM.

**65.** ***Sonia comstocki*** Clarke

a. Texas, Brewster County, Big Bend National Park, 20–21 August 1995, E. C. Knudson, slide DJW 3866, MGCL.
b. Texas, Brewster County, Big Bend National Park, 20 August 1995, E. C. Knudson, slide DJW 3867, MGCL.
c. Texas, Brewster County, Big Bend National Park, 9 August 1983, E. C. Knudson, slide ECK 695, MGCL.
d. Texas, Brewster County, Big Bend National Park, 20–21 August 1995, E. C. Knudson, slide DJW 3868, MGCL.
e. Texas, Brewster County, Big Bend National Park, 20–21 August 1995, E. C. Knudson, slide DJW 3865, MGCL.

**66.** ***Sonia filiana*** (Busck)

a. California, San Diego County, San Diego, 28 October 1908, W. S. Wright, slide DJW 3891, USNM.
b. California, San Diego County, San Diego, slide 70912, USNM.
c. California, San Diego County, Torrey Pines State Reserve, 18–25 September 2005, N. Bloomfield, slide DJW 3890, USNM.
d. California, San Diego County, San Diego, 12 October, G. H. Field, slide DJW 3892, USNM.
e. California, San Diego County, Torrey Pines State Reserve, 1–8 October 2005, N. Bloomfield, slide DJW 3893, USNM.

**67.** ***Sonia vovana*** (Kearfott)

a. Arizona, Mojave County, Beaver Dam Canyon, 26 September 2000, G. J. Balogh, slide DJW 3901, USNM.
b. California, Kern County, Carrizo Plains, 20 August 1935, D. Hann, slide 70907, USNM.
c. Arizona, Coconino County, Walnut Canyon, 12 August 1964, J. G. Franclemont, slide DJW 3900, USNM.
d. Colorado, Otero County, 15 mi S of La Junta, 27 August 2000, D. J. Wright, slide DJW 3895, USNM.
e. Kansas, Morton County, 7.5 mi N of Elkhart, 25 August 2000, D. J. Wright, slide DJW 3894, USNM.
f. Colorado, Otero County, 15 mi S of La Junta, 18 August 1997, D. J. Wright, slide DJW 3897, USNM.

**68.** ***Sonia hodgesi*** Wright and Gilligan

a. PTP. Arizona, Cochise County, Coronado National Forest, 14 August 2011, R. Brown and S. Lee, slide DJW 4024, MEM.
b. HTP. Arizona, Santa Rita Mountains, Madera Canyon, 5 August 1959, R. W. Hodges, slide DJW 4036, USNM.
c. PTP. Arizona, Cochise County, Chiricahua Mountains, SW Research Station, 15 August 2011, J. W. Brown, slide DJW 4037, USNM.
d. PTP. Arizona, Cochise County, Chiricahua Mountains, SW Research Station, 14 August 2011, R. Brown and S. Lee, slide DJW 4025, MEM.

**69.** ***Sonia gibsoni*** Wright and Gilligan

HTP. New Mexico, Grant County, 14 mi SW of Silver City, 2 August 2005, L. D. Gibson, slide DJW 4059, USNM.

**70.** ***Suleima helianthana*** (Riley)

a. Texas, Randall County, Palo Duro Canyon, 9 May 1959, J. and S. Burns, slide DJW 4627, EME.
b. Ohio, Wyandot County, Killdeer Plains WA, 16 August 2006, T. M. Gilligan, slide DJW 4349, TMG.
c. New Mexico, Roosevelt County, Oasis State Park, 17 May 1998, Acciavatti, slide DJW 4366, CMNH.
d. Ohio, Wyandot County, Killdeer Plains WA, 14 June 1991, L. D. Gibson, slide DJW 4352, LDG.
e. Kansas, Morton County, Cimarron National Grassland, 2 August 1999, D. J. Wright, slide 153270, USNM.
f. Kentucky, Campbell County, A. J. Jolly Park, 15 July 1982, L. D. Gibson, slide DJW 4340, LDG.
g. Texas, Briscoe County, Caprock Canyon State Park, 29 March 2015, J. D. Hooper, slide DJW 4529, USNM.
h. California, Riverside County, Lake Mathews, 1 September 1966, Powell, Rude, and Wolf, slide JAP 4835, EME.
i. Texas, Randall County, Palo Duro Canyon, 9 May 1959, J. and S. Burns, slide DJW 4470, USNM.
j. Arizona, Cochise County, Huachuca Mountains, 26 March 1986, R. S. Wielgus, slide DJW 4531, USNM.
k. Ohio, Wyandot County, Killdeer Plains WA, 14 June 1991, L. D. Gibson, slide DJW 4352, LDG.
l. Ohio, Adams County, 1 mi SE of Lynx, 25 July 1997, D. J. Wright, slide DJW 4530, USNM.
m. Illinois, Putnam County, 12 July 1965, M. O. Glenn, slide DJW 4421, INHS.
n. Kentucky, McCracken County, Paducah Zip Track, 12 August 2008, L. D. Gibson, slide DJW 4341, LDG.
o. Colorado, Arapahoe County, Deer Trail, 16 September 2007, T. M. Gilligan and C. E. Harp, slide DJW 4350, TMG.
p. Ohio, Wyandot County, Killdeer Plains WA, 16 August 2006, D. J. Wright, slide 153272, USNM.
q. California, Los Angeles County, Long Beach, 31 July 1952, C. Henne, slide DJW 4512, EME.
r. Texas, Pecos County, Sheffield, 4 September 1961, J. M. Burns, slide DJW 4513, USNM.
s. Kentucky, McCracken County, Paducah Zip Track, 12 August 2008, L. D. Gibson, slide DJW 4341, LDG.

**71.** ***Suleima lagopana*** (Walsingham)

a. California, Tulare County, Visalia, 7 July 1953, W. H. Lange, slide 70882, USNM.
b. LTP. California, Colusa County, 3 mi from Indian Valley, 28 June 1871, Walsingham, slide 11605, BMNH.
c. California, Yolo County, Davis, 12 August 1956, E. Jaycox, slide DJW 4471, EME.

**72.** ***Suleima powelli*** Wright and Gilligan

a. HTP. California, Imperial County, 17 mi NW of Glamis, 27 June 1978, J. A. Powell, slide DJW 4466, EME.
b. PTP. California, Imperial County, 17 mi NW of Glamis, 1 April 1978, J. A. Powell, slide DJW 4631, EME.
c. PTP. California, Imperial County, 17 mi NW of Glamis, 27 June 1978, J. A. Powell, slide DJW 4630, EME.
d. PTP. California, Imperial County, 17 mi NW of Glamis, 27 June 1978, J. A. Powell, slide DJW 4467, EME.
e. PTP. California, Imperial County, 17 mi NW of Glamis, 27 June 1978, J. A. Powell, slide JAP 4243, EME.
f. California, Imperial County, 17 mi NW of Glamis, 27 June 1978, J. A. Powell, slide DJW 4475, EME.
g. California, Imperial County, 17 mi NW of Glamis, 27 June 1978, J. A. Powell, slide JAP 4840, EME.
h. PTP. California, Imperial County, 17 mi NW of Glamis, 27 June 1978, J. A. Powell, slide DJW 4468, EME.
i. California, Imperial County, 17 mi NW of Glamis, 27 June 1978, J. A. Powell, slide DJW 4469, EME.
j. California, Imperial County, 17 mi NW of Glamis, 27 June 1978, J. A. Powell, slide DJW 4478, EME.

**73.** ***Suleima baracana*** (Kearfott)

a. Wyoming, Carbon County, Kortes Dam, 22 June 2004, S. Johnson, slide DJW 4325, SJ.
b. Colorado, Logan County, 1.2 mi W of Route 61, 8 August 2007, L. D. Gibson, slide DJW 4344, LDG.
c. Colorado, Weld County, 0.6 mi N of I-76, Exit 57, 7 August 1996, L. D. Gibson, slide DJW 4346, LDG.
d. Washington, Whitman County, Wilma, 12 April 1934, J. F. G. Clarke, slide DJW 4608, USNM.
e. California, Riverside County, Temecula, 13 May 1994, R. Leuschner, slide DJW 3971, EME.
f. Texas, Brown County, Lake Brownwood State Park, 21 April 1966, A. and M. E. Blanchard, slide DJW 4609, USNM.
g. Texas, Culberson County, Guadalupe Mountains, Cherry Canyon, 22 May 1973, R. W. Hodges, slide DJW 4576, USNM.
h. New Mexico, Dona Ana County, Las Cruces, 8 April 1979, R. Leuschner, slide DJW 4612, EME.
i. Arizona, Cochise County, Pueblo Del Sol, 22 April 1986, R. Wielgus, slide DJW 4611, USNM.
j. Arizona, September 1900, O. C. Poling, slide 70883, USNM.
k. Colorado, Weld County, N of I-76 at Exit 57, 7 August 1996, L. D. Gibson, slide DJW 4345, LDG.
l. Colorado, Weld County, E side of County Road 91, 26 July 1995, L. D. Gibson, slide DJW 4363, LDG.
m. Texas, Guadalupe Mountains, Frijoles, 24 May 1973, A. and M. E. Blanchard, slide DJW 4590, USNM.
n. Montana, Carter County, Medicine Rocks State Park, 4 September 2002, D. J. Wright, slide DJW 4614, USNM.
o. Arizona, Pima County, Redington, slide DJW 4618, USNM.

**74.** ***Suleima leeana*** Wright and Gilligan

a. HTP. Arizona, Cochise County, Paradise Road, 17 August 2015, R. L. Brown, slide DJW 4319, ASUHIC.
b. PTP. Arizona, Cochise County, Paradise Road, 17 August 2015, R. L. Brown, slide DJW 4320, ASUHIC.
c. PTP. Arizona, Cochise County, John Hands Campground, 20 August 2015, S. Lee, slide DJW 3998, ASUHIC.
d. PTP. New Mexico, Socorro County, Cibola National Forest, 15 August 2015, S. Johnson and J. Vargo, slide DJW 4624, SJ.
e. PTP. Arizona, Cochise County, John Hands Campground, 20 August 2015, S. Lee, slide DJW 3999, ASUHIC.
f. PTP, New Mexico, Grant County, 16 August 2015, J. Vargo, slide DJW 4634, USNM.

**75.** ***Suleima blanchardi*** Wright and Gilligan

a. HTP. Texas, Padre Island National Seashore, 29 September 1975, A. and M. E. Blanchard, slide DJW 4491, USNM.
b. PTP. Texas, Padre Island National Seashore, 5 June 1978, A. and M. E. Blanchard, slide DJW 4501, USNM.
c. PTP. Texas, Padre Island National Seashore, 29 September 1975, A. and M. E. Blanchard, slide DJW 4622, USNM.
d. PTP. Texas, Padre Island National Seashore, 5 June 1978, A. and M. E. Blanchard, slide DJW 4623, USNM.
e. PTP. Texas, Padre Island National Seashore, 19 July 1976, A. and M. E. Blanchard, slide DJW 4492, USNM.
f. PTP. Texas, Padre Island National Seashore, 20 May 1979, A. and M. E. Blanchard, slide 97546, USNM.

**76.** ***Suleima skinnerana*** Heinrich

a. New Mexico, Grant County, Gila National Forest, Emory Pass, 16 August 2015, Johnson and Vargo, slide DJW 4327, SJ.
b. PTP. Arizona, Cochise County, Huachuca Mountains, August 1905, H. Skinner, slide DJW 4480, USNM.
c. Texas, Jeff Davis County, Davis Mountains State Park, 20 August 1984, E. C. Knudson, slide DJW 4465, MGCL.
d. PTP. Arizona, Cochise County, Huachuca Mountains, August 1905, H. Skinner, slide DJW 4600, USNM.

**77. *Suleima flavidana* Wright and Gilligan**

a. HTP. Arizona, Cochise County, Chiricahua Mountains, 31 July–2 August 1964, D. R. Davis, slide DJW 4481, USNM.
b. PTP. Texas, Val Verde County, 15 mi W of Del Rio, 19 August 1995, E. C. Knudson, slide DJW 4464, MGCL.
c. PTP. Texas, Brewster County, BIBE Green Gulch, 16–18 May 1999, E. C. Knudson, slide DJW 4510, MGCL.
d. Arizona, Santa Rita Mountains, Madera Canyon, R. W. Hodges, slide DJW 4568, USNM.
e. Mexico, Puebla, October 2015, K. Mawela and L. Mukwevho, slide 146392, USNM.
f. PTP. Texas, Culberson County, Guadalupe Mountains National Park, 1 August 1992, E. C. Knudson, slide DJW 4463, MGCL.

**78. *Suleima daracana* (Kearfott)**

a. SNTP. California, Placer County, 1 May, A. H. Vachell, slide DJW 4483, USNM.
b. Oregon, Baker County, 28 mi SE of Joseph, 19 June 1970, J. F. G. Clarke, slide DJW 4487, USNM.
c. LTP. California, Placer County, 1 May, A. H. Vachell, slide DJW 4583, AMNH.
d. Washington, Whitman County, Union Flat, 15 May 1929, J. F. G. Clarke, slide DJW 4570, USNM.
e. Washington, Whitman County, Pullman, 12 May 1930, J. F. G. Clarke, slide DJW 4484, USNM.
f. Washington, Whitman County, Pullman, 4 October 1933, J. F. G. Clarke, slide DJW 4572, USNM. [reared from *Helianthella douglasii*]
g. Oregon, Baker County, 6.4 mi N of Richmond, 7 June 1970, J. F. G. Clarke, slide DJW 4574, USNM.
h. Oregon, Baker County, 28 mi SE of Joseph, 21 June 1970, J. F. G. Clarke, slide DJW 4506, USNM.
i. Oregon, Baker County, 28 mi SE of Joseph, 21 June 1970, J. F. G. Clarke, slide DJW 4508, USNM.
j. Washington, Asotin County, 1.5 mi N Anatone, 22 June 1970, J. F. G. Clarke, slide DJW 4606, USNM.
k. Oregon, Wallowa County, Wallowa Lake, 19 June 1970, J. F. G. Clarke, slide DJW 4485, USNM.
l. SNTP. California, Placer County, 20 April, A. H. Vachell, slide DJW 4569, USNM.
m. Washington, Whitman County, Pullman, 12 May 1930, J. F. G. Clarke, slide DJW 4598, USNM.
n. Oregon, Baker County, 9 mi E of Halfway, 7 June 1970, J. F. G. Clarke, slide DJW 4585, USNM.
o. Washington, Whitman County, Union Flat, 15 May 1929, J. F. G. Clarke, slide DJW 4571, USNM.
p. Washington, Whitman County, Pullman, 17 April 1934, J. F. G. Clarke, slide DJW 4597, USNM.
q. Oregon, Baker County, 28 mi SE of Joseph, 19 June 1970, J. F. G. Clarke, slide DJW 4488, USNM.
r. Washington, Whitman County, Union Flat, 15 May 1929, J. F. G. Clarke, slide DJW 4596, USNM.
s. Washington, Whitman County, Pullman, 11 November 1933, J. F. G. Clarke, slide DJW 4573, USNM.

**79. *Suleima cinerodorsana* Heinrich**

a. Kentucky, Owsley County, E side of County Road 1141, 20 July 1980, L. D. Gibson, slide DJW 4347, LDG.
b. Kentucky, Bracken County, Meldahl Dam, 1 September 1992, L. D. Gibson, slide DJW 4348, LDG.
c. Illinois, Putnam County, M. O. Glenn, 14 August 1966, slide DJW 4420, INHS.
d. Ohio, Hancock County, Liberty Township, 16 August 1991, T. M. Gilligan, slide DJW 4409, TMG.

**80. Unresolved specimens**

a. Colorado, Chaffee County, 4 mi SW of Buena Vista, 15 July 1982, R. W. Hodges, slide DJW 4490, USNM.
b. Colorado, Boulder County, Silver Saddle Motel, 25–26 June 2015, J. W. Brown, slide DJW 4489, USNM.

**81. *Suleima mendaciana* Blanchard and Knudson**

a. Texas, Cameron County, Southmost, 27 October 1986, E. C. Knudson, slide DJW 4460, MGCL.
b. Texas, Uvalde County, Garner State Park, 6 October 1984, E. C. Knudson, slide DJW 4499, MGCL.
c. Texas, Uvalde County, Garner State Park, 6 October 1984, E. C. Knudson, slide DJW 4461, MGCL.
d. Texas, Culberson County, Sierra Diablo WMA, 18 August 1984, E. C. Knudson, slide DJW 4462, MGCL.

**82. *Notocelia culminana* (Walsingham)**

a. California, Plumas County, 1 mi S of Meadow Valley, 9–16 September 1983, J. A. Powell, slide DJW 4150, EME.
b. Canada, British Columbia, Fraser River near Yale, 16 August 1973, R. Leuschner, slide DJW 4149, EME.
c. Michigan, Otsego County, Old State Road, 19 August 2007, L. D. Gibson, slide DJW 4153, LDG.
d. New York, Suffolk County, Long Island, Orient, 1 September 1947, R. Latham, slide DJW 4222, CUIC.
e. New Mexico, San Miguel County, Santa Fe National Forest, 17 August 2015, S. Johnson and J. Vargo, slide DJW 4328, SJ.
f. California, Contra Costa County, Pleasant Hill, 20 October 1960, W. E. Ferguson, slide DJW 4140, EME.
g. Maine, York County, Kennebunkport, August, C. H. Clapp, slide DJW 4337, CMNH.

**83. *Notocelia lobana* Wright and Gilligan**

a. PTP. Kentucky, Bullitt County, Route 480, 6.9 mi E of Route 61, 6 September 1989, L. D. Gibson, slide DJW 4248, LDG.
b. PTP. Ohio, Lake County, Hells Hollow Park, 24 August 2018, J. Wiedmann, slide DJW 4311, JW.
c. PTP. Ohio, Greene County, Wright-Patterson AFB, 28 August 1992, E. H. Metzler, slide DJW 4291, OSUC.
d. PTP. New York, Suffolk County, Shelter Island, 10 July 1946, R. Latham, slide DJW 4314, CUIC.
e. PTP. Ohio, Adams County, Lynx Prairie, 10 September 1988, L. D. Gibson, slide DJW 4247, LDG.
f. PTP. Ohio, Lake County, Hidden Lake Woods, 30 August 2017, J. Wiedmann, slide DJW 4312, JW.
g. PTP. New York, Suffolk County, Montauk Island, 10 June 1954, R. Latham, slide DJW 4313, CUIC.

**84. *Notocelia illotana*** (Walsingham)

a. Colorado, Jefferson County, Red Rocks, 5 June 1969, J. Scott, slide DJW 4147, EME.
b. California, Siskiyou County, Castle Lake, 12 July 1998, J. A. Powell, slide DJW 4146, EME.
c. Canada, Manitoba, Spruce Woods Provincial Park, 6 June 1972, J. R. Powers, slide DJW 4155, EME.
d. New York, Suffolk County, Long Island, Orient, 7 August 1934, R. Latham, slide DJW 4236, CUIC.
e. Washington, Whitman County, Kramer Prairie, 1 June 2004, R. Zack, slide DJW 4231, WSDA.
f. Idaho, Latah County, 6 mi NE of Moscow, 29 June 1975, Turner and Powell, slide DJW 4142, EME.
g. Pennsylvania, Allegheny County, Oak Station, 24 May 1912, F. Marloff, slide DJW 4143, EME.
h. Washington, Chelan County, Leavenworth, 9 July 2010, D. G. Holden, slide DJW 4233, WSDA.
i. New York, Suffolk County, Long Island, Orient, 1 July 1935, R. Latham, slide DJW 4224, CUIC.

**85. *Notocelia cynosbatella*** (Linnaeus)

a. Canada, British Columbia, Thetis Island, 3 September 1978, R. Guppy, slide DJW 4148, EME.
b. Canada, British Columbia, Thetis Island, 30 May 1978, R. Guppy, slide DJW 4156, EME.
c. Washington, King County, Seattle, 25 May 2020, T. M. Mustelin, slide DJW 4253, TMM.
d. Washington, Thurston County, 3 mi NW of Tenino, 6 March 2005, E. LaGasa, slide DJW 4229, WSDA.
e. Canada, British Columbia, Thetis Island, 10 May 1977, R. Guppy, slide DJW 4144, EME.
f. Canada, British Columbia, Thetis Island, 28 June 1976, R. Guppy, slide DJW 4145, EME.
g. Washington, King County, Seattle, 18 May 2018, T. M. Mustelin, slide DJW 4254, TMM.
h. Washington, Thurston County, 1 mi N of Olympia, 10 April 2005, E. LaGasa, slide DJW 4230, WSDA.

**86. *Notocelia purpurissatana*** (Heinrich)

a. Canada, Ontario, Renfrew County, 20 August 2016, J. J. and L. M. Dombroskie, slide DJW 4225, CUIC.
b. Canada, Alberta, Two Lakes PRA, 1 July 2006, J. Dombroskie and D. Lawrie, slide DJW 4218, CUIC.
c. Canada, Alberta, Kananaskis, 9 August 2010, J. Dombroskie, slide DJW 4219, CUIC.

**87. *Notocelia rosaecolana*** (Doubleday)

a. Canada, Nova Scotia, Kings County, Sheffield Mills, 26 June 2001, K. Neil, slide DJW 4141, EME.
b. New York, Tompkins County, Ithaca, Palmer Woods, 15 June 2018, J. Dombroskie, slide DJW 4221, CUIC.
c. Kentucky, Larue County, Thompson Creek Glade, 3 May 2006, L. D. Gibson, slide DJW 1716, LDG.
d. Kentucky, Gallatin County, Markland Dam, 20 May 2007, L. D. Gibson, slide DJW 4151, LDG.

# Animal Name Index

**Species numbers** and **genera** are listed in bold font; page numbers are listed in regular font; plate references are listed in parentheses with letters for adults and numbers for genitalia. Junior synonyms are listed in smaller font.

# Plant Name Index

**Species numbers** are listed in bold font; page numbers are listed in regular font.